Springer Transactions in Civil and Environmental Engineering

Springer Transactions in Civil and Environmental Engineering (STICEE) publishes the latest developments in Civil and Environmental Engineering. The intent is to cover all the main branches of Civil and Environmental Engineering, both theoretical and applied, including, but not limited to: Structural Mechanics, Steel Structures, Concrete Structures, Reinforced Cement Concrete, Civil Engineering Materials, Soil Mechanics, Ground Improvement, Geotechnical Engineering, Foundation Engineering, Earthquake Engineering, Structural Health and Monitoring, Water Resources Engineering, Engineering Hydrology, Solid Waste Engineering, Environmental Engineering, Wastewater Management, Transportation Engineering, Sustainable Civil Infrastructure, Fluid Mechanics, Pavement Engineering, Soil Dynamics, Rock Mechanics, Timber Engineering, Hazardous Waste Disposal Instrumentation and Monitoring, Construction Management, Civil Engineering Construction, Surveying and GIS Strength of Materials (Mechanics of Materials), Environmental Geotechnics, Concrete Engineering, Timber Structures.

Within the scopes of the series are monographs, professional books, graduate and undergraduate textbooks, edited volumes and handbooks devoted to the above subject areas.

More information about this series at http://www.springer.com/series/13593

Geetam Tiwari · Dinesh Mohan
Editors

Transport and Safety

Systems, Approaches, and Implementation

Springer

Editors
Geetam Tiwari
Department of Civil Engineering
Indian Institute of Technology Delhi
New Delhi, Delhi, India

Dinesh Mohan
Indian Institute of Technology Delhi
New Delhi, India

ISSN 2363-7633 ISSN 2363-7641 (electronic)
Springer Transactions in Civil and Environmental Engineering
ISBN 978-981-16-1114-8 ISBN 978-981-16-1115-5 (eBook)
https://doi.org/10.1007/978-981-16-1115-5

This Springer imprint is published by the registered company Springer Nature Singapore Pte Ltd.
The registered company address is: 152 Beach Road, #21-01/04 Gateway East, Singapore 189721, Singapore

Preface

Motorized road transport activities impact human health in many different ways, directly as road safety, and noise and air pollution issues and indirectly through reductions in active transport and social exclusion. Transport also contributes to global issues of climate change, and its operations are associated with urban equity, justice and poverty. Traffic safety remains one of the most important aspects of transport, directly or indirectly impacting specific issues across local and national boundaries. The select readings in this volume present transport safety in a broad global and societal context as well as in specific detail. Three quarters of the articles included here address issues on traffic safety policy—ranging from vision zero to modelling of head impact injuries. These readings will be of special interest to advanced researchers in engineering and planning disciplines working on traffic and transport safety as well as policy-makers concerned with setting up institutions and legislation for traffic safety.

Nine of the 16 chapters of this volume were presented as TRIPP Annual Lectures at the Indian Institute of Technology Delhi. The Transportation Research and Injury Prevention Programme (TRIPP) at the Indian Institute of Technology established an annual lecture series on sustainable transport in 2007 with an endowment from the Volvo Research and Educational Foundations with an objective to honour outstanding researchers who have made significant contributions in the field of sustainable transportation.

In 2015, TRIPP organized an International Symposium on Transportation Planning and Traffic Safety in Delhi to commemorate the completion of twenty years of TRIPP at IIT Delhi. Five chapters included in this volume are the state-of-the-art lectures delivered at this symposium by international experts. This volume also has two independent chapters written as position papers for International Council of Road Safety Research Independent (ICORSI) Symposium in 2018 and 2019 by the editors of this volume. Overall, the volume presents a wide range of topics dealing with safe and sustainable transport systems relevant specifically to low- and middle-income countries.

We acknowledge the support provided by Volvo Research and Education Foundations (VREF); the sponsors of the International Symposium on Transportation Planning and Traffic Safety 2015 at IIT Delhi; the ICORSI Symposium 2018 at Paris, France; and the ICORSI Symposium 2019 at Changsha, China.

New Delhi, India

Geetam Tiwari
Dinesh Mohan

Contents

Editors and Contributors

About the Editors

Geetam Tiwari is the MoUD Chair Professor for Transport Planning at the Indian Institute of Technology Delhi and was an Adlerbretska Guest Professor for sustainable urban transport at the Chalmers University of Technology, Sweden from 2007 to 2010. She obtained her B.Arch. from the University of Roorkee and Master of Urban Planning and Policy, and Ph.D. in Transport Planning and Policy, from the University of Illinois, Chicago. She has extensive research experience in dealing with transportation issues of special relevance to low income countries. These include development of bus systems and road designs that would make transportation efficient and safer. She has been working in the area of traffic and transport planning focusing on pedestrians, bicycles and public transport. She has published over 70 research papers on transportation planning and safety in national and international journals and peer reviewed seminar proceedings. She edited four books on transportation planning and road safety. She has received the International Velocity Falco Lecture Prize, Barcelona, Spain; the Stockholm Partnerships Award for local impact, innovative thinking and a potential for replication or transferability: a Centre for excellence grant from the Volvo Research and Educational Foundations (VREF); the IRTE & Prince Michaels Award for promoting road safety research and the LMA (Lucknow Management Association) Award for Woman Achiever, 2010. She is an Advisor to the Urban Age series of conferences coordinated by the London School of Economics since 2005.

Dinesh Mohan is an Honorary Professor at the Indian Institute of Technology (IIT) Delhi; he has been a Distinguished Professor at the Shiv Nadar University and continues as the Director of the Independent Council for Road Safety International. He serves on the Editorial Boards of two international journals dealing with safety. In the past he has worked at the Insurance Institute for Highway Safety, Washington DC (1975–1978) and the University of Michigan Transportation Research Institute (1971–1975). He has been a consultant on safety related matters to government departments in India, Nepal, Indonesia, Thailand, Bangladesh, Iraq and Libya; he has been a consultant to the automotive industries including TELCO, Ashok Leyland, Volvo Trucks, Eicher Motors Ltd., Escorts Ltd., Maruti Udyog Ltd., SIAM, Bajaj Auto Ltd. and to international organisations like the World Bank and WHO. His research focuses on transportation research (safety and pollution), human tolerance biomechanics, motorvehicle safety, road traffic injuries, childhood injuries, effectiveness of automobile safety equipment, evaluation of injuries to cyclists and motorcyclists and motorcycle helmet design. He has been the recipient of many awards including the IRCOBI Bertil Aldman Award, and the Distinguished Alumnus Awards from the University of Delaware (USA) and the Indian Institute of Technology (IIT) Bombay (India).

Contributors

Matts-Ake Belin has a long history within the Swedish government, and he works primarily with overall safety policies, strategies and collaboration with different stakeholders. In 2007–2009, he worked for World Health Organization where he participated in the development of global road safety strategies and global partnerships. He has also chaired the technical committee 3.1 on National Road Safety Policies and Programme, World Road Association. Currently, he is Director of Vision Zero Academy at the Swedish Transport Administration and he is also responsible for developing the programme for the third Global Ministerial Conference on Road Safety, held in Stockholm in 2020. He is also affiliated as Adjunct Professor at KTH Royal Technology Institute in Stockholm, Sweden.

Tony Bliss is Honorary Senior Fellow, Urban Transport, at the Transport, Health and Urban Design Research Hub, University of Melbourne, Australia; Director, Road Safety Management Limited; and an internationally recognized road safety management specialist. Current activities include research into city land use, transport and health governance issues; the assessment of safe sustainable mobility investment opportunities in the regional trade corridors of South Asia; and advising on the preparation of a Vision Zero Road Safety Strategy for Auckland city and its region. In his previous position as Lead Road Safety Specialist at the World Bank, he directed the development, promotion and implementation of multi-sectoral strategies designed to improve road safety performance in low- and middle-income countries throughout East Asia and the Pacific, Europe and Central Asia, Latin America and the Caribbean, the Middle East and North Africa, South Asia and Sub-Saharan Africa, working with World Bank sector teams and country, regional and global partners. He also established and managed the World Bank Global Road Safety Facility, a donor funding platform designed to support road safety management capacity strengthening in low- and middle-income countries.

Nicolas Bourdet is Mechanical Engineer. He is Scientific Researcher at the Institute of Fluid Mechanics and Solids, Strasbourg University, France. He defended his Ph.D. thesis on biomechanics at the Strasbourg University. He works on impact biomechanics.

Werner Brilon is Professor Emeritus for transportation and traffic engineering in the Department of Civil Engineering at the Ruhr-University Bochum, Germany. He studied civil engineering at Karlsruhe University (now KIT) during 1963–1969 and has since been involved in transportation research. He is Member of various national and international working committees and working groups. In 2001–2002, he worked as Chairman of the Scientific Advisory Board of the Federal Minister of Transport (BMVBW) and also worked with Research Association for Roads and Traffic (FGSV), and was appointed Chairman of the Research Advisory Board from 1997 to 2003 and Head of Working Group 2 (road design) in 1999–2007. Currently, he is Chairman of the committee to develop the next German Highway Capacity Manual (HBS 2015). He was appointed as Member of the TRB Committee on Highway Capacity and Quality of Service ("HCM Committee") from 1994 to 2005 and is also Member of the committee "Roundabouts" of the Transportation Research Board, Washin ton DC. He was awarded by several organizations for his professional achievements. Among these, he has been appointed as Honorary Professor by the Technical University of Cracow, Poland.

Caroline Deck is Ph.D. Researcher in biomechanics at the Institute of Mechanics of the Fluids and Solids (IMFS), ULP-CNRS, Strasbourg, France. She works on mechanics, shock wave interaction and identification of tissues, mechanical laws, modellization and head trauma numerical reconstructions, establishment of biomechanical criteria and head tolerance limits under impacts, conception, evaluation and optimization of head protective systems against biomechanical criteria, non-lethal weapon evaluation against biomechanical criteria and understanding of blast effects and its simulation.

Rune Elvik (born 1955) was educated as Political Scientist at the University of Oslo. He has worked as Road Safety Researcher at the Institute of Transport Economics since 1980. He obtained doctoral degrees from the University of Oslo in 1993 and 1999 and from Aalborg University in Denmark in 2007. He was Associate Editor of Accident Analysis and Prevention from 1997 to 2004 and Editor-in-Chief (together with Karl Kim) from 2005 to 2013. He has participated in many research projects funded by the European Commission and been Member of Transportation Research Board Committee on safety data, analysis and evaluation. He is Author of about 120 papers in scientific journals and many research reports.

Rahul Goel is Research Associate at the MRC Epidemiology Unit, University of Cambridge. He did his B.Tech. in civil engineering as well as his Ph.D. at the Indian Institute of Technology Delhi. His core area of research is transportation planning with a focus on the intersection of travel behaviour, built environment and health outcomes. In his Ph.D., he did interdisciplinary research integrating travel patterns with exposure to air pollution, traffic injuries and physical activity. He has published in peer-reviewed international research journals covering disciplines of transport planning, environment, traffic injuries, public health and public policy.

Christer Hydén is Professor Emeritus in traffic safety and has been working with traffic safety ever since 1971. He took his degree at Lund Institute of Technology, 1987, and became Full-Time Professor in 1993. First achievement was the development of the Swedish Traffic Conflict Technique, which was the basis for his thesis in 1987. Research on conflict theory and practice has been one of the main themes ever since. He has, in that context, been involved in several projects around the world where conflict studies have been the main issue. He led the first project on speed limiter in cars and was also responsible for a large-scale experiment on small roundabouts in the 1990s. He has been President of ICTCT, the International Co-operation on Theories and Concepts in Traffic, from 1979 to 2011. He has presented papers at a large number of national and international research conferences, and been lecturing at several international traffic safety courses organized by the Swedish Road Research Institute and Lund Institute of Technology, in Eastern Europe, Africa, Asia and South America. He received in 1991 the Volvo Traffic Safety Award for the development of the Swedish Traffic Conflict Technique and in 2014 the Frank Blackmore Award for Lifetime Contribution to Roundabouts, 2014, by TRB Committee on Roundabouts.

Hermann Knoflacher is Professor Emeritus, Institute for Traffic Planning and Traffic Engineering TU, Wien. From 1963 to 1970, he was involved in traffic safety research at the university. From 1970 to 1983, he was Head of the Traffic Safety Research Institute, Traffic Safety Board, Austria. From 1975 to 2008, he served as Professor and Head of the Institute for Traffic Planning and Traffic Engineering TU, Wien. Since 1963, he has been involved in practical work in civil engineering, transport planning, traffic safety and urban planning. His fields of research are as follows: design of transport elements; user behaviour; traffic infrastructure and mobility; sustainable development of cities and mobility; traffic safety; energy consumption; environment; and basic interdisciplinary research. He has published 12 books, more than 500 scientific publications, more than 500 research and planning projects, and lectures in the field of transport planning and traffic engineering worldwide (USA, Europe, Asia, Australia). He has more than 300 realized projects for cities, regions and national transport policy measures.

Murray Mackay (OBE, FREng) is Professor Emeritus of transport safety, University of Birmingham, England. He established the Birmingham Accident Research Centre at the University of Birmingham in 1964 and ran it until 1996. He has specialized in in-depth crash investigation, understanding the epidemiology of traffic crashes, the mechanisms of impact injury and the performance of safety systems and their limitations. He is Founding Director of the Parliamentary Advisory Council on Transport Safety in Britain and the European Transport Safety Council in Brussels. He was President of the International Research Council on the Biomechanics of Injury (IRCOBI) from 1990 to 2010. His awards include S.A.E.

Siegel Transportation Safety Award, N.H.T.S.A. Award for Safety Engineering Excellence, Joint Recipient of the Volvo International Traffic Safety Award, Safety Award in Mechanical Engineering from the Institution of Mechanical Engineers, London, and Award of Merit of the American Association for Automotive Medicine.

Robert B. Noland is Distinguished Professor at the Edward J. Bloustein School of Planning and Public Policy and serves as Director of the Alan M. Voorhees Transportation Center and Director of the Ph.D. programme in planning and public policy. He received his Ph.D. at the University of Pennsylvania in energy management and environmental policy. Prior to joining Rutgers University, he was Reader in Transport and Environmental Policy at Imperial College London and Policy Analyst at the US Environmental Protection Agency and also conducted postdoctoral research in the Economics Department at the University of California at Irvine. The focus of his research is the impacts of transportation planning and policy on both economic and environmental outcomes. He is currently Co-editor-in-Chief of *Transportation Research Part D* (Transport and Environment) and Associated Editor of the *International Journal of Sustainable Transportation* and was formerly Chair of the Transportation Research Board Special Task Force on Climate Change and Energy.

Yves Page is Project Manager of "Automated Vehicle Experimentation" at Renault since January 2016 after being Road Safety Expert from 2011 to 2015. He was previously Deputy Director of Advanced Security Projects at the Research, Advanced Studies and Research Department and in charge of the technological plan safety (2009–2010). He was also Deputy to Director of the Laboratory of Accidentology, Biomechanics and Human Behavior Studies (LAB), PSA Peugeot Citroën, from 2004 to 2008. He has been a member of the Scientific Committees of the AAAM, the French Foundation Road Safety, the Expert Committee to the National Council of Road Safety, the Belgian Institute of Road Safety and of the Ethics Committee of IFSTTAR. He has published, for over twenty years, more than 80 publications in automobile safety and road safety.

Ian Roberts is Professor of epidemiology and public health at the London School of Hygiene and Tropical Medicine. His main research interests are the prevention and treatment of traumatic injury and the links between energy use and health. He trained as a paediatrician in the UK and then in epidemiology at the University of Auckland, New Zealand, and at McGill University, Canada. He established and is the coordinating editor of the Cochrane Injuries Group, an international network of individuals that prepares and maintains systematic reviews of the effectiveness of interventions in the prevention, treatment and rehabilitation of injury.

David Satterthwaite is Senior Fellow at the International Institute for Environment and Development (IIED) in London and Visiting Professor at the University College London. The IIED is a global leader in policy research in sustainable development. He is also Editor of the international journal, Environment and Urbanization. In 2004, he was awarded the Volvo Environment Prize.

Most of his work has been on poverty reduction in urban areas in Africa, Asia and Latin America, undertaken with local teams. He has a particular interest in how organizations and federations of slum/shack dwellers have demonstrated more effective approaches to urban problems and where their partnerships with local governments have increased the scale and scope of what they can achieve. He has written and edited various books on urban issues, including Squatter Citizen (with Jorge E. Hardoy), Environmental Problems in an Urbanizing World (with Jorge E. Hardoy and Diana Mitlin) and Empowering Squatter Citizen (with Diana Mitlin), all published by Earthscan. He also co-authored two recently published books on urban poverty with Diana Mitlin—Urban Poverty in the Global South: Scale and Nature and Reducing Urban Poverty in the Global South, both published by Routledge.

Fred Wegman gained a M.Sc. in civil engineering, traffic engineering from the Delft University of Technology in November 1972. From 1974 to 1977, he was Traffic Engineer in the Municipality of Amsterdam. In 1977, he joined the (SWOV) Institute for Road Safety Research. Today, he divides his time between SWOV where he is Managing Director and the Delft University of Technology (TU Delft) at the Department of Transport and Planning at the faculty of civil engineering where he is Professor. He is President of the International Traffic Safety Data and Analysis Group and Chairman of the Core Advisory Group of the Global Road Safety Facility in the World Bank. He is Member of the Steering Committee of the Road Safety Performance Index (European Transport Safety Council). He is Author of several influential publications and is in the forefront of the "Safe Systems" development (which is a systems approach to road safety) around the world. He has initiated moves for co-operation with both national and international partners in furthering SWOV's vision of science and technology in the service of sustainable road safety.

Remy Willinger is Professor at the University of Strasbourg. His research ranges from biological tissue identification and modelling to human body characterization followed by lumped and distributed modelling. Once validated, the human body models are used for the simulation of real-world accidents in order to derive tolerance limits relative to specific injury mechanisms. Human models are also coupled to protective systems for the optimization in respect of biomechanical criteria. Most of his work addressed head and neck impact biomechanics that lead to model-based injury criteria.

Chapter 1
The Swedish Vision Zero—An Advanced Safety Culture Phenomenon

Matts-Åke Belin

1.1 Introduction

According to estimates of the World Health Organization (WHO 2015), more than 1.25 million people die each year on the world's roads. Also according to WHO, the difference in fatality rate is huge between low and high countries; the situation is worst in low-income countries.

Poor road safety has been highlighted by the UN and the WHO and together with the World Bank jointly published the World Report on Traffic Injury Prevention in 2004 (Peden et al. 2004). The World Report is to road safety, what the Brundtland Report (World Commission on Environment and Development 1987) is to the environmental sector. The World Report on Traffic Injury Prevention was launched in conjunction with the World Health Day, April 7, 2004, which was dedicated to road safety. A week later, on April 14, 2004, the first road safety debate was held in the UN General Assembly and a historic United Nations resolution on road safety was adopted (United Nations General Assembly 2004). Another important milestone was the UN General Assembly's adoption of a global plan for road safety, ("The Decade of Action for Road Safety 2011–2020") which was adopted in order to save lives and stop the expected negative trend in the number of traffic accidents in the world (United Nations General Assembly 2010).

At the UN summit on September 25–27, 2015, world leaders decided on 17 new global sustainable development goals that would guide the international development cooperation over the next 15 years (https://sustainabledevelopment.un.org/?menu=1300). Although several targets have a bearing on road traffic injuries, targets 3.6 and 11.2 are most clearly connected to road safety.

M.-Å. Belin (✉)
Swedish Transport Administration, Vision Zero Academy, KTH Royal Institute of Technology, Borlänge, Stockholm, Sweden
e-mail: matts-ake.belin@trafikverket.se

G. Tiwari and D. Mohan (eds.), *Transport and Safety*, Springer Transactions in Civil and Environmental Engineering,
https://doi.org/10.1007/978-981-16-1115-5_1

Goal 3. Ensure healthy lives and promote well-being for all at all ages.

- Target 3.6: By 2020, halve the number of global deaths and injuries from road traffic accidents.

Goal 11. Make cities and human settlements inclusive, safe, resilient and sustainable.

- Target 11.2: By 2030, provide access to safe, affordable, accessible and sustainable transport systems for all, improving road safety, notably by expanding public transport, with special attention to the needs of those in vulnerable situations, such as women, children, persons with disabilities and older persons.

In November 2015, there was a meeting in Brazil where a road safety declaration was adopted, describing more clearly the focus of the global road safety work (https://www.who.int/violence_injury_prevention/road_traffic/Brasilia_Declaration/en/).

Sweden has 2.8 fatalities per 100,000 inhabitants in 2015 which could be compared with India's 16.6 fatalities per 100,000 inhabitants. 2.8 is the lowest fatality rate in the global status report on road safety 2015 (WHO 2015) and it is not an exaggeration to state that Sweden is a leading country when it comes to road safety.

However, to become the world's leading country, it has been a long and winding journey, and in this paper, it will be explored, from a macrolevel, as to how a country like Sweden has developed its public policies and strategies over the years, and how Swedish society has organized its road safety work in accordance with new challenges and strategies. This paper will in particular explore Vision Zero as a policy innovation and how it influences road safety work.

1.2 Safety Culture—A Concept to Explore How Road Safety Strategies Are Evolved Over the years

In 1986, a catastrophic accident occurred in a reactor at the Chernobyl nuclear power station in the former Soviet Union. After the disaster, the International Atomic Energy Agency (IAEA) carried out several investigations and concerns about the safety culture within the Soviet nuclear power industry (IAEA 1986). This was the first time the concept of safety culture was explicitly coined and made available for a large audience as an important concept in safety. Based on this new interest, an expansion of research into safety culture, especially within organizations, occurred. One important conclusion from this research is that safety culture is not a binary concept, that either you have a safe culture or you do not. Safety culture is an evolutionary concept, and it seems like organizations improve their safety culture through different stages. According to Westrum (2004), safety culture is defined as the organization's pattern of response to the problems and opportunities it encounters. Westrum (2004) identified three dominant types of responses—the pathological, the bureaucratic and the generative. Based on this initial idea, the model was developed and

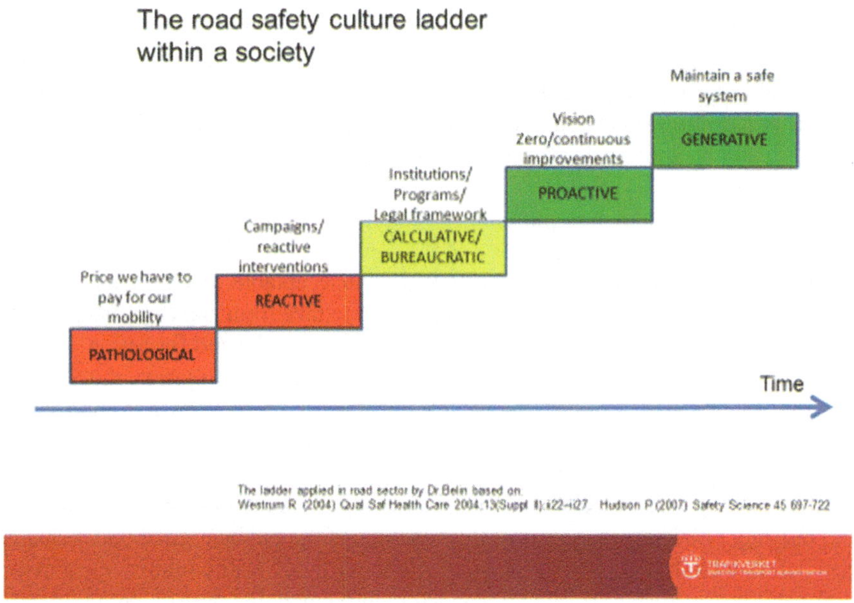

Fig. 1.1 Road safety culture ladder within a society

extended (Hudson 2007) from three to five stages in a sequence by introducing the reactive and the proactive stages and also replacing the label bureaucratic with the calculative. In Fig. 1.1 you will see an example of how this safety culture theory can be applied in the field of national road safety.

Although the concept of safety culture has focused mainly on safety culture within organizations, in this paper, these ideas are applied to Sweden as a society in order to explore how Sweden has developed its policies and strategies over the years and how the Swedish Vision Zero fits in.

1.3 Swedish Road Safety—From Pathological, Reactive to Calculative

Before the Second World War, Sweden was a rather poor country but after the war, Sweden witnessed a tremendous economic growth, fast motorization and urbanization. The popularity of the automobile took off and the road transport system was developed rapidly. Unfortunately, there was also a negative side to this development: The greater the volume of motor traffic, the more people were killed and seriously injured in traffic accidents see Fig. 1.2 (Swedish Official Statistics). In the mid-60s, Sweden had some 17 fatalities per 100,000 inhabitants annually on the roads. This is similar to what middle-income countries around the world are facing nowadays

(18.4 fatalities per 100,000 inhabitants) according to World Health Organization estimates (WHO 2015). Motorization and urbanization were definitely factors that contributed to economic growth, and these side effects in terms of injuries were in general perceived by decision makers and the general public as the price we have to pay for better mobility and economic growth. In light of safety culture, this phase could, more or less, be characterized as a **pathological phase** and the main responsibility for the problem was put almost completely on the victims, the road users themselves.

The situation during the 1950s and 1960s was unacceptable, and it correlated badly with the modern welfare state that was beginning to take form and especially among the medical experts; there was a growing frustration and demand for action. Parallel with this growing awareness to do more to reduce road traffic injuries, the Swedish government prepared a rather unique reform, namely to go from left-hand traffic to right-hand traffic. The rationale for this reform was that Sweden's Scandinavian neighbors were driving on the right side of the road as did most of the others in Europe. Furthermore, most Swedish cars also had left-hand steering. However, there was a strong public opinion against this reform and the public argued that a change from left hand to right could increase the number of road traffic injuries even more. However, the Swedish government decided to implement the reform, but in order to take on board these public fears and to make sure that the reform could be carried out without increasing the number of road traffic injuries, the government set up a special organization, the so-called Högertrafik kommissionen (the right-hand traffic commission). This commission consisted of several experts within different areas of expertise such as road, human factor, and vehicle design. The commission planned and implemented massive information campaigns before and during the change in 1967, and the reform was a great success. Figure 1.2 shows that the change was successful from a road safety point of view. Instead of increasing the road traffic

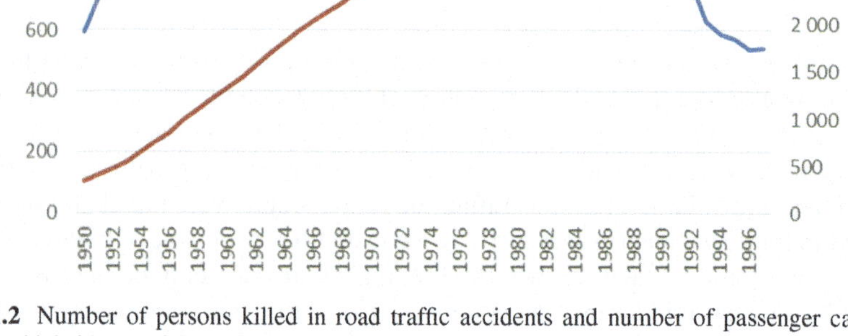

Fig. 1.2 Number of persons killed in road traffic accidents and number of passenger cars per thousand inhabitants 1950–1996 in Sweden

deaths, which had been the worst fear among critics of the reform, the number of deaths in road traffic decreased the next year, but in the years that followed, the number went up again.

However, during the middle of the 60s, a seed had been sown for a comprehensive and systematic road safety work through Ralph Nader's book "Unsafe at Any Speed" (Nader 1965). In the USA, this book helped spur the passage of the National Traffic and Motor Vehicle Safety Act in 1966 and the creation of several predecessor agencies which would eventually become the US National Highway Traffic Safety Administration NHTSA (Graham 1989).

At the same time, a former Swedish Prime Minister, Olof Palme, who was at the time the Swedish Transport Minister, was deeply affected by Ralph Nader and his book. He even arranged for the book (Nader 1967) to be translated into Swedish and he also took the initiative, based on an American model, to set up a special authority for road safety issues: the Swedish Road Safety Authority.

The establishment of the Road Safety Authority can be said to be the starting point for systematic road safety activities in Sweden. This work was successful during the 1970s and the number of people killed on the roads dropped from 21.3 killed per 100 000 inhabitants in 1966 to 12.9 killed per 100 000 inhabitants in 1982—a decrease of over 40% (Trafikanalys 2015). Sweden was definitely moving into the calculative/bureaucratic phase where the road safety work was institutionalized and characterized by a planning approach with goals, targets and the use of cost-effectiveness methods to prioritize effective countermeasures. The calculative/bureaucratic phase is, from a road safety point of view, a good phase during which the number of fatalities usually decreased by approximately 50%. There was also a risk that sooner or later a plateau would be reached and that is what happened in Sweden as well.

During the 1980s, the positive trend was broken and traffic growth and road injury figures began to follow each other: the more car traffic, the more people were killed on the roads. In 1989, Sweden had 10.6 fatalities per 100.000 inhabitants and Sweden was, once again, approaching four-figure numbers in road deaths. A sense of lost control was spreading in society and together with the political pressure to do something more radical it eventually (in 1993) led to the closure of the Swedish Road Safety Agency and to an enhanced role for the Swedish Road Administration.

Parallel with this process to change the institutional prerequisites for the national road safety work, Sweden was facing a severe economic recession in the first half of the 90s. During the period 1990–1993, Swedish GDP dropped almost 5% and the unemployment level increased dramatically. From a road safety point of view, at least in the short run, we know that economic recessions might be good for safety and this was also the case in the beginning of the 1990s. The number of fatalities dropped between 1989 and 1996 by more than 40%.

With the negative trend in deaths and injuries on the roads in the 1980s freshly in mind, together with the knowledge that the recession would gradually change into a more normal pattern of economic development, road safety strategists in Sweden shared a common insight that something different was needed in the Swedish work on road safety. Business as usual was not a long sustainable strategy. Heavily influenced by the new liberal wave that was sweeping across the world at the time, there was a substantial lack of trust in rules, regulations, monitoring and the over-protective, "nanny-state" attitude of many authorities. Instead of regulating safety in the spirit of the Traffic Safety Authority, ideas were aroused that the real potential for greater road safety lay in a dialogue with the road users. The most important thing was to influence the road safety culture among road users, which in turn would have a long-term impact on their behavior in traffic. However, this strategy turned out to be a historical bracket and a step backward to a more reactive approach.

1.4 Vision Zero—Adopted by the Swedish Parliament 1997—A Proactive Approach to Safety

In the autumn of 1994, Professor Claes Tingvall was appointed as the new Road Safety Director of the Swedish Road Administration. It soon became apparent that Professor Tingvall had a different idea of how our road safety work should be conducted in Sweden. Professor Tingvall and a small team at the Swedish Road Administration started to develop ideas for a new long-term strategy for traffic safety. Using existing scientific evidence, they created a new model based on the biomechanical tolerances of human beings, which guided them in identifying interventions that could help mitigate the creation of kinetic energy during crashes. Thus, Vision Zero was born. The results of this development work were documented in a memorandum entitled "Vision Zero—An idea for a road transport system without health losses" (Vägverket 1996). This policy memorandum quickly attracted political interest, and the Minister of Transport and Communications at the time, Ms. Ines Uusmann, initiated a policy-preparation process which culminated in the Swedish Parliament giving its firm backing to the Social Democrats Government's proposal for adopting a new direction in traffic safety work in October 1997 (Swedish Parliament 1997).

"Madam Speaker! This is a white paper bill. The government has tabled a proposal for an entirely new way of thinking within the sphere of road safety work. The bill contains no sign of a concrete proposal, which we shall be forced to return to later. As things stand at present, it is a matter of adopting the new way of thinking, giving it a firm foundation among our citizens and making preparations—by means of investigations and other methods—for future decision-making." (Swedish Parliament 1997).

This quotation is from the Social Democrat MP involved in the parliamentary debate, Ms. Lena Sandelin, pending the decision to adopt Vision Zero. The decision was a directional or alignment decision, with no direct concrete proposals for measures.

1.5 Vision Zero a Policy Innovation

A public policy (Parson 1995; Anderson 2000; Peters and Pierre 2006) like Vision Zero can basically be defined as a policy theory, a program theory or an intervention theory (Hoogerwerf 1990; Vedung 1997; Patton 2002; Leeuw 2003; Mickwitz 2003; Rossi 2004). From that perspective, the decision to adopt Vision Zero was a decision to adopt a new strategy which aims to influence the concrete work on road safety directly, but also—more indirectly—the institutional preconditions and approaches, which in turn also have an impact on the actions of various players so that they take action to increase the safety of the road transport system.

Vision Zero differs from a traditional road safety policy in a number of ways (Belin et al. 2011), and it is probably not an exaggeration to state that from a safety culture perspective, Vision Zero belongs to the more advanced phases in the safety culture ladder.

1.6 Definition of the Road Safety Problem—Traditional Versus Vision Zero

Knowledge based on investigations of actual traffic accidents that answer questions about why accidents happen points sharply in the direction of the fact that it is the individual road user who is the missing link in the road transport system. The traditional road safety activities are to a significant extent based on behavioral science research which draws the conclusion that 90% of all road traffic accidents can be explained by the human factor. In the traditional safety work, the principal challenge is to prevent conscious and subconscious faulty human action (Swedish government 1940). Vision Zero instead accepts, as a basic starting point, that human beings make conscious and subconscious mistakes. That is why accidents occur, and the safety work must in the first instance be directed at those factors which can prevent accidents leading to death and serious injury. Accidents in themselves can be accepted, but not their serious consequences.

According to Vision Zero, the principal cause as to why people die and are seriously injured is that the energy to which people are exposed in a traffic accident is excessive in relation to the energy that the human frame can withstand. Vision Zero is, among other things, based on the research that the famous American road safety expert William Haddon conducted in the 1960s (Haddon 1968, 1970, 1972, 1973, 1980). Knowledge of energy and tolerance has to a great extent served as a basis for the development we have seen of the passive safety characteristics of vehicles and for the development of different protection systems such as child safety seats, helmets, seat belts, etc. One important consequence of Vision Zero as a general policy for safety work is that the view of knowledge which has served as a basis for the development of a sub-component in the road transport system, namely the vehicle, has also become a general principle for the entire road transport system.

1.7 Perspective on Responsibility—Traditional Versus Vision Zero

In the traditional safety work, ultimate responsibility for safety rests with the individual. According to a traditional view, it is the individual road user who ultimately controls and manages the risks that may occur when traveling on the road transport system. The regulations surrounding the road transport system are clear and unambiguous on this point. If a road traffic accident occurs, it is possible in most cases to hold a certain road user liable for the deficient observance of regulations. Even if, for example, a road authority has made a mistake in the design of a road, it is the responsibility of the road user, through the general requirements for caution that are built into the traffic legislation, to at the same time provide compensation through his/her behavior for such shortfalls. According to Vision Zero, it is not the individual road user who has the ultimate responsibility but rather the so-called system designers.

The responsibility for safety is thus split between the motorists and the system designers (i.e., infrastructure builders and administrators, the vehicle industry, the haulage sector, taxi companies and all the organizations that use the road transport system professionally), on the basis of the principles that:

- the system designers have ultimate responsibility for the design, upkeep and use of the road transport system, and are thus responsible for the safety level of the entire system;
- as before, the road users are still responsible for showing consideration, judgment and responsibility in traffic and for following the traffic regulations;
- if the road users do not take their share of the responsibility, for example, due to a lack of knowledge or competence, or if personal injuries occur or for other reasons are likely to risk occurring, the system designers must take further measures to prevent people from being killed or seriously injured.

In Vision Zero, the responsibility for safety is a chain of responsibility that both begins and ends with the system designers (Belin et al. 2016).

1.8 What Should Be Achieved in the Long Run—Traditional Versus Vision Zero

A more traditional approach to people getting killed and seriously injured as a consequence of road traffic accidents has been the utilitarian philosophical approach (Bowen 2012; Belin et al. 2012). Utilitarianism, as it has come to be applied within the road traffic sector, means that safety has to be weighed against other types of benefits. In theory, and to a large extent in practice, this approach means that those killed and seriously injured is a price that society has to pay for the mobility of the road transport system, and that a nonzero number of deaths and serious injuries is acceptable. Safety can be gradually be improved, but only to the extent that it is socioeconomically advantageous. In addition, the traditional road safety work is to a large extent based on the idea that people are willing to take risks and that it is part of human nature.

The long-term objective of Vision Zero is to create a road transport system in which nobody is killed or seriously injured as the result of a traffic accident. Thus, Vision Zero aims to create a safe road transport system. The justification for this absolute and uncompromising attitude is what philosophers would attribute to deontological ethics (Bowen 2012; Belin et al. 2012), i.e., nobody should need to be killed or seriously injured when moving via the transport system from Point A to Point B. Road transportation can be regarded as a type of production. Just as little as society can accept people killing or seriously injuring themselves as a consequence of producing goods and services within an industry, Vision Zero can accept it when transportation is produced. According to Vision Zero, mobility is therefore subordinate to safety, at least in the long term. If it is impossible to otherwise create a safe system, it should inexorably have consequences for mobility. Furthermore, Vision Zero is based on the fact that people do not want to die or get seriously injured as the result of a road traffic accident, and therefore, each person has his or her own Vision Zero. Vision Zero and a traditional safety policy thus differ from each other when it comes to the long-term objective of the safety work. Traditional safety work is to a large extent based on the notion that individuals and society largely speaking do not ask for safety.

There are other values that are given a higher priority, such as accessibility and personal freedom. Traditional traffic safety strategies are thus to a large extent based on the "unwilling road user" and society who must be forced into giving consideration to safety (Johnston et al. 2014). Vision Zero is instead based on individuals and society demanding safety. The fact that people sometimes act as though they do not require safety has, according to Vision Zero, rather more to do with inability, ignorance and a lack of social support than a lack of will.

The evolution from a traditional approach to Vision Zero can be summarized in Fig. 1.3.

1.9 Vision Zero and Implementation

Vision Zero is the kind of theory which can only be appreciated fully when it is implemented. However, to go from policy intentions to sound policy outcomes in a complex world is difficult, as experiences of evaluation and implementation research have shown (Sabatier and Mazmanian 1979; Hill and Hupe 2009; Vedung 1997). Therefore, since the 1970s, a number of studies have been carried out in an attempt to pin down what has been called the missing link in policy research (Nakamura and Smallwood 1980; Hill and Hupe 2002). This research has contributed to a skeptical perspective on the opportunities to get policy intentions implemented in society, and it has even been portrayed as the "misery research" of the social sciences (Rothstein 1998). Implementation of road safety interventions is no exception. For example,

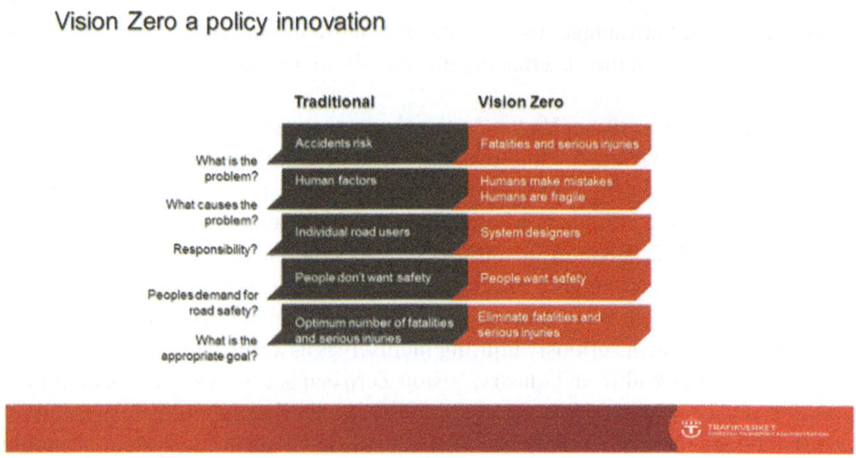

Fig. 1.3 Traditional road safety policy versus Vision Zero

according to Bax et al. (2014), studies of actual priorities in road safety policy, in particular in the Scandinavian countries, suggest that these priorities are inefficient.

Fortunately, Swedish society, to a great extent, has been able to go from words to action and to base the daily road safety work on Vision Zero (Belin and Tillgren 2012). This process can be described as in the model below where Vision Zero has influenced our methods to make things happen including some direct interventions which have eventually influenced the number of fatalities and serious injuries in the road transport system (Fig. 1.4).

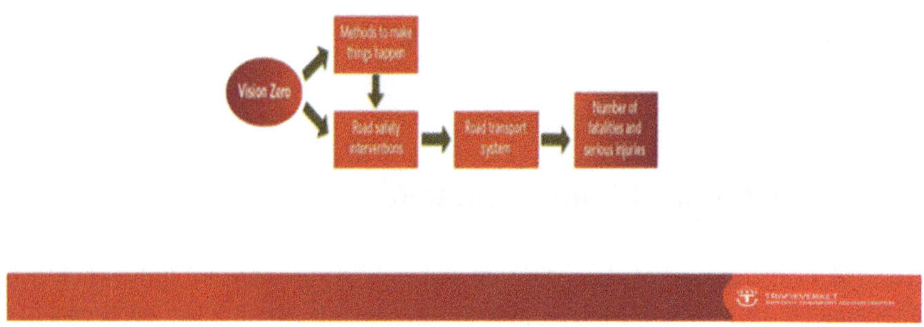

Fig. 1.4 Vision Zero—implementation process

1.10 Vision Zero—Road Safety Interventions: A Few Concept Examples

Vision Zero challenges our traditional way of looking upon road safety problems and the ways in which we solve them. Let us take a few concrete road safety problems, both from an urban and a rural environment, and exemplify how Vision Zero changes the ways in which these problems can be solved.

1.11 Unprotected Road Users in Urban Areas

According to WHO, 22% of all road deaths in the world are pedestrians (WHO 2015). Some traditional interventions to solve this problem are to lower the speed to 50 km/h in urban areas, to regulate the way pedestrians are allowed to cross the street through zebra crossings, sometimes together with traffic lights. These interventions are usually combined with information and education (especially for young children), and sometimes, these regulations are enforced by the police. The drivers are also educated to take extra care to avoid collisions with vulnerable road users. In general, it is up to the individual road users to negotiate and handle risks. These interventions are based on a traditional approach where the perception is that road traffic injuries are caused mainly by individual road users' lack of ability or will to follow the rules. Thus, the strategy is to step-by-step reduce the risks by aiming to shape the road users' behavior to excellence. The ultimate goal is to create "the perfect human behavior." Many of these interventions do have road safety benefits and will reduce the number of fatalities and injuries, and these interventions fit well into the calculated/bureaucratic phase's incremental approach, but these interventions might not in the long run eliminate severe injuries. Some of these interventions (e.g., zebra crossings) might even, contrary to common sense, increase the number of injured pedestrians (Elvik et al. 2009).

Vision Zero provides an alternative approach and will influence the kind of interventions that need to be implemented. First, Vision Zero's ultimate goal is to create a safe system. The aim is not to, step by step, reduce the risks and make it a little better. Second, Vision Zero does not make any claim that it is possible to create "the perfect road user." On the contrary, Vision Zero is based on the idea that people are people, who will make mistakes and sometimes break the law. Instead of wishing for the perfect road users, Vision Zero faces the fact that people are not perfect and that the challenge is to so design the road system that it can accommodate these human errors. This brings us to one of the core elements in Vision Zero, namely human tolerance against external violence. This is a crucial part of Vision Zero: to design a road transport system which does not exceed people's tolerance against external violence. For the first time in the history of road safety, Vision Zero forced us to start discussing human tolerance levels. This is common when we discuss tolerance levels of environmental factors. In an urban area where we mostly deal with conflicts between protected and unprotected road users, the most important factor will be the unprotected road users' tolerance level. Although our knowledge is tentative and when new research is published we need to adjust our recommended tolerance levels, it seems that 30 km/h for unprotected road users is an appropriate tolerance level. Consequently, we have an important designing element for our road safety work. If we plan for conflicts between a protected and unprotected road user, we need to strive for energy levels less than 30 km/h. This tolerance level is universal and it does not matter if it is in Sweden or India.

This brings us to two main types of interventions within urban areas, namely physical separation and speed controlling. Physical separation through sidewalks, bridges, tunnels, special areas only for pedestrians, basically eliminates conflicts between pedestrians and cars. This type of intervention is appropriate when the motorized traffic demands high mobility. In areas where pedestrian and cyclist mobility is prioritized and in areas where pedestrians must be mixed with cars, the interventions need to control speeds below 30 km/h. There are several traffic calming interventions which are effective, such as speed bumps and roundabouts, among others. There are also promising vehicle technologies such as intelligent speed adaptation systems and auto brake systems. Compared to a more traditional approach, these interventions have a clear target with an emphasis on the design of the system (Photo 1.1).

1.12 Head-On Collisions on Rural Roads

Sweden has a fairly large rural road network. Parts of the road network with high traffic volume have over the years been rebuilt as modern motorways. But Sweden has a small population, and therefore, 13-m-wide road type was built, a kind of cheap motorway, during the 1970s and the 1980s. This road type was straight with high visibility and the risk for accidents calculated per kilometer driven was quite low. The roads were also built with grade-separated intersections. Although it was not justified to build a motorway, these roads accommodated quite a heavy traffic volume and the speed limit was 110 km/h. Even though the risk of accidents was small on this kind of road, those which happened were always severe. These roads were commonly referred to as "death roads." Head-on collisions due to unsafe overtaking and high speeds were the major causes of these catastrophes. Traditional approach to try to solve this problem was public information about the risks, driving education and to some extent police enforcement. The perception was that a change of behavior would solve the problem but in reality, these interventions were not effective.

When Vision Zero was adopted, these parts of the road network came into focus. A small proportion of the road network represented a large proportion of the number of fatalities and serious injuries. Something needed to be done. A motorway would of course solve the problem with head-on collisions, but it was too large a network which would be too expensive to rebuild. In that time in Sweden, it was estimated that a motorway costs approximately 1 billion SEK per saved life. Motorways are an effective but very expensive road safety intervention. Another alternative would

Photo 1.1 Speed reduction and separation measures in urban areas

be to lower the speed. Just as unprotected road users have a certain tolerance against external violence, we also know quite well how much external violence a belted passenger in a modern car can tolerate, namely around 80 km/h. So it would be a solution to lower the speeds which also is good from an environment perspective. However, it would probably be impractical to get the road users to comply with 80 km/h on these high-standard roads. This forced Sweden to come up with a new solution, an innovation which is called 2 + 1 roads. Basically, this road is a three-lane road, which consists of two lanes in one direction and one lane in the other, alternating this every few kilometers, separated usually with a steel cable barrier. The speed limit on these roads is usually 100 km/h. This was, especially in the beginning, a very cost-effective solution to a major road safety problem. In the beginning, this solution was estimated to cost 30 million SEK per saved life (Photo 1.2).

After some initial resistance, especially among the professional road builders, a 2 + 1 road program was carried out, and now, Sweden has more than 3,000 km of 2 + 1 roads, and according to evaluations, the number of fatalities has been reduced by 79% (Carlsson 2009). This is a good example of a Vision Zero approach in its practical

Photo 1.2 2 + 1 road

application. However, it also shows that even in this environment, when you eliminate head-on collisions, fatal crashes could occur and new countermeasures need to be developed and implemented that address both single crashes in the side areas and rear-end crashes (Fig. 1.5) (Swedish Transport Administration 2015).

1.13 Alcohol-Related Crashes

It is a well-known fact that alcohol and driving is a deadly combination. Sweden has a long tradition to combat this problem, and in 1939, Sweden adopted a limit on how much alcohol a driver can have in the bloodstream before it will be defined as a criminal act. These alcohol limits have been adjusted over the years from 0.8 per mille in 1939 to the existing legislation of 0.2 per mille. These regulations have also been backed up with information, education and police enforcement. A quite traditional approach definitely contributed to the social norm in the Swedish society that it is not acceptable to drink and drive. According to estimations made in Sweden,

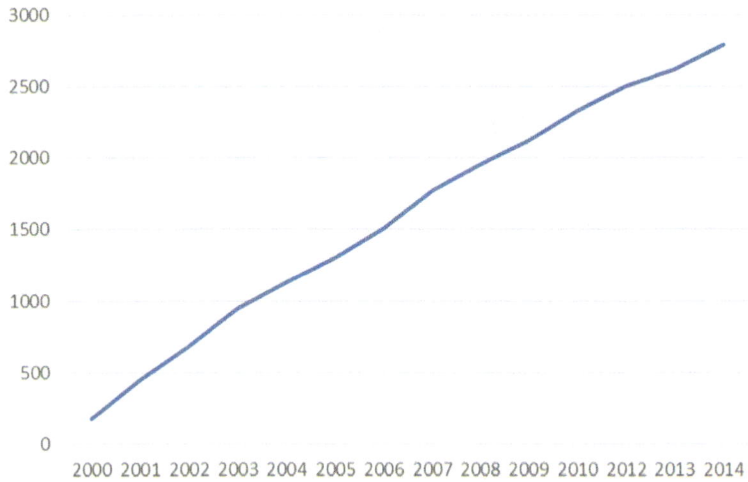

Fig. 1.5 2 + 1 roads 2000–2014 (tens of km at year's end) (Swedish Transport Administration 2015)

99.78% of the traffic is sober. Therefore, approximately only two drivers per 1,000 are influenced by alcohol. However, according to in-depth studies of fatal crashes in 2014, 20% of all killed drivers of passenger cars were influenced by alcohol. Although we do not have a social norm problem in Sweden, we still have a road safety problem.

Also, this well-known drink and drive problem could be analyzed from a Vision Zero point of view. The traditional perspective on moral responsibility is so embedded in our strategies that we may not see the forest for the trees. Alcohol definitely increases the risk for people to have a crash but the major problem is that this event happens in an inherently unsafe system. Advanced road safety strategies force us to focus beyond individual deficits to how this problem can be prevented through a systemic perspective. Therefore, from a Vision Zero perspective, the top priority is to design a system in which people will not be killed or seriously injured regardless of what caused the crash. First of all, the 2 + 1 road, which already has been mentioned, is probably one of the most effective interventions to prevent people from getting killed in alcohol-related crashes (or crashes due to fatigue or distraction among others) (Fig. 1.6) (Swedish Transport Administration 2015).

Second, there are several promising vehicle technology solutions on the market. These interventions are indifferent to what causes the drivers' incapability; instead, they focus on prevention. For example, a lane departure system which forces the car to keep its position in a lane could prevent the car from moving out of it. Most of the technologies on the market today are warning systems, but in the future, these will probably become (as in automated driving) controlling systems. Third, the most common technological solutions are the so-called alcolocks. An alcolock is a technical product that prevents a vehicle from being started if the driver has been drinking. The alcolock is an immobilizer which is connected to the vehicle, and

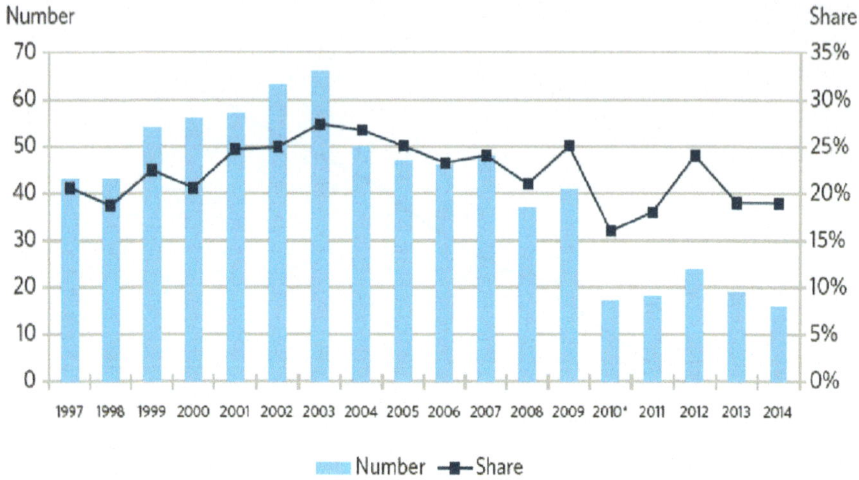

Figure 20. Share of drunk driver (alcohol > 0.2 mg/ml) fatalities among all fatalities and number of drunk driver fatalities, 1997-2014. Source: The Swedish Transport Administration's in-depth studies.
**From 2010 excluding suicide.*

Fig. 1.6 Alcohol-related crashes (Swedish Road Administration 2015)

before the vehicle is started, the driver blows into the alcolock. If the driver is under the influence of alcohol (over the legal limit of 0.2 parts per thousand), the ignition is locked so that the vehicle cannot be started (www.mhf.se/sv-SE/sakrare-i-trafiken/bast-pa-alkolas-alkomatare/alkolas/alcolocks-in-english/about-alcolocks/).

1.14 Speed and Vision Zero

From a Vision Zero perspective, speed is the most important factor to control. That is because speed is another word for kinetic energy which, as has been mentioned before, is the principal cause of injuries. Therefore, when we talk about speed, we need to discuss three important aspects; the designed speed, the posted speed and the operation speed. All these aspects have been influenced by Vision Zero, but let us first discuss the more traditional approach to speed.

In year 1907, the Swedish government launched the first road traffic regulation for automobiles. Among other things, the regulation stipulated that motor vehicles were not allowed to drive faster than 15 km/h in urban areas and 25 km/h in rural areas. During the period 1910–1930, the maximum speed limit was increased to 35 km/h in urban areas and 45 km/h in rural areas. The use of the automobile was heavily regulated mainly because it was seen as an unwelcome element in a transport system which mainly consisted of horse transports. In the 1930s, a strong opinion was raised against these static speed limits. The advocators argued that the vehicles and the roads had a higher standard and therefore were designed to allow a much higher speed. It was better, according to the advocators, to put the whole responsibility on the individual to adjust their speed according to the situation. Therefore, a new speed regime with free speed, both in urban and rural areas, and with a big proportion of self-responsibility was introduced in 1936.

After the Second World War when the number of cars increased rapidly and the number of fatalities increased dramatically, the epidemic situation forced the Swedish government to take different steps to improve the road safety situation. The experts were not sure that the freedom for the drivers to choose their own speed was a good idea. Besides, it was difficult for the police to enforce an inappropriate choice of speed and the police needed clearer guidelines regarding which speed to allow. The politicians responded to that request and the first step was to re-regulate the speed in urban areas. In year 1955, a new default speed that stipulated 50 km/h in urban areas (this speed limit is still in place) was introduced. During the years 1960–1967, temporary speed limits for the rural roads were introduced—especially during holidays. The speed limits were 80, 90 or 100 km/h. In 1968, a trial with general differentiated speed limits was introduced and the idea was to allow higher speeds on roads with higher standards. In 1971, a default 70 km/h speed limit for rural roads was introduced. The debate about speed limits or not vanished from the agenda and was replaced by a discussion of which criteria the speed limits should be based on (e.g., on what road should responsible authorities allow 90 or 110 km/h).

One of the most important criteria when the speed limits were first discussed was the drivers' acceptance. The advice was that the speed should be around 85th percentile, which means the speed 85% of the vehicles are not exceeding.

Soon it became obvious that road alignment, passing sight distance and accident rate needed to be considered before a speed was decided, and these accident-related criteria have dominated since the 1970s. In the 1980s, the experts advocated that speed limits should be set from a cost–benefit perspective. The idea was that you could calculate an optimal speed limit for different road environments. This method has never been implemented in reality, though.

Speed limit is one of the most important steps that has been taken by the government to improve the road safety in the road transport system. According to the handbook of road safety measures (Elvik et al. 2009), a large number of studies have evaluated the effects of changes in traffic speed on the number and severity of accidents. According to the handbook (Elvi et al. 2009), the relationship between speed changes and changes in the numbers of accidents and injuries can be described as a power function.

Based on these studies, the handbook (Elvik et al. 2009) also provides these exponents. The conclusion that can be drawn from this is that, other things being equal, small changes of speed have a great impact on the outcome of accidents (Fig. 1.7).

The designed and posted speed limits are one thing, but the compliance of these limits is something different. Traditionally, enforcement, sanctions, information and

Best estimate	Accident/Injury severity
4.5	Fatalities
3.0	Serious injuries
1.0	Property-damage-only Accidents

Fig. 1.7 Exponents

education have been major strategies to make the road users comply with the stipulated speed limits. In Sweden, the number of those who comply with the speed has been rather stable, and the share of the traffic volume within the speed limits is 48% on the national road network and 63% on the municipality roads (Swedish Transport Administration 2015). There are therefore large opportunities for improvements.

Vision Zero forces us to really consider the basic motives and criteria for the speed for which we actually design our road transport system and for the speed limits that should be posted. This issue has already been touched upon when we discussed interventions in urban and rural areas but in this section, these aspects will be explored in more detail.

According to Vision Zero, road users' tolerance against external violence should be the basic design parameter for the speed. Below is an example of how this principle is put in practice (Tingvall and Haworth 1999). The table shows possible long-term maximum travel speeds related to the infrastructure, given best practice in vehicle design and 100% restraint use (Fig. 1.8).

Possible travel speed (km/h)	Type of infrastructure and traffic
30	Locations with possible conflicts between pedestrians and cars
50	Intersections with possible side impacts between cars
70	Roads with possible frontal impacts between cars
100+	Roads with no possibility of a side impact or frontal impact (only impact with the infrastructure)

Fig. 1.8 Long-term maximum travel speeds from (Tingvall and Haworth 1999)

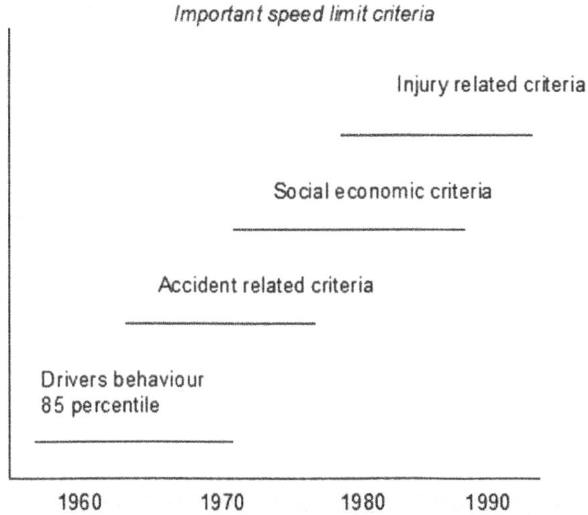

Fig. 1.9 Speed limit evolution in Sweden

In Fig. 1.9, the different approaches to setting speed over the years in Sweden are summarized (Belin 2007; Global Road Safety Partnership 2008).

From a Vision Zero perspective, it is the speed we should focus on, not necessarily the speed limits itself in order to achieve a safe system. The actual speed could be influenced in different ways, and it seems like our strategies, also in this case, have evolved over time. In the model below, this evolution is presented. It starts with awareness, which then goes to regulation and traditional manual enforcement. Experience from Sweden indicates that with these strategies, we can expect to achieve in general 50% compliance. To achieve better performance, which is necessary to achieve a safe system, we need to change our strategies and focus more on technology. The question is should this transition be influenced by our traditional strategies or should it be based on a Vision Zero approach. In the paper (Belin et al. 2010), these differences are explored through a comparative analysis between Victoria, Australia (the traditional approach) and Sweden (the Vision Zero approach). In Sweden, the speed cameras are named traffic safety cameras (Fig. 1.10).

The program that was implemented in Victoria in 1991 was intended to solve the problem of a large proportion of drivers who continuously exceeded speed limits and thereby created road safety problems. The purpose was to achieve a massive increase in the likelihood of detection accompanied by a high-level advertising/public education campaign to inform drivers of the risks of speeding and of the risks of getting caught (Government of Victoria, Department of Justice 1993). The Victoria approach is clearly focused on driver behavior, and the speed camera program attempts to create a feeling among drivers that speeding can be detected at any time and in any place in the whole road system. To achieve this feeling of continued surveillance, the intention is to catch a large proportion of the drivers that exceed the speed limit so that they experience the consequences and avoid re-offending (specific deterrence) and, in their turn, tell other drivers that they have been caught (general deterrence). The

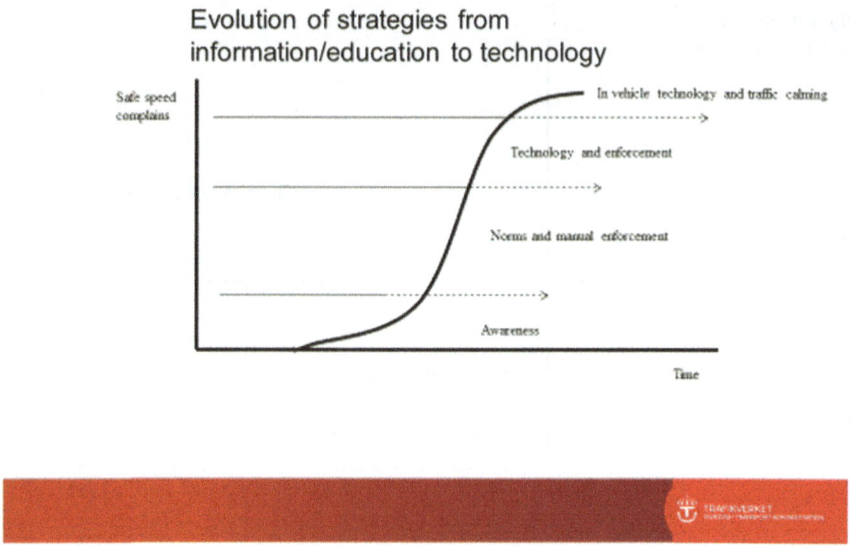

Fig. 1.10 Evolution of strategies to reach 100% behavior compliance

road safety effect in the Victoria system is supposed to be achieved by the fact that the perceived risk of being detected and fined prevents a large proportion of the drivers from speeding, and therefore, the average driving speed in the whole system will decrease, as too will the number of fatalities and injuries.

In the Swedish program implemented in 2006, on the other hand, the focus is more on dangerous environments. The main road safety problem is an imbalance in the situation between the road network and its safety level, and the speed at which people choose to travel.

The reason for using automatic road safety control is to achieve a preventive effect in places which have been assessed as dangerous and where there is poor speed adaptation and no other countermeasures are planned (Swedish government 2004). The basic problem is therefore the inherently dangerous design of the environment. The cameras are supposed to support and create a new social norm among drivers that is easier and better to follow than those stipulated by posted speed limits. This is very close to a nudge approach (Sunstein and Thaler 2012) to safety. Drivers who are caught by the system are regarded as an undesirable consequence. Everyone should know where the cameras are located and voluntarily adjust their speed accordingly. In the Swedish system, information on the existence of cameras will help most of the drivers to voluntary respect the speed limits, which means that both the average speed and the number of fatalities and injuries will decrease.

These two different intervention theories can be summarized according to this table

The different components in the program theory	Victoria (1991)	Sweden (2006)
What main problem does the speed camera program try to solve?	The problem is that a large proportion of drivers continuously exceed the speed limit and create road safety problems	The main problem is that the speed, on a large proportion of the road network, exceeds the speed level which the roads, from a safety point of view, are designed for
What does the program try to achieve?	The main purpose is to create a feeling among these drivers that speeding can be detected at any time and in any place in the whole road system	The main purpose is to support and create a new social norm among drivers that it is easier and better to follow the speed limit
What does the intervention mechanism look like?	The main chain of influence is to catch a large proportion of those drivers who exceed the speed limit so that they experience the consequences and avoid re-offending and, in their turn, tell other drivers that they have been caught	The main chain of influence is to inform (through signs and open cameras) the drivers that a large proportion of the traffic network (large proportion of the traffic) is covered by cameras
How is the program intended to achieve road safety effects?	A large number of drivers will be deterred from speeding. Excessive speeds and the average speed will decrease and the number of fatalities and injuries will decrease in the whole system	It will prevent most of the drivers from speeding. The average speed will decrease, as too will the number of fatalities and injuries

Clearly, both Victoria, Australia, and Sweden have defined speed as an important factor in the road safety work. They also recognize the value of using new technology in order to prevent people from speeding. Therefore, from a functional perspective, the two speed camera systems could be classified as a similar kind of intervention. However, the approach applied in Victoria is based on the idea that speeding is a deliberate offense in which a rational individual wants to drive as fast as possible and consciously calculates the costs and benefits of his behavior. According to this theory, decisions to speed would be part of what Kahneman (2011) calls our System 2. The underlying aim of the Victorian intervention is to increase the perceived costs of committing an offense while decreasing the perceived benefits, so that the former outweigh the latter. A potential traffic offender thus needs to assess the perceived risk of being caught, the fear of being caught and the fear of a likely penalty and punishment (Zaal 1994). In consequence, the speed camera system in Victoria is intended to influence driver's behavior through two different processes: specific deterrence and general deterrence. Specific deterrence is a process of encouraging an apprehended offender, through his actual experience of detection and its consequences, to avoid re-offending. General deterrence is a process of influencing a potential traffic law offender, through his fear of detection and its consequences, to avoid offending (Cameron and Sanderson 1982).

The Swedish approach, on the other hand, appears to be based on a belief that road safety is of the utmost priority for road users, and that one of the reasons for speeding is due to lack of information. The Swedish approach thus places less trust in the capabilities of individual road users to make decisions and to calculate the benefits and costs of speeding. It also appears to be based on the idea that other road users, and their choice of speed, have a significant influence on the individual's choice of speed (Åberg et al. 1997; Haglund and Åberg 2000) and the decision according to Kahneman (2011) is more automatic and operated through humans System 1. The traffic safety camera system in Sweden is therefore supposed to increase the level of information in order to support drivers in making a safe speed choice and, through a change in speed behavior among a large proportion of the traffic (at least the local traffic near cameras), to create a new social norm with respect to what is an appropriate speed.

Despite different intervention mechanisms, both the Australian and the Swedish approaches appear to deliver substantial road safety benefits (Cameron et al. 1992; Bergdahl 2007; Swedish Road Administration 2015). However, the approach employed in Victoria seems capable of delivering greater road safety benefits at less cost than the Swedish approach. But the Victorian approach does raise deep concerns about the relations between the government and the citizens. According to Delaney et al. (2005), there are certain specific recurring controversies that tend to arise whenever cameras are used. These comprise: fine revenue: opponents claim that the aim of cameras is to raise revenue rather than to increase safety; fairness: some opponents believe that cameras are unfair due to such factors as failure to identify the driver, failure to notify the offender on the spot, lack of witnesses, lack of an opportunity to explain the circumstances, cameras located where it is safe to speed or that speed limits are set too low in these locations, speeding not perceived as a safety problem, reliability of speed cameras and speedometers and privacy: photo enforcement allows police to act as a "Big Brother." All these concerns have an impact on how many cameras are tolerated in society. It is very difficult to develop this approach. Also, politically, one would be continually questioned.

Due to a perception of fairness, the Swedish system has not been questioned and therefore this system can be developed step by step. The number of safety camera stations has increased from 600 to 1,100 and Sweden plans a yearly addition of 200 safety camera stations until 2020. In the end of year 2020, Sweden will have 2,200 safety cameras (Swedish Transport Administration 2015). This is probably one of the largest safety camera programs per capita in the world.

The step from manual enforcement to technology is crucial, but as the previous section has shown, it is important to know if the interventions are based on a traditional approach or Vision Zero. Furthermore Vision Zero influences not only our safety cameras but also the next step in the evolution which is traffic calming technologies and vehicle technologies. Traffic calming technologies such as speed bumps, roundabouts and other solutions are especially important in urban areas. Intelligent speed adaptation systems are also important in vehicle technology.

1.15 Vision Zero—How to Make Things Happen

Over the years, a comprehensive knowledge base has been developed regarding the severity and incidence of traffic injuries. This is also true of important risk factors and effective countermeasures (Elvik et al. 2009; Peden et al. 2004). However, there are still large gaps in the knowledge regarding the road transport system where different components, vehicles, infrastructure and users interact and communicate with each other. Large investments, predominantly industrial, are being made for this type of research worldwide. In addition, non-fatal injuries are becoming increasingly important to highlight.

Deaths and serious injuries in road traffic accidents are ultimately a result of the road transport system's design and function. The road transport system can be defined as a phenomenon consisting of users, vehicles and the traffic environment that interact with each other in order to satisfy society's need for road transport (Larsson et al. 2010). These components and their interactions can be seen as a result of how the various actors individually and jointly act in a road transport system. In turn, these individual actors are influenced by how social actors individually and jointly act and thus affect road transports. An area of knowledge that is greatly understudied is how societies, local communities and markets work from a safety perspective. Such knowledge would contribute to how safety issues can be solved most effectively and appropriately and thus be of practical benefit to both the individual and society. There is therefore a lack of understanding of the dynamic process that aims to formulate and implement road safety policy and understand how road safety measures are effectively disseminated (Racioppi 2004; Wegman and Hagenzieker 2010).

The state, parliament, government and particularly the Swedish Road Administration, played a central role, both directly and indirectly, in the implementation of Vision Zero. With regards to its role as an influencer of different stakeholders, the state must decide whether to intervene or whether to let the market and society take care of itself (Bemelmans-Videc et al. 1998).

Historically, traffic safety work has been focused on management by rules (Graham 1989; Friedland et al. 1990; Belin et al. 2014). In the work to realize Vision Zero, management by rules still exists but occupies less space than in the past. New measures and new approaches have been introduced, in which networking and management by objectives are fundamental approaches. Although we need more systematic knowledge about these implementation and innovation processes, this fact has not hindered the Swedish government to, mainly through a trial and error process, implement several new methods.

The work to make roads safer, to achieve Vision Zero, has been primarily characterized by four forms of governance which have shown to be effective and have driven the work forward:

Management by objectives and results. In order to create long-term sustainability and a systematic way of working, the Swedish road safety efforts are based on management by objectives. The work includes setting objectives for various indicators judged to have the greatest impact on road safety, for example, adherence to the speed limit, safe roads, safe vehicles and the use of helmets. Changes to the various indicators are measured, monitored and discussed at annual results conferences.

Network governance. Collaboration between various parties has been and is a key factor in the work toward achieving Vision Zero. This has led to different parties working together in the same direction through coordination and harmonization. Especially important is the collaboration between the vehicle industry and road authorities.

Benchmarking. The market's demand is the mechanism that governs development and results. This way of working has had good results in terms of safer cars such as the Euro new car assessment program.

Evidence-based work to improve road safety. Implementing new road safety measures requires fundamental information about the problems that need to be solved. There are two sources of statistics and knowledge about road deaths in Sweden: STRADA and the Swedish Transport Administration's in-depth studies of fatal accidents. Swedish Traffic Accident Data Acquisition (STRADA) is an information system containing data about injuries (collected through hospitals) and accidents that have occurred in the road transport system. Since 1997, in-depth studies have been conducted into all fatal accidents on the roads in order to document what has happened before, during and after the accident.

All these methods are examples of processes which are vital to make things happen.

1.16 New Institutional Conditions

In 2006, the Moderate Party formed a majority government together with the Center Party, Liberal People's Party and the Christian Democrats following the election. The Social Democrats were overthrown after twelve years in power. In 2010, this Alliance lost its absolute majority in the parliament but continued to govern as a minority government until the election in 2014.

During 2009 and 2010, the Alliance government made major changes in the government structure within the transport sector. Among other things, the Swedish Road Administration and the Swedish Railway Administration were merged together into the Swedish Transport Administration. A new regulating government body was also set up—the Swedish Transport Agency. However, the government did not decide which authority should be the leading agency and responsible for maintaining and developing Vision Zero. In 2014, the Social Democrats came back to power and together with the Green Party they formed a minority government. In 2016, the Swedish government made a renewed commitment to Vision Zero (www.regeringen.se/informationsmaterial/2016/09/information-material-renewed-commitment-to-vision-zero/) and they also decided that the Swedish Transport Administration is responsible for managing and coordinating the overall road safety work in Sweden.

1.17 Vision Zero and the Road Safety Trend

In the diagram below, you see how the number of fatalities per million inhabitants has developed over the years. In 1999, the Swedish government launched an 11-point program for Vision Zero and based on this several interventions have been implemented: traffic calming in urban area, 2 + 1 roads, new speed limit system and traffic safety camera program among other things (Strandroth 2015). The number of fatalities per million inhabitants has been reduced from 67 fatalities in year 2000 to 28 fatalities in 2010, a more than 50% reduction (Trafikanalys 2015). However, in recent years, there are some signs that Sweden has reached a plateau (Swedish Transport Administration 2015), and therefore, there are great expectations that the Swedish government will start new an initiative.

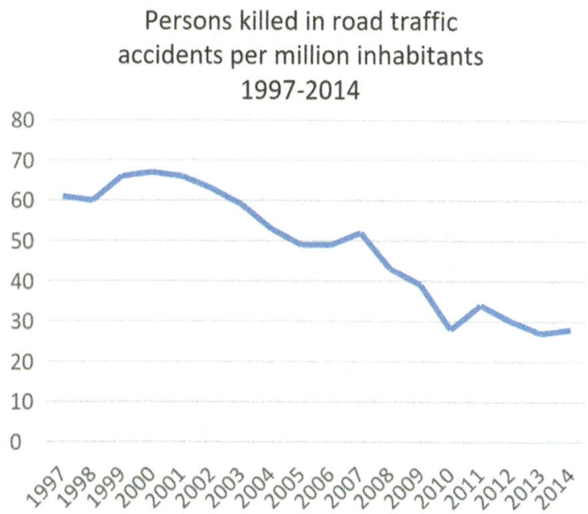

1.18 Conclusions

This chapter gives an overview of how the road safety work in Sweden had developed over the years. Guided by theories about safety culture, Sweden has, during a 50 year period, developed its road safety policies and strategies. The first important milestone was the establishing of a special national agency for road safety in 1968. This institutional change was probably one of the most important factors to enable Sweden to take the step to a more systematic road safety work, with other words to reach the calculative/bureaucratic safety culture phase. A very important step which contributed to a more than 50% reduction of fatalities. This institutional change also helped to increase the demand for knowledge and capacity to work with safety, until road safety became a specific area of expertise. On the other hand, experience from 1980 shows that these kinds of achievements could, in a relatively short time, be reversed. An economic recession in the beginning of 1990 gave Sweden the necessary respite and time to more thoroughly reconsider its long-term road safety strategies. Except for a short time in the middle of the 90 s, when the Swedish government seemed to have abdicated its responsibility for safety, the Swedish Parliament's adoption of Vision Zero 1997 constitutes a new era in the Swedish road safety work. The adoption of Vision Zero is, from a safety culture perspective, a landmark which shows that Sweden has climbed on to the safety culture ladder.

Vision Zero as a public road safety policy differ from a more traditional road safety policy (in its in many is different in several important respects). Specifically, in the formulation of the problem, the perception of responsibility, the demand for safety from road users, as well as its recognition of the main aim of road safety work (Belin et al. 2011). It is therefore no exaggeration to claim that Vision Zero represents a kind of conceptual paradigm shift. This conceptual change has affected analysis, research and development (e.g., in-depth studies, STRADA, analysis models, etc.), in policy and implementation (e.g., performance management, benchmarking, etc.) and in the adoption of specific road safety measures (e.g., 2 + 1 roads, safety cameras, speed systems, vehicle engineering solutions).

Vision Zero is a policy innovation and interventions based on this have contributed to a reduction in the fatalities of more than 50% during a ten-year-period. This is nothing less than an outstanding achievement, especially for a country which was already performing well. This is very promising because it shows that continuous improvements of both strategies and interventions can help us do better.

However, there is still a long way to go before we reach a safe road transport system where we will have eliminated fatalities and serious injuries and then the challenge would be to maintain a safe system. Hopefully Sweden, after a few years of stagnation, will renew the Vision Zero and initiate a new era in traffic safety work.

References

Åberg L, Larsen L, Glad A, Beilinsson L (1997) Observed vehicle speed and drivers' perceived speed of others. Appl Psychol Int Rev 46(3):287–302

Anderson JE (2000) Public policymaking an introduction. Houghton Mifflin Company, Boston, New York

Bax C, Leroy P et al (2014) Road safety knowledge and policy: A historical institutional analysis of the Netherlands. Transp Res F Traffic Psychol Behav 25(Part B):127–136

Belin M (2007) Trends in setting speed limits—Sweden as a case study. Swedish Road Administration, Borlänge, Sweden

Belin M-Å, Tillgren P, Vedung E, Cameron M, Tingvall C (2010) Speed cameras in Sweden and Victoria, Australia—a case study. Accident Anal. Prevention 42(6):2165–2170

Belin M-Å, Tillgren P, Vedung E (2011) Vision Zero—a road safety policy innovation. Int J Injury Control Safety Promotion, 1–9

Belin M-Å, Tillgren P (2012) Vision Zero—how a policy innovation is dashed by interest conflicts, but may prevail in the end. Scand J Public Adm 16(3)

Belin M-Å, Vedung E, Meleckidzedeck K, Tingvall C (2014) Carrots, sticks and sermons: state policy tools for influencing adoption and acceptance of new vehicle safety systems. In: Regan MA, Horberry T, Stevens A (eds) Driver acceptance of new technology: theory, measurement and optimisation. Ashgate, 241ff

Bemelmans-Videc M-L, Rist RC, Vedung E (eds) (1998) Carrots, sticks and sermons policy instruments and their evaluation. Transaction Publishers, New Brunswick, NJ

Bergdahl F (2007) Automatic speed cameras - A study of the effects on vehicle speeds, examensarbete 20 p. Teknisk-naturvetenskaplig fakultet. Uppsala Universitet, Uppasala

Bowen SA (2012) Moral philosophy. Sage, Thousand Oaks, USA

Carlsson A (2009) Uppföljning av mötesfria vägar, slutrapport. VTI rapport 636. VTI, Linköping

Delaney A, Ward H, Cameron MH, Williams AF (2005) Controversies and speed cameras: lessons learnt internationally. J Public Health Policy 26(4):404–415

Elvik R, Vaa T et al (2009) The handbook of road safety measures, 2nd edn. Emerald Group Publishing Limited

Friedland ML, Trebilcock MJ, Roach K (1990) Regulating traffic safety. University of Toronto Press

Global Road Safety Partnership (2008) Speed management: a road safety manual for decision-makers and practitioners. Global Road Safety Partnership, Geneva

Government of Victoria, Department of Justice, Bourne MG, Cooke RC (1993) Victoria's speed camera program. Crime prevention Studies 1

Graham JD (1989) Auto safety: assessing America's performance. Auburn House

Haddon W (1968) Changing approach to epidemiology prevention and ameliroation of trauma—transition to approaches etiologically rather than descriptively based. Am J Public Health Nation's Health 58(8):1431–2000

Haddon W (1970) Escape of tigers—an ecologic note. Am J Public Health Nation's Health 60(12):2229–3000

Haddon W (1972) Logical framework for categorizing highway safety phenomena and activity. J Trauma 12(3):193–200

Haddon W (1973) Energy damage and 10 countermeasure strategies. Hum Factors 15(4):355–366

Haddon W (1980) Advance in the epidemiology of injuries as a basis for public-policy. Public Health Rep 95(5):411–421

Haglund M, Åberg L (2000) Speed choice in relation to speed limit and influences from other drivers. Transport Res F: Traffic Psychol Behav 3(1):39–51

Hill M, Hupe P (2009) Implementing public policy, 2nd edn. Sage, Los Angeles, CA

Hoogerwerf A (1990) Reconstructing policy theory. Eval Program Plann 13:285–291

Hudson P (2007) Implementing a safety culture in a major multi-national. Saf Sci 45:697–722

International Atomic Energy Agency, IAEA (1986) INSAG's summary report on the post-accident review meeting on the Chernobyl, Accident Safety Series No.75-INSAG-1

Johnston IR, Muir C, Howard EW (2014) Eliminating serious injuries and death from road transport—a crisis of complacency. Taylor and Francis Group LLC, Boca Ration

Kahneman D (2011) Thinking, fast and slow. Farrar, Straus and Giroux

Leeuw FL (2003) Reconstructing program theories: methods available and problems to be solved. Am J Eval 24(1):5–20

Mickwitz P (2003) A framework for evaluating environmental policy instruments. Evaluation 9(4):415–436

Nader R (1965) Unsafe at any speed: the designed—in dangers of the American Automobile. Grossman Publisher, New York

Nader R (1967) Den livsfarliga bilen. Swedish translation. Stig Björklund, Unsafe at any speed: the designed- In Dangers of the American Automobile

Nakamura RT, Smallwood F (1980) The politics of policy implementation. St. Martin's Press, New York

Parsons, D. W. (1995). Public policy: an introduction to the theory and practice of policy analysis. Edward Elgar Pub., Aldershot, Brookfield, VT

Patton MQ (2002) Qualitative research and evaluation methods, 3rd edn. Sage, Thousand Oaks, CA

Peden M et al (2004) World report on road traffic injury prevention. World Health Organization, Geneva
Peters BG, Pierre J (2006) Handbook of public policy. Sage
Racioppi F (ed) (2004) Preventing road traffic injury: a public health perspective for Europe. World Health Organization Regional Office for Europe
Rothstein B (1998) Just institutions matter—the moral and political logic of the universal welfare state. Cambridge University Press
Sabatier P, Mazmanian D (1979) The conditions of effective implementation: a guide to accomplishing policy objectives. Policy Anal 5(4):481–504
Strandroth J (2015) Identifying the potential of combined road safety interventions: a method to evaluate future effects of integrated road and vehicle safety technologies. Chalmers University of Technology, Göteborg
Sunstein CR, Thaler RH (2012) Nudge: improving decisions about health, wealth and happiness. Penguin Books Limited
Swedish Government (1940) Ministry of Transport and Communications. Principbetänkandet I Trafiksäkerhetsfrågan SOU 1940:33
Swedish Government (1997) Bill 1996/97:137: Vision Zero and the traffic safety society
Swedish Government (2004) Bill 2003/04:160, Fortsatt arbete för en säker vägtrafik
Swedish Parliament, Committee on Transport and Communications (1997) Committee report 1997/98:TU4. Protocol 1997/98:13
Swedish Transport Administration (2015) Analysis of road safety trends 2014. Management by objectives for road safety work towards the 2020 interim targets. Publication nr. 2015:103. Ineko
Swedish Transport Administration (2015) ATK Annual Report 2015. Publication Number: 2016:097
Tingvall C, Haworth N (1999) Paper presented to the 6th ITE International Conference Road Safety & Traffic Enforcement: Beyond 2000, Melbourne, 6–7 Sept 1999
Trafikanalys (2015) Road traffic injuries 2015
United Nations General Assembly (2004) Improving global road safety A/RES/58/289
United Nations General Assembly (2010) Improving global road safety, A/RES/64/255
Vedung E (1997) Public policy and program evaluation. Transaction Publishers, New Bruswick, NJ
Vägverket (1996) Nollvisionen – En idé om ett vägtransportsystem utan hälsoförluster
Wegman F, Hagenzieker M (2010) Editorial safety science special issue road safety management. Saf Sci 48(9):1081–1084
Westrum R (2004) A typology of organizational cultures. Qual Saf Health Care 13:ii22–ii27
World Commission on Environment and Development (1987) Our common future
World Health Organization, WHO (2015) Global status report on road safety 2015
Zaal D (1994) Traffic law enforcement: a review of the literature. Monash University Accident Research Centre

Chapter 2
Sustainable Safety: The Dutch Example of a Safe System Approach

Fred Wegman

2.1 Introduction

The Global Status Report on Road Safety 2018 (WHO 2018) estimates that worldwide about 1.35 million people in 2016 are killed in traffic. This is based on official figures and WHO estimates, but this figure is most probably an underestimation. A large majority (80%) of road fatalities occur in middle-income countries that have 76% of the world's population and 59% of all registered motorized vehicles. A further 13% of road fatalities occur in low-income countries (with only 1% of the world's motor vehicles), leaving 7% to high-income countries. The overall global road traffic fatality rate (mortality rate) amounts to 18.2 road deaths per 100 000 inhabitants, while the rate for high-income countries is 8.3, and 27.5 for low-income countries. Not only are people killed in road crashes, they are often injured, sometimes seriously. We don't have precise figures, but it is estimated the annual number of non-fatal road injuries as being between 20 and 50 million. These road crashes, fatalities and injuries result in considerable economic costs, estimated to be between 0.5 and 6% of the GDPs of high-income countries (Wijnen and Stipdonk 2016), with values ranges between 1.1 and 2.9% for low- and middle-income countries. A major part of costs is related to injuries (an average share of 50% for both HICs and LMICs).

This issue can be considered from another perspective than that of economic costs: deaths and injuries have an enormous impact on families and communities. Economically disadvantaged families are hardest hit by both the direct medical costs and the indirect costs such as the loss of wages resulting from injuries. Road traffic injuries are estimated to be the eighth leading cause of death globally. They are the leading cause of death for children and young adults aged 5–29 years (WHO 2018).

F. Wegman (✉)
Delft University of Technology, Delft, The Netherlands
e-mail: f.c.m.wegman@tudelft.nl

G. Tiwari and D. Mohan (eds.), *Transport and Safety*, Springer Transactions in Civil and Environmental Engineering,
https://doi.org/10.1007/978-981-16-1115-5_2

This means that road casualties must be considered a very serious public health problem.

More than half of all road traffic deaths occur among pedestrians, cyclists and motorized 2–3 wheelers, the so-called vulnerable road users. Different traffic mixes in different countries result in different distributions of casualties over different transport modes. For example, in many South-East Asian countries, motorized two and three wheelers are very popular which is reflected in a high proportion of fatalities using this transport mode (43%).

Many, if not all, countries in the world wish to reduce the number of people killed and injured on their roads. This is a reasonable conclusion which may be derived from the "Stockholm Declaration" of the Third Global Ministerial Conference on Road Safety: Achieving Global Goals 2030 (www.roadsafetysweden.com). This Declaration calls upon UN Member States to contribute to reducing road traffic deaths by at least 50% from 2020 to 2030. Furthermore, the Declaration invites to set targets to reduce fatalities and serious injuries for all groups of road users, and especially vulnerable road users such as pedestrians, cyclists and motorcyclists and users of public transport. Delegates of 140 countries have adopted the Stockholm Declaration and this document will now go forward to the United Nations to be negotiated and result in a resolution of the UN General Assembly. The European Commission welcomes the Stockholm Declaration. The Declaration reinforces the ambition of the European Union at halving the number of fatalities and serious injuries on European roads by 2030, as a milestone on the way to "zero fatalities and injuries" by 2050. Many countries worldwide share the ambition to improve road safety and work with targets, as is illustrated by, for example, in the Road Safety Annual Report 2019 of the IRTAD-group (OECD/ITF 2019).

The number of road casualties per country in a certain period of time (a couple of years for example) is changing. In a substantial number of countries, an increase has been observed, whereas some other countries show a decrease. The Global Status report (WHO 2018) presented a decline in 48 countries between 2013 and 2016 in middle- and high-income countries, but no reductions were observed in any low-income country. The same report indicated an increase in the number of fatalities in 104 countries.

Sometimes, it is stated that an increase in economic growth, resulting in increased motorization and an increase in kilometres travelled, will inevitably result in an increase in the number of road casualties. This correlation may have been the case for some countries during a certain period of their history. However, in general terms, we have no evidence that such a correlation between economic growth, motorization, exposure and casualties (or mortality rates or fatality rates) exist. This correlation is dependent on several other factors. And for many countries in the world, we have evidence that if, over a period of time, fatality rates go down and if the percentage of increase in exposure (kilometres travelled) is lower than the percentage of decrease of fatality rate, a decrease of the number of fatalities will be the result, given the simple relationship:

$$[\text{Fatalities}] = [\text{exposure}] \times [\text{fatality rates}]$$
$$[\text{Fatalities}] = [\text{kilometres}] \times [\text{fatalities/kilometre}]$$

The basic question is why does the fatality rate decrease over time? The answer to this question is not a simple one. Even if there are strong indications which factors and developments make a contribution to this decline, they do not give a full explanation.

This chapter is an updated version of my TRIPP Lecture from 2013 (Wegman 2013).

2.2 Crashes with Fatalities and Injuries

Sometimes, people believe that road crashes are caused by poor human behaviour and that the causes are dominated by intentional errors only, such as violations and traffic offences. And many studies tell us that in almost all crashes the human being is to blame, and only a minority of crashes can be attributed to roads and vehicles. A very illustrative picture came to me when I studied road safety problems in the State of South Australia (Wegman 2012).

South Australian crash statistics showed that, in one-third of all fatal crashes, drivers had a BAC above the legal limit; did not wear a seat belt, or were reported to have driven at excessive speed. This picture was made even more extreme: quite a proportion of drivers involved in a crash had a previous crash history (in which they had been at fault), had a previous history of driving offences, or even a criminal history. A new English word entered my English vocabulary: hoon driving. This is defined as:

> Hoon driving causes a vehicle to travel at very high speed or in a manner that produces burnouts and doughnuts. Dangerous driving, careless driving, failure to have proper control of the vehicle and causing the vehicle to make excessive noise or smoke are also considered hoon-related offences if they are committed in circumstances involving the improper use of a motor vehicle.

Those who were involved in a crash were criminalized. I learned that the dominant opinion was to blame, if not criminalize, the driver involved in a crash. Two conclusions were presented to me (Wegman 2012): the first was that road safety can be improved by better detection of hoon behaviour and tackling it, and the most powerful tool being that of more severe punishments. The second conclusion: the road safety problem is only related to this selected group of drivers and not to anyone else.

My understanding of road crashes is different from the view presented above. I thought that there could be several explanations for this unexpected view, so my first inclination was to ask researchers from the Centre for Automotive Safety Research CASR to help me understand the nature of road crashes in South Australia and to explain why these causes were different from what I had learned in other parts of the world, such as Europe and North America. Not unexpectedly, the (CASR) researchers confirmed my expectations (Wundersitz and Baldock 2011). They had used a data set compiled from the Coroner's reports of fatal crashes in 2008 containing information from their own in-depth crash investigations of non-fatal crashes. The researchers had categorized road crashes into two groups: the first, in which extreme behaviour could be identified (high level speeding and drink-driving); and the second in which extreme behaviour could not be found. The latter group consisted of ordinary road users, having more or less ordinary behaviour. This group was called "system failure", in which well-intentioned road users have a crash because of some sort of human error. The researchers also created a third category which they called illegal system failure. This included illegal behaviour, such as travelling above the speed limit or an illegal BAC, but not at what is regarded as extreme levels.

The study showed that the often expressed opinion that crashes are only, or mainly, caused by antisocial road users who grossly disregard all rules, is not confirmed by South Australian data (Table 2.1). The comparison with Australian findings led to another important result. If we use the same categorization for injury crashes, a similar, but slightly different picture emerges. The distribution is just under 10% extreme behaviour and almost 90% ordinary behaviour. In other words, injury crashes tell a different story than fatal crashes. This means that we cannot rely on fatal crash data only when formulating road safety strategies. We need to include injury crash analysis as well. As I recommended paying more attention to injury crashes in South Australian policies in the future, this focus may open new avenues for policymaking.

We seldom witness a road crash, and, to understand the causes of a crash, we have to rely on (silent) witnesses, for example, by carrying out a crash reconstruction. Causes of crashes are not easy to find in this manner. In the past, we relied on police information which was written down in crash report forms. These forms focused on violations, in order to assist in police activities, as they are supposed to do. However, based on road safety studies all over the world, more specifically in-depth studies, we have learned that, in the majority of crashes, more than one cause plays a role. These studies always result in the conclusion that the human being failed in one way or another. This leads us to three questions: which factors increase risk, how can we influence these factors in order to reduce risk and how cost-effective are these interventions?

Table 2.1 Proportion of crashes based on extreme behaviour or system failure (Wundersitz and Baldock 2011)

Data source	Extreme behaviour (%)	Illegal system failure (%)	System failure (%)
Fatal crashes 2008	43.4	22.9	33.7
Non-fatal metropolitan injury crashes 2002–2005	3.3	9.9	86.8
Non-fatal rural crashes 1998–2000	9.4	16.6	74.0

Based on this analysis, two main approaches for further improvement can be defined:

- Elimination of extreme behaviour
- Creation of a Safe System in which human errors are considerably reduced, if not eliminated.

The second approach has a positive impact on the first. Assume that in the future we will build cars with a seat belt lock and an intelligent speed adaptor, in which it will be impossible to drive without proper use of a seat belt and without opportunity to drive faster than the prevailing speed limit. We will then not have to rely on the traditional ways of improving driving behaviour, such as police enforcement. However, as we do not have these locks (yet) and we are not sure whether they will ever become a standard feature in the future cars, we cannot eliminate the first approach.

Yet, we can reach one very fundamental conclusion based on the South Australia data, and the picture for other countries will not be entirely different. Fatal crashes seem to be subject to other dominant causes than crashes that result in injury only and this seems to be more so in the case of urban crashes than rural crashes. The large majority of injury crashes is caused by system failures; fatal crashes are caused equally by extreme behaviour and system failures. If we were to use only fatal crashes as a basis, we would not use the complete image of road traffic crashes. In addition, if we realize that in many highly motorized countries the decline in fatalities is much stronger than that in the number of road injuries (IRTAD 2019).

Another important conclusion arises: policy focusing on a reduction of fatal crashes has been more successful than policy aimed at reducing injury crashes. And we may not deem it a matter of course that the number of serious road injuries will decrease to the same extent as the number of fatalities if the present road safety policies were to be continued. Finally, if we realize that from a cost perspective road injuries are responsible for a substantial share of the costs (more than 50% in the Netherlands), it is obvious that road injuries deserve more attention, both in the analyses and in finding solutions. In other words: road safety policy requires a different approach for the reduction of road injuries than the approach that was used to reduce the number of fatalities.

2.3 Road Safety Problems: High Risks and Low-Hanging Fruit

When analysing road safety, two types of problem can be identified (Wegman 2010): *generic* problems and *specific* problems. Specific problems are those safety problems which are concentrated on specific locations, specific road user groups, specific behaviour or specific vehicles.

Generic problems are caused by the fact that road traffic is inherently unsafe: ordinary people are killed in crashes under normal circumstances. This means that everybody can be involved in a crash at any particular time and many of us will be involved in a lifetime, because road traffic has not been designed with safety as an important requirement for the design of the system. To identify generic problems, we first of all need general knowledge about why crashes occur and which factors influence the seriousness of their outcome, and secondly, we need to identify which conditions/circumstances are inherently dangerous and eliminate, or at least "treat" or tackle these conditions/circumstances.

Road safety policy has a long history in identifying risk-increasing factors and a long tradition in bringing down these *specific risks*. An important assumption behind this approach is perhaps the assumption that a concentration of risks increases the chance of cost-effective measures, because only a small sample of a population, locations, etc., need to be treated. Also, in public health, this is a well-known and widely supported approach: cure those who are ill and identify and treat high risk groups or circumstances.

Much of the past road safety policies were based on high risks, high numbers and frequent causes, and on well-identified patterns in crashes. Risks, for example, were determined and divided into age groups, which showed that the young and the elderly have increased risks. The answer that policy has come up with, is the effort to reduce these high risks: "smoothing the peaks in distributions". Analysis of road safety was aimed at the detection of peaks (in distributions), explaining them, and finding measures to overcome them. This approach requires high-quality road safety data, not only data of crashes recorded by the police, but also "exposure" data, data on road user behaviour, such as seat belt use in traffic, and data on safety quality of roads and vehicles.

Looking back—as I did in my Westminster Lecture, organized by PACTS (Wegman 2010), how did we identify high risk locations, crash-prone drivers and near-wrecks? As was mentioned above, we have a long history of identifying and treating locations with a disproportionate number of crashes. Certain individuals were characterized as crash-prone drivers and the policy was aimed at removing them from traffic or, alternatively, teaching them to be better road users. Some vehicles were found to be in such a sorry state that they could not but contribute to crashes happening.

And indeed, crashes are not randomly distributed across road networks. There are locations with a concentration of crashes, and in the past, we tried to improve these locations. This was motivated by the idea that circumstances that were specific for a location were partly responsible for the high number of crashes. Location-specific, infrastructural measures would then be taken to decrease the number of crashes, to put a stop to the concentration of crashes at that location. In their Road Safety Handbook, the Norwegian researchers Elvik et al. (2009) concluded that this can be an effective approach. Their meta-analysis predicts an 18% reduction of crashes with serious casualties on this type of location.

In the Netherlands, we investigated how this approach had developed in recent years. The conclusion was that it was successful policy. In the period 1987–1989,

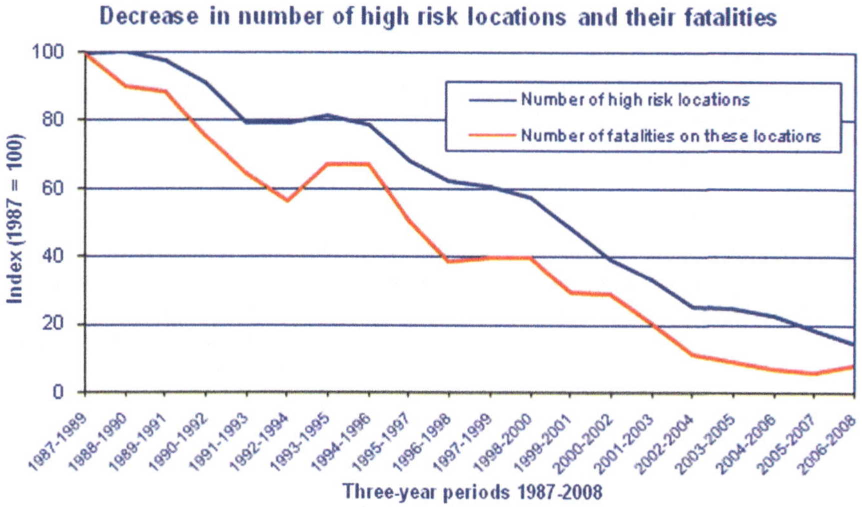

Fig. 2.1 Developments of number and fatalities on high risk locations in the Netherlands (SWOV 2010)

10% of the fatalities were from crashes on locations that can be labelled "high risk locations", their number decreased to 1.8% in the period 2006–2008 (see Fig. 2.1).

Therefore, we can say that the least safe locations have been successfully dealt with. But it is hardly possible for such an approach to have a positive effect in the future years. One could say that the approach has become a victim of its own success and will barely make a further contribution to the reduction in the number of road crash casualties in the Netherlands.

Next, there are the crash-prone drivers. They too give rise to the idea that part of the road safety problem is concentrated in a small part of the population which is responsible for a disproportionally high share of the crashes. Lately, this crash-prone driver theory has ceased to be a starting point in road safety policy. And justly so, in my opinion, the first question that could be asked: who are these crash-prone drivers anyway, and how are they identified? Crash-prone drivers are not those who have already been involved in a crash. Having been involved in a serious crash has proved to be a bad predictor of the involvement in another crash. Must crash-prone drivers then be defined as those who have committed many traffic offences? This may be the case: some traffic offenders do break the rules frequently and they are involved in crashes more frequently than non-offenders, but the predictive value at an individual level is low. Therefore, fighting recidivism makes sense, but expecting to identify offence-prone drivers is an unlikely achievement.

And finally, the near-wrecks: bad vehicles with defects. These defects contribute to the occurrence of crashes or their severity. In the Netherlands, serious vehicle defects, especially defective brakes and bold/poor tires, are seen as either a main or a contributory factor only in a few crashes. It is often said that Periodic Technical Inspection will pick out defective cars, and that the vehicle owner will then have the defects repaired. This sounds logical, but as yet has not been scientifically proved. Vehicle testing can and will trace part of the defects before a crash occurs. But which proportion? If we estimate this to be dozens of percentage points, it will only prevent a few per cent of the crashes. There is nothing against this approach, it may just not be very cost-effective.

These three examples (hazardous locations, dangerous road users and technical inspection of vehicles) are reason for me to conclude that we certainly had "peaks" or "spikes" in the distributions, and that they have (partly) been eliminated. In theory, those peaks could still be eliminated entirely. If we measure the success of any policy or investment by the extent to which it reduced the number of casualties, this approach of identifying specific high risks and tackling them, has been and can still be a rather effective and efficient approach. If it is simple to identify safety problems and if it is simple to cure and to eliminate these problems, we speak of low-hanging fruits, especially if the cure is relatively cheap.

Road users run different risks. Sometimes, the risk is higher, on top of the risk that everybody has: *risk factors or risk-increasing factors* play a role. These (human-related) risk-increasing factors are: lack of experience, psycho-active substances: alcohol and drugs, illnesses and ailments, emotion and aggression, fatigue and distraction (Wegman and Aarts 2006). Certain road and traffic conditions carry higher risks; we certainly have knowledge about risk-increasing road and traffic conditions.

A significant example of a *risk-increasing factor* is drink-driving. This topic has received considerable attention in many countries and also in the Netherlands. Drink-driving is involved in less than 1% of all kilometres travelled; drink-driving is seen as socially unacceptable by the vast majority of the population, which results in only a very small percentage of the drivers being above the legal limit. However, still a substantial proportion of traffic fatalities are in alcohol-involved crashes and we face a small group of hard-core drink-driving offenders with a rather high BAC. Strategies have been successful in reducing drink-driving, but the remaining problem is stubborn. The time has come for a specific policy to identify the "hard-core" offenders and change their behaviour. Strategies need to be different from those in the past that were successful.

Allow me to introduce another risk-increasing factor: distraction. Some years ago, I was alarmed by the results of the 100-car study in the USA. In this study, drivers were followed for a year by observation systems installed in their cars: a black box, small cameras (Dingus et al. 2006). The idea was to observe their everyday behaviour. A wealth of data resulted of which I find the following result the most striking. Nearly 80% of all crashes and 65% of all near-crashes involved driver inattention just prior to the onset of the traffic conflict. This is rarely found on police registration forms, because who would tell the police that a cigarette fell to the floor just prior to the crash and that in a state of some panic he or she was trying to retrieve it? Therefore, I have begun to think that the idea many people have, including road safety professionals and decision makers, of crashes being caused only by traffic offences, as we also rather frequently see on police registration forms must present a distorted picture of the truth.

So, *both (generic) basic factors and risk-increasing factors have been and will be relevant in improving road safety*. Worldwide we have substantially increased our knowledge about the extent to which certain behaviour and specific circumstances affect road safety and we have been successful in reducing specific risk-increasing factors in many countries all over the world. As the nature of the problem will be different, interventions will need to be different in the future than those in the past.

One example to illustrate this: if a jurisdiction reached 95% seat belt wearing rates by legislation, publicity, and by increasing general deterrence by police enforcement, the policies to accomplish a 100% wearing rate will most probably be completely different. Furthermore, I expect that, following the 20–80 rules, traditional policies to tackle "hard-core" problems will inevitably become less effective. How to reach 100% compliance and how to fight "hard-core" offenders are questions that have not yet been satisfactorily answered.

When a country has made considerable progress and has taken the most obvious measures, the traditional approach of reducing the peaks will become increasingly less effective and become less efficient. The traditional policies will lose their effectiveness. The change in thinking that is required as a first step in achieving further road safety improvement could be based on economic considerations in terms of whether profitable investments can be realized.

The next step is not to label individual locations, individual persons or individual vehicles as relatively hazardous. Instead the conditions involving high risks need to be identified*: we move from specific risks to generic or inherent risks.* We move from risk-increasing factors to basic risk factors (Wegman and Aarts 2006). A reasonable case can be made for the relative importance of generic problems having increased over time, while the importance of specific problems has decreased.

Future road safety policies in countries such as the Netherlands will therefore increasingly need to be aimed at the generic character of road safety, and less at the specific aspects. The next phase in policymaking has arrived. This also supports the saying that past success is no guarantee for the future and also gives rise to a question about the efficiency of policy and whether it is not subject to the law of diminishing returns.

However, it must be noted that every country needs to analyse its own road safety problems and its own potential for solutions. For example, it is too simple to expect that Dutch problems and their solutions can be copied on a one-to-one basis to the Indian situation. The so-called SUNflower countries (Sweden, UK and the Netherlands) are three countries with a mature road safety history and performance record and are the three safest countries in the world. When we compared their road safety, the conclusion was that the three countries have similar levels of safety as a result of continued improvements over recent decades. Policy areas targeted have been similar. But implemented policies differ at the level of detail. And, interestingly enough, although they are the safest countries worldwide, each of the three countries identified room for further improvement in well-established safety fields and they saw possibilities to learn from each other (Koornstra et al. 2002). International experts are only able to mirror thoughts and ideas from local road safety professionals as a starting point for policy development. International communication and cooperation is nothing more and nothing less than the exchange of "evidences" assuming that a recipient country has sufficient capacity to transfer that information to local conditions.

2.4 A Safe System Approach

There are two good reasons for the departure from more traditional paradigms on road safety. The first reason lies in the fact that serious road crashes will occur as long as the inherent unsafe conditions, the generic road safety problems are not dealt with. Our traffic system is designed in such a way that it cannot (sufficiently) prevent crashes and serious injuries. The most dangerous traffic issues are the large differences in speed and mass together with the human being who has to deal with these differences. The human being is physically vulnerable and, moreover, makes errors and commits offences. And the system is not designed to deal safely with these errors and offences.

A second good reason lies in the fact that our traditional policies have become both less effective and less efficient; this is also the result of the core characteristics of our road safety problems not yet having been addressed fundamentally.

In the Netherlands, these two reasons triggered a paradigm shift and resulted in *Sustainable Safety* being developed, the Dutch version of a Safe System approach. The increasingly diffuse character of the road safety problem requires a different approach from that of the past. With the Sustainable Safety vision, SWOV seems to have found a suitable answer; the main lines of this vision will be explained from here on. More details can be found in *Advancing Sustainable Safety* (Wegman and Aarts 2006) and in an OECD/ITF report on the Safe System approach (2008).

For the Dutch Sustainable Safety approach Letty Aarts and myself identified the following key aspects:

- An ethical approach
  - The next generation should not have to face a traffic system with the current casualty levels and the number of casualties must be reduced considerably: *towards zero*.
- A proactive approach
  - There is no need to wait for crashes to happen before acting; sufficient knowledge is available that can be applied.
- An integral/holistic approach
  - Man, vehicle and road are to be integrated into one safe system.
  - The whole network, all vehicles, all road users are to be covered.
  - Road safety should be aligned with other policy areas: infrastructure, planning, health, environment, etc.

- Man is the measure of all things

  Human capabilities and limitations are the guiding factors.
  "Don't blame the victim".

- Reduction/elimination of latent errors in the system (system gaps)
  - Which means we are not fully dependent on whether a road user makes a mistake or error in preventing a crash
- Criterion of preventable injuries
  - The criterion of preventable injuries must be used to identify which interventions are most effective and cost-effective.

The Sustainable Safety approach, as developed in the Netherlands (Koornstra et al. 1992; Wegman and Aarts 2006), is the first example of a so-called Safe System approach (OECD 2008). Sweden has its Vision Zero and Australia later developed its own Safe System approach (2011). But all three developed a similar approach trying to fit into the local conditions.

To introduce this approach, it may help to answer the question why still so many crashes occur? An intentional or unintentional human error plays a role in almost every crash. No matter how well educated and motivated, man makes errors and does not always abide by the rules. Many studies of road traffic crashes wrongly indicate that the factors "road" and "vehicle" play only a minor role.

In present day, road traffic has not been designed with safety in mind. For avoiding crashes, we now are almost completely dependent on the extent to which man is capable of (and sometimes willing) correcting his own errors. And, in its turn, errors are also made in doing so.

Both intentional errors and unintentional errors are made. Intentional errors are committed by the "unwilling" man; unintentional errors are committed by the "incapable" man. We do not yet know sufficient about human behaviour to know which of these two error types is dominant in crashes. We assume that both types of error occur frequently and therefore both error types deserve our attention. The often heard opinion that all errors are intentional and that eliminating these errors (often traffic offences) would therefore be sufficient to considerably improve road safety, is not true to reality, as has been clearly illustrated by the work of the South Australian researchers Wundersitz and Baldock (2011).

Unintentional errors, inadvertence, slackening attention, distraction from the driving task are all very frequent and sometimes lead to crashes. Road safety can be improved by limiting the possibility of unintentional errors, and in the Sustainable Safety vision SWOV has detailed how this can be done.

Furthermore, man also makes intentional errors and traffic offences. It has been sufficiently established that traffic offences increase the number of crashes. It is equally well known that a road user who has been fined for an offence adapts his behaviour for some time, usually a few months. Traffic enforcement supports and improves traffic behaviour. But research has also taught us that the positive effect of enforcement is not so much the result of behavioural changes of those who have been fined—that effect is limited—but of the preventive effect of the possibility of being fined and the punishment that comes with it: the general deterrence effect of traffic enforcement (Goldenbeld 2005).

A crash is rarely caused by one single unsafe action; it is usually preceded by a whole chain of poorly attuned occurrences. This means that it is not only one or a series of unsafe road user actions that cause a crash; also hiatuses in the traffic system contribute to the fact that unsafe road user actions can in certain situations result in a crash. These hiatuses are also called latent errors (Reason 1990). In summary: crashes occur when latent errors in the traffic system and unsafe actions during traffic participation coincide in a sequence of time and place (Fig. 2.2).

As unsafe actions can never entirely be prevented, the Sustainable Safety vision aims at banishing the latent errors from traffic: the road traffic system must be *forgiving* with respect to unsafe actions by road users, so that these unsafe actions cannot result in crashes. The sustainable character of measures mainly lies in the fact that actions during traffic participation are made less dependent on momentary and individual choices. Such choices may be less than optimal and can therefore be risk-increasing.

Adjusting the environment to the abilities and limitations of the human being is derived from cognitive ergonomics, which in the early 1980s made its entry coming from aviation and the processing industry. In all types of transport other than road traffic, this approach has already resulted in a widespread safety culture. Further

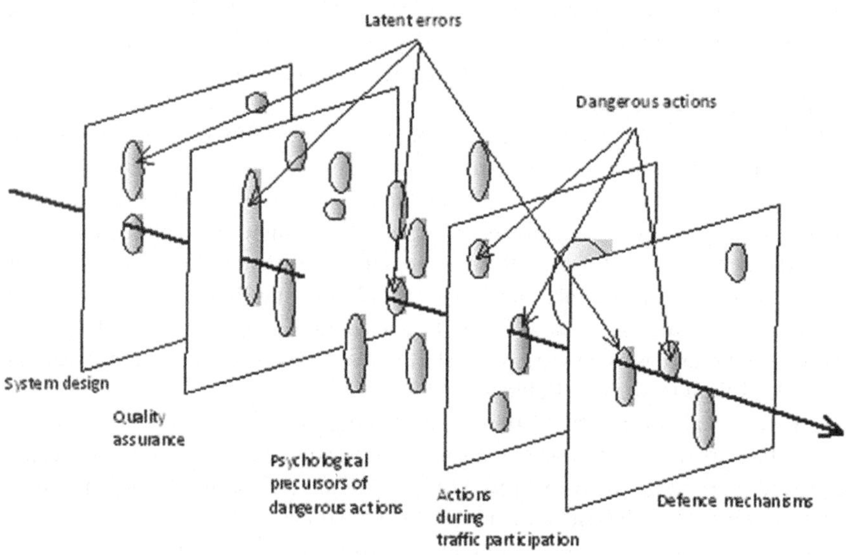

Fig. 2.2 Diagram showing the development of a crash (bold arrow) as a result of latent errors and unsafe actions in the different elements composing road traffic (based on Reason 1990). If the arrow encounters "resistance" at any moment, no crash will develop

incorporation of the Sustainable Safety vision should eventually lead to road traffic that can be considered as "inherently safe" as the result of such an approach.

2.5 Sustainable Safety in the Netherlands: A Safe System Example

The Dutch approach has been presented in many documents and at many events, for example, in the Twentieth Westminster Lecture on Transport Safety (Wegman 2010). The objective of Sustainable Safety is to prevent road crashes from happening, and where this is not possible, to reduce the severity of injuries as much as possible. This can be achieved by a proactive approach in which human characteristics are used as the starting point: *a user-centric system approach.* These characteristics refer on the one hand to human physical vulnerability, and on the other hand to human (cognitive) capacities and limitations. People regularly make errors unintentionally and are not always able to perform their tasks as they should. Furthermore, people are not always willing to comply with rules and violate them intentionally. By tailoring the environment (e.g. the road or the vehicle) to human characteristics, and by preparing the road user for traffic tasks (by training and education), we can achieve an inherently safe road traffic system.

The most important features of inherently or Sustainably Safe traffic are that latent errors in the traffic system (gaps in the system, as indicated by Reason (1990), that result in human errors or traffic violations causing crashes) are, as far as possible, prevented and that road safety depends as little as possible on individual road user decisions. The responsibility for safe road use should not be placed solely on the shoulders of road users but also on those who are responsible for the design and operation of the various elements of the traffic system (such as infrastructure, vehicles and education). Given the fact that people make errors, not always abide by the rules, and are also vulnerable, it is of the utmost importance that latent errors in the traffic system are prevented so that they cannot be a breeding ground for the occurrence of crashes. To prevent serious unintended errors, the Sustainable Safety vision as much as possible adjusts the traffic environment and the traffic demands that go with it to what the majority of road users can deal with. This to a large extent evokes the required behaviour, tells the road user what is to be expected and ensures that possible errors are counterbalanced by a forgiving environment. In addition, the breeding ground for intentional and unintentional offences becomes fertile. For as far as offending behaviour can already be detected before traffic participation (e.g. use of alcohol or driving without a licence), the prevention of taking part in traffic fits into a sustainably safe traffic.

Road users must be well-informed and well-trained to participate in traffic. And in situations where their skills do not meet the environment's demands, specific measures must be taken to induce safe behaviour. Road users, being capable of correctly assessing their situation-dependent condition and the resulting driving

Table 2.2 The five sustainable safety principles (Wegman and Aarts 2006; Wegman et al. 2008)

Sustainable safety principle	Description
Functionality of roads	Mono-functionality of roads as either through roads, distributor roads or access roads in a hierarchical road network
Homogeneity of mass and/or speed and direction	Equality in speed, direction and mass at moderate and high speeds
Forgivingness of the environment and of road users	Injury limitation through a forgiving road environment and anticipation of road user behaviour
Predictability of road course and road user behaviour by a recognizable road design	Road environment and road user behaviour that support road user expectations through consistency and continuity in road design
State awareness by the road user	Ability to assess one's task capability to handle the driving task

skills, is essential for making adequate decisions that can prevent a possible crash. However, different road users have different skills, and the more experienced road users should be asked to consciously use safe traffic behaviour which is aimed at facilitating the less skilled road users. Thus, by making traffic a social system, a forgiving driving style can prevent the occurrence of a crash caused by other road users' errors.

In traffic, the vulnerable human being needs to be protected by his environment, either by physical constructions that prevent conflicts, or by constructions that absorb the kinetic energy which is released in a crash. For this purpose, the road users' vehicles sharing the same space should have similar masses. If this cannot be achieved, speeds must be reduced. On the one hand, this system is embedded in a traffic-related taxonomy of high-speed traffic streams, and destination and residence on the other. Between these two extremes, traffic must be guided along correct, sustainably safe roads.

This brings us to the *five central principles* of a Sustainably Safe road traffic, as identified in 2006 (Wegman and Aarts 2006): functionality, homogeneity, recognizability, forgivingness and state of awareness. A brief explanation of these principles can be found in Table 2.2.

These principles have all been based on scientific theories and research methods arising from disciplines such as traffic planning and engineering, psychology and biomechanics.

2.5.1 Traffic Planning: Functionality of Roads

Traffic flow organization manifests itself in many ways and with various and different objectives. As long ago as the 1970s, a functional road categorization system had been introduced which formed the basis for the Sustainable Safety *functionality principle*. This principle starts from the premise that roads can only have a single function (*mono-functionality*) and that they must be used in keeping with that function. The road function can on the one hand be "to facilitate traffic flow" (associated with "through roads"), and on the other hand "to provide access to destinations" (associated with "access roads"). In order to provide a proper transition between "giving access" and "facilitating traffic flow", a third category of function was defined: the "distributor road".

2.5.2 Homogeneity: Dealing with Physical Vulnerability

If road users perform dangerous actions that lead to crashes, the human body's integrity is jeopardized. This vulnerability results from the release of kinetic energy and the body's biomechanical properties.

To deal with the issue of vulnerability in a proactive fashion, Sustainable Safety requires that controls are placed on factors that may intensify the severity of a crash: differences in speed, direction and mass. This forms the foundation of the *homogeneity principle*. This principle states that where vehicles or road users with great differences in mass have to use the same road space, speeds will have to be so low that, should a crash take place, the most vulnerable road users involved should not sustain fatal injuries. In addition, where traffic is moving at high speeds, road users should be separated spatially. This principle resulted, for example, in the recommendation (#8) of the Academic Expert Group for the Third Global Ministerial Conference on Road Safety (Swedish Transport Administration 2019) that a maximum road traffic speed limit of 30 km/ be mandated in urban areas unless strong evidence exists that higher speeds are safe.

2.5.3 Forgivingness: Physical and Social

The principle of *forgivingness* (a forgiving roadside) can contribute to reducing injury severity in crashes. Forgiving surroundings ensure that the consequences of errors remain limited, for example, when vehicles leave the road and enter safe road sides: for example, safe (matted) shoulders, obstacle-free zones or collision-friendly obstacle protection.

Apart from the physical aspects of forgivingness, this principle also has social implications which are more relevant for preventing dangerous actions. Traffic is a social system in which crash causes can partially be traced to the interaction between road users. Social forgivingness has more recently been defined as "the willingness to anticipate a potentially unsafe action of another road user, and to act in such a manner that negative consequences of this potentially unsafe action are prevented or in any case limited" (Houtenbos 2008).

Forgiving road behaviour, particularly, of the more competent road users, could allow for the less competent road users, such as children and elderly road users. In other words, road users can, by means of forgiving driving behaviour (in terms of being anticipative or defensive), increase the room for manoeuvre of less competent road users. Errors should still be regarded as errors by the less competent in order that they can learn but a forgiving approach should lead to fewer or less serious crashes.

2.5.4 Predictability

People can perform tasks at different levels of control: skill-based, rule-based or knowledge-based (Rasmussen 1983). Generally speaking, the longer people are trained in performing a task, the more automatic their behaviour. The benefit is that task execution takes less time and attention, and that fewer (serious) errors are made (Reason 1990). To prevent dangerous actions, Sustainable Safety strives to avoid knowledge-based task performance in particular. People have to be sufficiently capable and experienced to take part in traffic but they also need to perceive what is expected from them and what they can expect from other road users. This is manifest in the *predictability principle*, the benefits of which can be delivered, according to the Sustainable Safety vision, by *consistency and continuity* in road design. This means that the design needs to support the user's expectations of the road, and that all components of the design need to be in line with these expectations.

People not only act dangerously because they make errors unintentionally, they can also exhibit dangerous behaviour by intentionally violating traffic rules. In situations where the road environment does not stimulate proper behaviour, a sustainably safe road traffic system benefits from road users who spontaneously obey traffic rules from a normative point of view. To achieve this, traffic regulations have to fit with the environment, and people have to be educated about the logic and usefulness of rules. Where people still fail to comply with the rules, policing to a level where a reasonable chance of being caught is perceived is the usual measure to enforce compliance.

Another element in the updated vision is that traffic has to be sustainably safe for *everybody*, and not just for "the average road user". Fuller's task-capability interface model (Fuller 2005) supplies a theoretical framework here. Fuller's model states that road users' task capability is the sum of their capacities less the sum of their impairments caused by their present condition (e.g. because of fatigue or use of alcohol). For safe road use, the task capability has to be large enough to meet the task requirements. These task requirements are primarily dictated by the environment, but they can also be altered by the road user, for instance, by increasing or decreasing driving speed.

2.5.5 State Awareness

This principle requires that road users should be able to assess their own task capability for participating in traffic. Task capability can be insufficient due to a lack of competence (e.g. because of a lack of driving experience), or because of—or aggravated by—a state of mind that temporarily reduces the task capability (e.g. because of fatigue, or the use of alcohol or drugs).

Since task capability differs from individual to individual (e.g. inexperienced and elderly road users with underdeveloped or diminished competencies, respectively, and also fatigued "average" road users or road users under the influence of alcohol or drugs), generic road safety measures are a necessity for safe traffic. However, for the group of road users with a lower task capability, in particular, these measures are not sufficient for safe participation in traffic. Therefore, *generic measures* have to be supplemented with *specific measures* aimed at these groups or situations involving them. Specific measures can be found in areas such as regulation, education, enforcement (e.g. banning drivers under the influence of alcohol or drugs) and intelligent transport systems (ITS).

2.6 Sustainable Safety: Implementation and Effects

The Sustainable Safety approach uses as a starting point the idea that the present traffic system is inherently hazardous (that serious crashes can happen anywhere and at any time) and that all possible solutions are considered in an integral and rational manner. There is no a-priori preference for improving roads, vehicles or changing behaviour. The rationality should not be restricted to road safety only, but wider deliberations are desirable (congestion, environment, scenery, economic development, health care, etc.). The purpose of Sustainable Safety is to offer the road user such an environment that the risk of errors and violations is limited considerably. This implies a fundamentally different approach than that used in the past: proactive adaptation of the environment of the road user considered as a system, following the five principles of Sustainable Safety to meet the characteristics defined by human physical vulnerability and of human (cognitive) capacities and limitations.

The new approach required a system of priority setting. The concept of "avoidable crashes" was developed for rational decision making (Wegman 2000). Avoidable crashes are those crashes where we know what caused them, of which we know how to prevent them, where the prevention costs are socially cost-effective and which fit into the Sustainable Safety vision. This approach implies that limited (financial) resources ensure that only cost-beneficial measures are taken and that cost-effectiveness defines the priority setting. This is in line with the Dutch Government's decision making about investments. This concept differs, for example, from the Vision Zero approach, which judges cost/benefit considerations as not relevant.

An exception to this "avoidable crashes approach" is conceivable if in the framework of management and maintenance the existing road infrastructure is improved and road safety can be included. In those circumstances, cost-effectiveness is not a major consideration from a road safety perspective, but a quality assurance approach can be used to give guarantees for a sufficient road safety level.

The essence of Sustainable Safety is that it is a system approach, which means that measures should not only be taken on those locations where crashes have occurred, but that it assumes that crashes can and will happen anywhere. Hence, the inherently unsafe character of the system must be adapted (to "the human measure"). Putting this idea into practice requires political courage, especially from road authorities and from those who work on improving vehicle safety. In the areas of legislation, traffic education and traffic enforcement, no serious problems concerning realization are to be expected; in the Netherlands, it is rather a matter of gaining sufficient support within society, followed by gaining political support, and then choosing the implementation methods that will make a real contribution to the reduction in the number of casualties. Of course, road users and consumers will need to aspire to a higher safety level. Mobilizing the (latent) demand for a higher road safety level and making it manifest is an important task for social organizations.

As a matter of fact, there are no real obstructions that stand in the way of implementing Sustainable Safety; it was and is a matter of political leadership to realize it. It was a key adventure to get all key stakeholders behind the implementation and to organize implementation not for one year only, but for a considerable number of years. Both support from stakeholders and funding has been obtained.

Another very important aspect was to reach an agreement with all Dutch road authorities and to invite them to come to mutual agreements on implementation. As implementing the Sustainable Safety vision requires their support, this turned out to be a crucial step: in the Netherlands road authorities do, to a large extent, have an autonomous position and a considerable amount of freedom for decision making. As was mentioned earlier, the ambition is to achieve uniform solutions resulting in a predictable road course and layout due to consistency and continuity in road design, a predictable road course results in predictable road user behaviour and lower risks. Local and regional road authorities were found to have invested substantial amounts in the improvement of infrastructural safety.

Are the other components of Sustainable Safety affordable? The costs of traffic education and police enforcement are modest in comparison with the investments required for safety improvements of roads and vehicle, and therefore, their financing does not need to be a major problem. In addition, in the Netherlands, the costs of enforcement are paid back by the proceeds from traffic fines. Safer vehicles are purchased by the consumer, and therefore, the market controls itself. For the remaining components, however, it is unclear why parties other than the government should make the required investments. The "bulk" of the financing of the prevention of road traffic crashes will need to be made by the government.

To recapitulate this implementation aspect of Sustainable Safety, we may conclude that the introduction of Sustainable Safety has concentrated on the financing of a

Sustainably Safe infrastructure. This is a problem which is felt by all road authorities, especially by the local and regional road authorities. It involves a political choice. There are clear indications that in the Netherlands financing was in the first instance not made available from specific safety budgets, but that other budgets for management and maintenance were used. Additional safety assets, however, can be of assistance. Yet, it is more important to create conditions which allow safe use of these budgets for management and maintenance.

An assessment was carried out to learn if the implementation of Sustainable Safety met the high expectations. In an evaluation study, SWOV investigated how Sustainable Safety had been implemented in recent years, and what results had been achieved for road safety (Weijermars and van Schagen 2009; Weijermars and Wegman 2011). One preliminary observation is in place. Of course, we don't know exactly what would have happened if Sustainable Safety had not been implemented. But by making some assumptions, we can reach some conclusions. To do this, we have estimated the safety effects of a large number of measures that had been implemented in the period 1998–2007. The policy efforts that were made in the areas of infrastructural investments and traffic enforcement can be called considerable. SWOV has been able to make it plausible that these efforts have resulted in fewer traffic deaths and also, to a lesser extent, in fewer serious road injuries. The fatality rate in the period that was studied, the period 1998–2007, showed an annual decrease of 5%, which is considerably higher than the almost 2% per year in the period 1988–1997. Two scenarios were calculated to determine to what extent the measures that were implemented made a contribution to the decrease in the number of traffic deaths. These two scenarios make different assumptions about what would have happened during the period 1998–2007 if Sustainable Safety had not been implemented, a situation, of course, which we cannot know. In the pessimistic—perhaps, more realistic—scenario 300 road fatalities are saved on a yearly basis, in the more optimistic one the saving amounts to 400 fatalities. These are reduction percentages of more than 30% and 40%, respectively. Comparison of the costs of the investments versus the benefits of the fatalities, injured and crashes saved shows that the investments have been socially cost-effective. The cost–benefit ratio is 4:1.

It is also interesting to see that the reduction in the number of fatalities has been higher than the reduction that was expected at the beginning of the implementation of this policy. An important reason seems to be that the investments that were made by local and regional road authorities like municipalities and provinces were higher than estimated in advance. In addition, the increase in the risk of being caught by intensified police enforcement has also surpassed our expectations.

Therefore, the conclusion was that the high expectations had been met. And, furthermore, it was recommended to continue the implementation and to explore opportunities for higher quality implementation.

SWOV published the 3rd edition of Sustainable Safety in 2018 (www.sustainablesafety.nl). The first edition has been published in 1992 (Koornstra et al. 1992) and the second one in 2006 (Wegman and Aarts 2006). The ultimate ambition of the 3rd edition is being a casualty-free traffic system.

2.7 Lessons Learned

There is not one country in the world for which road safety improvement is an easy task. Lack of high-quality data or competent staff sometimes makes a thorough analysis of a country's own situation impossible. Road safety is a complicated issue characterized by very different opinions on the causes of crashes or the best measures to prevent them. These measures are often the subject of political discussion and difference of opinion. Often these debates revolve around how these measures are to be financed. Add to this the fact that road safety improvement is not a "real top priority" in policy anywhere, and the conclusion appears to be that reducing the number of road casualties is not that simple. This is even more so in countries with a rapid economic growth that comes with rapidly growing (motorized) mobility. This issue can become even more demanding if motorized mobility not at all or only to a limited extent takes account of vulnerable road users like pedestrians, cyclists and motorized two wheelers. This illustrates that road safety improvement is not by any means an easy task.

However, when designing road safety policies and when coming to questions which measures to include in these policies, it is interesting to look at the experiences with road safety improvement in a country like the Netherlands. Not, as was mentioned earlier, with the intention of copying the approach, but as a source of inspiration. The experiences with road safety improvement in the Netherlands can roughly be summarized as follows:

- Although road safety has never been a top policy priority in the Netherlands, The Dutch Parliament has always shown much interest in the subject. Members of Parliament can place topics on the agenda and make tools available for the implementation of policy. Sometimes, a "champion" can be identified. A real champion can truly make a difference and speed up road safety improvement.
- Road safety improvement is a "shared" responsibility of citizens/road users and the designers and operators of the road traffic system. Therefore, it is necessary that all stakeholders feel a responsibility for the road safety problem and want to contribute to its improvement. In order to commit stakeholders to road safety improvement, it is important that politics at the national, regional and local levels clearly indicate that they consider road safety improvement to be an important subject. In addition, it is necessary that top management makes clear to each stakeholder that it is committed to make a contribution. As many different organizations (e.g. road authorities, police, legal authorities, legislators, vehicle industry, schools) are required to contribute, it is advisable to make it attractive for organizations to participate, to cooperate, and sometimes to also be willing to be coordinated by a "lead agency" (without their responsibilities being taken away). It is important that the stakeholders' achievements are being monitored to make sure they "deliver".
- Research indicates that setting realistic and ambitious quantitative targets leads to better policy and more rapid road safety improvement (Allsop et al. 2011). Of course, such a quantitative target can only function if a high-quality data system is

available. Furthermore, it is important that the process and therefore its progress is closely monitored. Mechanisms are required (e.g. an annual report on road safety) to establish whether one is on the right track.

- Using a vision like Sustainable Safety has a number of important advantages. Developing such a vision, which addresses specific local problems and which is based on a good insight into what is effective and efficient in a country/district/town should therefore be considered. The Safe System approach, of which the Dutch Sustainable Safety vision is a good example, can be used to this effect. In the Netherlands, the vision has mainly been used by the professional world of researchers, policy makers, practitioners, elected politicians, etc. and not so much in the communication with road users. The advantage of the vision was that all stakeholders were invited to commit themselves to the vision and to use the vision as a guideline in their work. In the Netherlands, this resulted in a synergy, and therefore an increase in the effectiveness and efficiency of interventions. Another advantage was that it created a long-term perspective and that longer term financial sources for road safety policy were made available.
- It is important to put effort into gaining more political interest in road safety policy and to create support for measures among the population and road users. The media has an important role here and it is advisable to consider the media as an important partner in road safety policy.
- Evidence-based interventions should be at the heart of each road safety programme. It is therefore to be recommended that a trusted knowledge organization establishes, ex-ante and ex-post, whether interventions are evidence-based. In the Netherlands, the independent research institute SWOV has such a function. Furthermore, knowledge should form the basis for all advocacy work within a country. It is important to be aware of the fact that road safety improvement is not only a matter of reducing the number of road fatalities, but also of reducing the number of (serious) road injuries. Policy to reduce the number of road injuries will (need to) be different from the aimed at reducing road fatalities.

This brings us to the question whether a Safe System approach is a bridge too far for India or could it be a source of inspiration. I believe that a Safe System could also be of importance for India. Not, of course, in the sense that the measures, as they have been implemented in the Netherlands or in Sweden, could be copied on a one-to-one basis for the Indian situation; the countries differ too much. The vision could, however, be used as a source of inspiration and as a background for the analysis of risks in road traffic in India. My view is that the Sustainable Safety principles and the implementation of these principles in interventions can be applied in India if well adapted to the Indian (road traffic) conditions and society.

References

Allsop RE, Sze NN, Wong SC (2011) An update on the association between setting quantified road safety targets and road fatality reduction. Accid Anal Prev 43(3):1279–1283

Dingus TA, Klauer SG, Neale VL, Petersen A, Lee SE, Sudweeks J, Perez MA, Hankey J, Ramsey D, Gupta S, Bucher C, Doerzaph ZR, Jermeland J, Knipling RR (2006) The 100-Car Naturalistic Driving Study, Phase II—Results of the 100-Car Field Experiment DOT HS 810 593

Elvik R, Vaa T, Hoye A, Erke A, Sorenson M (2009) The handbook of road safety measures, 2nd revised. Emerald Group Publishing, Bingley

Fuller R (2005) Towards a general theory of driver behaviour. Accid Anal Prev 37(3):461–472

Goldenbeld Ch (2005) Verkeershandhaving in Nederland; Inventarisatie van kennis en kennisbehoeften [Traffic enforcement in the Netherlands. An inventory of knowledge and knowledge needs]. (in Dutch). R-2005-15. SWOV, Leidschendam

Houtenbos M (2008) Expecting the unexpected; A study of interactive driving behaviour at intersections. Ph.D. thesis at Delft University of Technology. TRAIL thesis series, Delft

Koornstra MJ, Mathijssen MPM, Mulder JAG, Roszbach R, Wegman FCM (eds) (1992) Naar een duurzaam veilig wegverkeer; Nationale Verkeersveiligheidsverkenning voor de jaren 1990/2010 [Towards sustainably safe road traffic; National road safety outlook for 1990/2010] (in Dutch). SWOV, Leidschendam

Koornstra MJ, Lynam D, Nilsson G, Noordzij PC, Petterson H-E, Wegman FCM, Wouters PJ (2002) SUNflower: a comparative study of the development of road safety in Sweden, the United Kingdom, and the Netherlands. SWOV, Leidschendam

OECD/ITF (2008) Towards zero: ambitious road safety targets and the safe system approach. OECD/ITF, Paris

OECD/ITF (2019) Road safety annual report 2019. OECD/ITF, Paris

Rasmussen J (1983) Skills, rules and knowledge; signals, signs and symbols and other distinctions in human performance models. In: IEEE Transactions on Systems, Man and Cybernatics, SMC-13, pp 257–266

Reason J (1990) Human error. Cambridge University Press

Swedish Transport Administration (2019) Saving lives beyond 2020: the next steps. Recommendations of the Academic Expert Group for the 3rd Global Ministerial Conference on Road Safety. STA, Borlänge

SWOV (2010) The high risk location approach. SWOV Fact sheet, January 2010. www.swov.nl/publicatie/high-risk-location-approach

Wegman F (2000) Hoe ervoor te zorgen dat vermijdbare ongevallen niet meer gebeuren [Road Safety: how to ensure that avoidable accidents no longer occur] (in Dutch) D-2000-10. SWOV, Leidschendam

Wegman FCM, Aarts LT (eds) (2006) Advancing Sustainable safety: national road safety outlook for 2005–2020. SWOV, Leidschendam

Wegman F, Aarts L, Bax Ch (2008) Advancing Sustainable Safety: national road safety outlook for the Netherlands for 2005–2020. Saf Sci 46(2):323–343

Wegman F (2010) Putting people at the centre: how to improve road safety in the 21st century? The 20th Westminster lecture on Transport Safety. The Parliamentary Advisory Council for Transport Safety, London

Wegman F (2012) Driving down the road toll by building a Safe System. Adelaide Thinker in Residence 2011–2012. Government of South Australia, Adelaide

Wegman F (2013) Road safety in India: a systems approach. 5th TRIPP Annual Lecture, TRIPP RP13-01. Transportation Research & Injury Prevention Programme, Indian Institute of Technology, New Dehli

Weijermars WAM, van Schagen INLG (eds) (2009) Ten years of Sustainable Safety; Road safety assessment 1998–2007. R-2009–14. SWOV, Leidschendam

Weijermars WAM, Wegman FCM (2011) Ten years of sustainable safety in the Netherlands; an assessment. Transp Res Rec 2213:1–8

World Health Organization (2018) Global status report on road safety. Supporting a decade of action. WHO, Geneva

Wundersitz LN, Baldock MRJ (2011) The relative contribution of system failures and extreme behaviour in South Australian crashes (CASR092). Centre for Automotive Research, Adelaide

Wijnen W, Stipdonk H (2016) Social costs of road crashes: an international analysis. Accid Anal Prev 94:97–106

Chapter 3
Traffic Safety: The Top Ten Issues

Tony Bliss

3.1 Introduction

Traffic safety is a vital dimension of sustainable transport explored through multiple disciplinary and interdisciplinary lenses. The growing accumulation of scientific evidence concerning the effectiveness and efficiency of interventions provides the bedrock for ongoing global improvements. Traffic safety management practices in high-income countries have successfully leveraged this scientific evidence base and in turn provide a sharper focus on issues concerning the delivery, or implementation, of improved traffic safety.

This focus on the implementation of traffic safety improvements is relevant to India and other low and middle-income countries experiencing growth in avoidable road crash deaths and injuries. While there have been sustained global efforts to promote good practice safety interventions, less attention has been paid to strengthening the management capacity necessary to deliver them. Traffic safety research has a role to play in addressing this, and for the next generation of researchers, ten priority issues have been identified for further consideration. For each issue, questions are posed, without answers, to extend this exploratory process.

T. Bliss (✉)
Faculty of Architecture, Building and Planning, Transport, Health and Urban Design Research Lab, Melbourne University, Melbourne, Australia
e-mail: tonybliss@roadsafetymanagement.com

Road Safety Management Limited, Wellington, New Zealand

G. Tiwari and D. Mohan (eds.), *Transport and Safety*, Springer Transactions in Civil and Environmental Engineering,
https://doi.org/10.1007/978-981-16-1115-5_3

3.2 Issue 1: Production

The first issue concerns how little attention we pay to the way traffic safety is actually produced. It is often sold as something you just take off the shelf and use like a magical potion, with an informed mixture of incantations and exhortations assuring success. Do the right thing in the right place at the right time. An abundance of this form of advice is offered to and by the global road safety community, which leaves the impression that quick success can be simply conjured up.

The reality is more prosaic and much more demanding. Traffic safety is produced, just like other goods and services, and its quality in terms of final results achieved is determined by the quality of the production process. Hence, in traffic safety management terms, it is important to understand the nature of this production process and its multi-faceted, systemic elements.

We can visualize the traffic safety management system as having three key elements, stacked up in pyramid form, comprising institutional management functions that produce interventions that in turn produce results (Bliss and Breen 2009).

At the bottom of the pyramid, we can identify seven institutional management functions, with the most important of these being what we term "results focus." This is a short-hand way of expressing what it is that we want to produce and our strategy for doing so in terms of interventions and targeted results. Six related management functions concern how we coordinate partnership and stakeholder activities to achieve the desired results; how we legislate to achieve the desired results; how we fund and allocate resources to achieve the desired results; how we promote activities to achieve the desired results; how we monitor and evaluate activities to achieve the desired results; and how we undertake research and development and knowledge transfer to achieve the desired results. It is this institutional "engine room" of the production process that so often gets neglected in discussions concerning how to improve national traffic safety performance.

In the middle of the pyramid, we have three categories of targeted interventions: interventions that concern the planning, design, operation, use and maintenance of the road network; interventions that concern the entry and exit of drivers, vehicles and operators to and from the road network; and interventions that concern the recovery of crash victims from the road network and their rehabilitation. These three intervention categories can be further disaggregated into two types: those that set and implement safety standards and rules, and those concerning related compliance regimes which comprise combinations of education, enforcement and incentives.

Finally, at the top of the pyramid, we have the results we wish to achieve which can be measured in terms of outputs, intermediate outcomes, final outcomes and social cost. In low and middle-income countries where final outcomes data quality is poor, sampled intermediate outcome measures, such as network speeds, helmet and safety belt wearing rates, safety rating of core network roads and vehicle safety ratings, can provide reliable indicators of overall safety performance and related trends.

Even with this simplified safety management system framework, we can see the complexity of the traffic safety production process which must operationalize the three identified elements, each of which necessarily breaks down into greater levels of detail. We must also take into account the dynamic nature of this production process and the evolution of its results focus over time.

Through to the 1960s, the results focus of the production process was preoccupied with improving road user attitudes and behavior, what we now term a "victim-blaming" approach (Rumar 1999). During the 1970s and 80s, this shifted to a "matrix" approach, with a more systemic results focus on pre-crash, in-crash and post-crash events engaging road users, vehicles and the road environment (Haddon 1968). By the 1990s, what subsequently become termed the "Safe System" approach was beginning to surface with the Dutch *Sustainable Safety* and Swedish *Vision Zero* strategies (Wegman and Elsenaar 1997; Tingvall 1995).

The Safe System approach aims to manage the impacts of injurious crash forces at the interfaces between road users, vehicles and the road environment, with a results focus seeking to eliminate crash fatalities and serious injuries. This zero-harm goal has become the norm in high-income country traffic safety strategies and is also being promoted to low and middle-income countries.

While high-income countries are struggling to address this level of ambition—with Sweden and The Netherlands leading the way—a new results focus is emerging. In our work at Melbourne University, we have termed this the "Complex System" approach, which encapsulates the Safe System approach within a more holistic set of measures seeking to address new forms of sustainable urban mobility arising from the convergence of information, communications and transport technologies and the exigencies of global climate change policies targeting large-scale atmospheric carbon reductions (Bliss 2015).

The traffic safety production process must be understood and acted upon within a wider set of transport goals seeking improved productivity, competiveness and prosperity, integrated with related goals concerning the environment, energy security, urbanization and public health. We are moving into an era of growing complexity and longer planning and programming horizons that hinge on more highly valuing costs and benefits accruing to future generations. Yet at the same time, we are experiencing powerful resistance to the state playing a strong leadership role in managing fundamental shifts in policy and practice required for sustainable success.

This issue of how road safety is produced raises questions that merit further consideration:

- *Why do we insist on traffic safety as being something simple to deliver?*
- *Why do we focus on interventions alone?*
- *Why do we cling to past traffic safety paradigms?*
- *Why do we view traffic safety as an imposition of the "nanny state"?*

3.3 Issue 2: Sequencing

Like all effective and efficient production processes, the sustained delivery of traffic safety requires proper sequencing. Foundations must be built before later stages of production can be operationalized (e.g., driver and vehicle entry and exit controls and registries underpin effective compliance regimes; high-capacity judicial systems and publicly acceptable administrative penalty procedures underpin feasible speed camera operations; and so on).

It is like building a house where key tasks have a logical ordering. There is little value in specifying the fitting of the roof, without considering the nature of the foundations and walls, let alone their construction. Yet ongoing calls for action to address traffic safety priorities often fail to recognize this reality. Talk is cheap, but the costs of not attending to fundamental requirements for feasible action are high.

Good practice, high-income countries took seventy years of motorization to build up sufficient capacity to be able to implement a more scientific management model, and it has taken a further four decades or so to achieve more acceptable traffic safety outcomes. The basic building blocks for success are visible in high-income countries, and they highlight the long planning and programming horizons required for the delivery of sustainable results.

Achieving comparable success more quickly in India and elsewhere will require a properly sequenced pathway of action to be followed. The next 15 years set for the achievement of the Sustainable Development Goals (United Nations 2015) provides a challenging but achievable time frame for this, providing a rapid start on building country management capacity can be assured. Experience with the delivery of the Millennium Development Goals underscores the importance of moving quickly to build a solid foundation for scaled-up action, if targets are to be achieved.

This issue concerning the sequencing of road safety delivery raises questions that merit further consideration:

- *Why do we say the delivery of substantially improved traffic safety can be achieved quickly?*
- *Why do governments and donors demand immediate success in terms of final results?*

3.4 Issue 3: Inequality

Aside from addressing climate change, inequality is perhaps globally the biggest "big" issue of our times, given the growing disparities in wealth and opportunities. For example, the work of Thomas Piketty focuses critical attention on growing income inequality in high-income countries and inequalities in the global distribution of wealth (Piketty 2014). It cannot be assumed that winners will compensate losers, or that principles of social justice, which point to assisting those worst off first, will prevail in contemporary, macroeconomic regimes, especially those promoting the virtues of austerity. Remarkable complacency about growing global inequality is evident.

We can see many of the contours of inequality manifested in the delivery of traffic safety measures and their performance. The growing performance gap between high-income countries and low and middle-income countries underpins the call for a Decade of Action for Road Safety and related Sustainable Development Goals. Within countries, worldwide safety inequalities are also evident between urban and rural areas, road types, vehicle types and road users.

Awareness and the unacceptability of these inequalities have been heightened by the promotion of the Safe System approach, with its goal of ensuring safety for all users of the road transport system. Improvements in analytical tools are highlighting the prevalence of unequal outcomes and the systemic means of addressing them. There are growing concerns that rapid technological change in the vehicle fleet and its communications with the road environment could create new inequalities with unintended safety consequences.

A major system failure continues to haunt us. As highlighted by the Global Burden of Disease findings (Murray and Lopez 1996), road crashes rank highly with suicide, drugs and interpersonal violence as a leading cause of death for our young people. Between the ages 15–34 years, this picture is evident across the developed and developing world and there has been little change in the rankings over the last two decades. After more than a hundred years of motorization, traffic safety risks for young people are still not being effectively managed (Bliss 2014).

This issue concerning road safety delivery inequalities raises questions that merit further consideration:

- *Why have we failed our young and our vulnerable road users?*
- *Why do we treat certain roads with differing levels of protection provided for users as being equal, in terms of setting speed limits?*
- *Why does the global donor community continue to ignore traffic deaths and injuries as a development priority in low and middle-income countries?*
- *Why has equity or fairness become subordinate to efficiency?*

3.5 Issue 4: Limits to Performance

Limits to traffic safety performance are set by technical production frontiers and institutional management capacity, with the latter often constraining achievement of the former. For any given road system, technical limits to the level of safety are governed by the protective quality of the infrastructure, vehicles, and safety clothing and helmets, at prevailing speed limits, and the degree of compliant behavior by system users. However, often the management capacity to achieve feasible levels of traffic safety is lacking. Surpassing current safety performance outcomes requires both the technical means and the managerial capacity to deliver it.

Technical production frontiers are determined by safety standards and rules, and desired shifts in results focus may require these standards and rules and related compliance regimes to be recalibrated and management practices adjusted accordingly to achieve improved safety. Yet implementation of the Safe System approach in high-income countries still provides examples of performance ambition exceeding both technical boundaries and institutional delivery capacity. For example, in Australasian jurisdictions, the desired goal of achieving zero harm will be impossible to attain with prevailing infrastructure designs and related speed limits, although there is little official acknowledgment of this.

This issue concerning road safety performance limits raises questions that merit further consideration:

- *Why do we avoid informed discussion or understanding of what traffic safety management systems are capable of in performance terms?*
- *Why do we call for safety performance way beyond the technical and institutional capacity to produce it?*

3.6 Issue 5: The Road User

Taking a speculative stance and looking to the future, there is value in going beyond the bounds of our contemporary traffic safety dialogue and evidence base. While our paradigmatic shift has de-emphasized victim blaming, we need to reconsider the capacity and agency of road users, in terms of their safety behaviors, and what we can realistically expect of them.

The Safe System approach requires road users to share responsibility for their safety, by complying with system safety standards and rules. However, while the dynamic nature of the traffic safety production process has resulted in sustained improvements in vehicle and infrastructure safety performance, our understanding and expectations of the road user have changed very little, beyond perhaps a sharper recognition of gender, age, physical vulnerability and so on.

We do have ideas concerning "culture" and "consumer behavior" and their potential influences on traffic safety outcomes, though these are loosely shaped and poorly accounted for in evidential terms. We also acknowledge the importance of human factors science, given the rapid rate of technological change and emerging complexities in human–machine interfaces, in terms of the cognitive load our brains can be expected to handle. But overall we still treat the road user as a constant in these matters and are not considering that perhaps a new type of human could be emerging or being shaped.

We must recognize that the human–machine interface is evolving rapidly and artificial intelligence is making progress. Where are we heading as human subjects in this process? Useful account could be taken of the insights provided by Marshall McLuhan on media as extensions of human capacities and identities (McLuhan 2003), and Michel Foucault on the mode of being of the human subject shaped by regimes of knowledge, power and self-formation (Foucault 2005). We must revisit our understanding of road users and their inherent capacities and actions.

This issue concerning the role of road users in the traffic safety management system raises questions that merit further consideration:

- *Why do we treat road users in terms of their human capacities as a constant?*
- *Why do we view information and communications technologies and their convergence with transport systems as neutral enablers of change?*
- *Why do we continue to pin our hopes on improving road user attitudes, behaviors and cultures as the means of improving safety performance?*

3.7 Issue 6: Design

Safe design is integral to achieving safe performance, something that was already becoming well understood in industrial safety a half-century ago. If you want to stop workers being injured by the machines they use, you make the machines safer. If you want to lose less lives in mining disasters, you make the mine systemically safer, rather than viewing and treating mine workers as having suicidal tendencies. This perspective seems self-evident to many traffic safety practitioners today.

Steady progress is being made in vehicle safety improvements, especially for vehicle occupants, yet in the case of road infrastructure, traffic safety has been lagging in its design focus, even with the Safe System approach bringing safe design to the forefront. In particular, the Dutch *Sustainable Safety* strategy sets out rigorous safety design principles and globally more attention should be paid to their application (Wegman and Aarts 2006).

In some instances, the goal of eliminating crash deaths and serious injuries is proactively shaping infrastructure design solutions, replacing reactive measures being taken only when system failures are of sufficient concern to merit intervention. The shift to roundabouts for safer junctions is a good example of this. Likewise, the adoption of low-cost wire rope barriers on high-speed interurban roads is eliminating deaths and injuries from head-on crashes. Aside from this, overall progress in safety design is slow.

Given the professional inertia cloaking current safe infrastructure design practices, there is still a long way to go before speed and safe design are addressed integrally to eliminate predictable road crash deaths and injuries. However, the emergence of a Complex System approach may demand safety performance requirements that designers must comply with, rather than trading safety off and continuing to produce infrastructure designs with a known kill-rate built into them. This requirement is already the case with rail and air transport systems and the growing complexity of road transport systems may dictate a similar ethos.

This issue concerning safe design raises questions that merit further consideration:

- *Why do we tolerate road infrastructure designs that are clearly failing?*
- *Why do we promote safety audit and inspection as the panacea for achieving safe road design, rather than hold operators accountable for safe infrastructure and empower and require infrastructure designers to proactively deliver it?*

- *Why do we resist designing for safe speeds in the road network?*

3.8 Issue 7: Development

Economic development and its sustainability goals provide the broader context for the consideration and promotion of traffic safety in low, middle and high-income countries. In terms of development impacts, we need address the issue of avoidable deaths and disabilities, and the associated health losses and economic impacts, let alone the pain and suffering of crash survivors and their families.

Deepening levels of inquiry exploring the linkages between economic development and human wellbeing are evident. Critical biopolitical perspectives are uncovering and elaborating the primacy of population health in preserving and enhancing life in the formation of modern states and their citizens (Foucault 2007, 2008). Reinforcing this is the macroeconomic view that improved health generates income growth and productivity, a reversal of the conventional understanding that income growth results in improved health (Bloom and Canning 2000). From this perspective, health creates wealth, which changes how we view health losses and investments made to improve health.

These insights resonate with a rethinking of country development priorities that have shifted from a narrow focus on income and spending to paying increased attention to the provision of accessible education and health, and ensuring social, cultural and national inclusiveness and political participation. Development aims to promote higher living standards for all, with an emphasis on improved health, education and peoples' ability to participate in the economy and society. Viewed within the twin pillar framework promoted by Nicholas Stern during his time at the World Bank, it seeks to foster an investment climate conducive to increased growth, productivity and employment, and to empower and invest in people to include them in the development process and ensure that they share its benefits (Stern et al. 2005). Yet the overwhelming impacts of population health and wealth losses arising from road crashes undermine this necessary inclusiveness and reinforce the case being made for their prevention to become a higher development priority.

We must confront the issue that some lives matter more than others: lives that the state is more willing to give away. This is an ethical disposition that Richard Allsop memorably termed "the scandal of tolerance" in his reflections on the findings of the SUNflower project that reviewed road safety development in three of the world's best-performing countries, Sweden, the UK and the Netherlands (Allsop 2002).

Over the first 30 years of the twenty-first century it is projected that the global vehicle fleet will at least double, with more than half of these vehicles entering the road networks of low and middle-income countries (Dargay et al. 2007). These networks include unprecedented numbers of vulnerable road users fated on current trends to become road crash victims. The resulting carnage will be huge. Global road deaths are projected to rise to around 2 million people a year by 2020 and continue growing, unless substantial new initiatives are taken. Making the bold assumption that the Decade of Action goal of halving 2020 fatalities can be achieved—whereas in reality, halfway through the Decade, time has already run out on this—and accounting for growing road transport demand, more than 50 million deaths and 500 million serious injuries on the world's roads can reasonably be anticipated over the first fifty years of this century (Bhalla et al. 2008).

Historically, only comparable eras of war or genocide have delivered such sustained violence. For example, in the 45 years following World War II, based on estimates made by Robert McNamara, around 40 million people were killed in national and regional conflicts, with 70% of the victims being civilians (McNamara 1992). In the modern parlance of twenty-first-century war, civilian victims have since become euphemistically described as "collateral damage." By way of comparison, the first three decades of road crashes in this century look set to generate fatalities on this scale, with equivalent collateral damage in terms of impacts on vulnerable road users. This is not accounting for the higher injury toll associated with road crashes, which result in at least ten serious injuries per fatality (and much more in road environments like India), a ratio only now being approached for combatants in modern warfare where improved body armor, victim recovery and trauma care services are keeping more battlefield victims alive and often left seriously disabled (Goldberg 2010).

Taking a longer-term view, this picture becomes even bleaker, with roads in certain ways resembling modern battle zones. Urgent and sustained traffic safety measures are required to stem the rising tide of crash fatalities and injuries arising from growing demand for road transport services and related infrastructure investment. Otherwise, sustained violence on the world's roads looks to be inevitable.

This issue concerning road safety as a sustainable development priority raises questions that merit further consideration:

- *Why do road deaths and disabilities on a mass scale go relatively unnoticed as a contradictory by-product of development?*
- *Why do we accept the giving away of lives and the heavy burden of injury resulting from our road transport system?*
- *What is holding us back from urgent, sustained action to address this avoidable carnage?*

3.9 Issue 8: Innovation

Hopes are increasingly being pinned on innovation to address global traffic safety concerns. However, with some notable high-income country exceptions, associated research and development budgets do not reflect this enthusiasm. We are living off the findings of past research and diminishing returns are setting in. New research and development programs are now needed to anticipate and address the growing complexity of urban mobility and related traffic safety issues. We must now commit to the long-term process of building a new evidence base for the twenty-first century.

The successes of traffic safety programs in good practice countries that began to accumulate toward the end of last century were built on the findings of decades of sustained research and development. While it takes considerable time to create and effectively translate an evidence base into action, the benefits achieved justify the research and development investments. Innovation, particularly in terms of piloting and evaluating new measures that theory suggests promise safety benefits, is integral to this process.

Certain traffic safety priorities call for innovative action. These include improving the safety of vulnerable road users in mixed mass, mixed speed environments like those found in India and its neighboring countries. Emerging mobility scenarios of a complex nature must also be addressed, such as vehicle-to-vehicle, vehicle-to-infrastructure and vehicle-to-road user communications, and autonomous driving, to ensure network safety is paramount. This in turn calls for a commitment to long-term research and development (Bliss 2014). In low and middle-income countries, with some possible exceptions, little attention is being paid to investing in these traffic safety priorities.

This issue concerning the promise of road safety innovation raises questions that merit further consideration:

- *Why do we just see innovation as technical applications in search of problems they can solve and related market or business opportunities?*
- *Why are we not seeking innovative infrastructure and vehicle design solutions for unique traffic safety issues in low and middle-income road network environments?*
- *Why have we lost the will for long-term investment in traffic safety research and development?*

3.10 Issue 9: Investment

We now get to a crunch issue that continues to be neglected, perhaps because it concerns finding sources for safety funding. Improving national traffic safety performance requires substantial and sustained investment. Though difficult to quantify—given that safety budgets are often embedded in larger infrastructure, enforcement and regulatory budgets—funding requirements are huge. In public management contexts that require transparent output and outcome linkages, it is possible to gain some insights into the scale of these requirements.

For example, traffic safety enforcement in Australasian jurisdictions requires around 20% or more of total policing budgets. This is at least an order of magnitude higher than enforcement budgets in low and middle-income countries. It is also investment of a proactive nature targeting measurable performance outcomes, rather than being diffused and diluted in traffic management and incident responses, as is usually the case in low and middle-income countries. Much more could be said about this, across the broad spectrum of traffic safety interventions, but the key point is that the successful performance of high-income countries in effect makes a clear business case for what is needed in investment terms. The complexity of their institutional arrangements alone can also be viewed as a surrogate indicator of success and commitment to sustained investment, and their longer-term goal of fatality and serious injury elimination points to the net benefits that are anticipated.

Significant investment will be required to achieve the Decade of Action targets of saving 5 million lives and avoiding 50 million serious injuries. It was estimated that this would result in gross benefits of more than USD$3 trillion, using a value of statistical life based on International Road Assessment Program estimation procedures (Guria 2009). Assuming that investments made to achieve these fatality and injury targets had an overall benefit-cost ratio of 10:1, which is optimistic though plausible, dividing US$3 trillion by 10 indicates the magnitude of investment that must be mobilized.

On this basis, it could be argued that low and middle-income countries cannot afford to invest in traffic safety, but this is not the case when the sheer scale of road transport investment is taken into consideration. Over the next four decades, global passenger and freight travel is expected to double from 2010 levels, with non-OECD countries accounting for nearly 90% of travel increases. Estimated network length will increase by 60%, with 90% of associated road infrastructure investment being in non-OECD countries. Global road capital construction, reconstruction and operations and maintenance costs are estimated to reach up to US$1.1 trillion annually over the next 20 years, dropping to around US$700 billion a year by 2050 (Dulac 2013). Justifiable savings could be achieved by shifting to more sustainable modes, but even so the overall level of investment in road transport will continue unabated on a huge scale.

From this perspective, it is specious to argue that improved traffic safety is unaffordable for low and middle-income countries. As is the case with high-income countries, the long-term funding required to sustainably improve traffic safety performance could reasonably be sourced and absorbed within projected mobility investments, providing the institutional road safety management capacity is built to deliver this. But in the immediate term, what is urgently required is sufficient international catalytic funding to accelerate this capacity-building process, to help unlock sustainable domestic funding sources targeted to deliver improved traffic safety over the coming decades.

This issue concerning the scaling up of road safety investment raises questions that merit further consideration:

- *Why are we so coy about the level of investment required to achieve sustainable traffic safety?*
- *Why do we promote the view that improving traffic safety is cheap and easy?*
- *Why is the global traffic safety community comfortable with derisory levels of catalytic funding that do not address county capacity-building requirements and are incapable of achieving measurable and sustainable results?*

3.11 Issue 10: Management

The final issue concerns traffic safety management and the associated issues of responsibility and accountability for performance. This brings us back full circle to the first issue of producing safety and the nature of this production process. Taking a "managing for results" perspective, what is measured is managed, and application of the scientific evidence base is fundamental to success.

Management must address production complexities and take responsibility for all elements of the management system comprising institutional functions that produce targeted interventions which in turn produce results. Management must also address all the other issues previously outlined: those of sequencing, inequality, limits to performance, the road user, design, development, innovation and investment. This is a formidable set of tasks.

Management responsibilities also include the less visible tasks of sustaining agency creativity, building teams and partnerships, and ensuring the ongoing creation of traffic safety knowledge necessary to achieve continual improvements in performance. Strategic and visionary leadership are vital to addressing these management priorities, and their general absence at country, regional and global levels remains a fundamental brake on progress.

While a focus on management is prioritized as the first pillar in the Global Plan for the Decade of Action and permeates much of the plan's content, in the main, it is neglected in global road safety dialogue and action. More attention must be paid to developing robust management capacity in low and middle-income countries. Strengthening peer-to-peer relationships with high-income countries would be an important first step in this development process, to help achieve the diffusion and sharing of knowledge vital to improving traffic safety productivity.

This issue concerning road safety management raises questions that merit further consideration:

- *Why do we assume that sufficient management capacity already exists in low and middle-income countries, just waiting to be mobilized?*
- *Why are many governments and their agencies, and international donors, missing in action, when it comes to showing leadership and supporting the strengthening of traffic safety management capacity in low and middle-income countries?*

3.12 Concluding Remarks

To sum up, there are many issues and questions remaining to be addressed, which all serve to illustrate what a rich and rewarding field we are privileged to be engaged in. These issues and questions also underscore the immense challenges we must face and the societal benefits we can help create—the value we can potentially add—if we commit to the pursuit of long-term success. It is appropriate at this point to conclude with the researchers' lament (Bliss 2014):

We do not know enough….

And we do not do enough with what we know.

References

Allsop R (2002) Closing address of SUNflower congress. In: Koornstra M et al (eds) SUNflower: a comparative study of the development of road safety in Sweden, the United Kingdom and the Netherlands. Appendix A. SWOV, Dutch Institute for Road Safety, Leidschendam

Bhalla K, Shahraz S, Naghavi M, Murray C (2008) Estimating the potential impact of safety policies on road traffic death rates in developing countries. In: Poster presentation at 9th World conference on injury prevention and safety promotion, Merida, Mexico

Bliss T (2014) Road safety in the 21st century—public expectations of government. In: 25th Westminster lecture on transport safety, Terrace Pavilion, Palace of Westminster, PACTS, Parliamentary Advisory Council for Transport Safety, London

Bliss T (2015) Road safety in the 21st century: innovation and national strategy. National Road Safety Forum, Sydney

Bliss T, Breen J (2009) Implementing the recommendations of the World report on road traffic injury prevention. Country guidelines for the conduct of road safety management capacity reviews and the specification of lead agency reforms, investment strategies and safe system projects. World Bank Global Road Safety Facility, Washington DC

Bloom D, Canning D (2000) The health and wealth of nations. Science 287

Dargay J, Gately G, Sommer M (2007) Vehicle ownership and income growth, worldwide: 1960–2030. Energy J 28(4)

Dulac J (2013) Global land transport infrastructure requirements: estimating road and railway infrastructure capacity and costs to 2050. International Energy Agency, Paris

Foucault M (2005) The hermeneutics of the subject. Lectures at the College de France 1981–1982. Palgrave Macmillan, New York

Foucault M (2007) Security, territory, population. Lectures at the College de France 1977–1978. Palgrave Macmillan, New York

Foucault M (2008) The birth of biopolitics. Lectures at the College de France 1978–1979. Palgrave Macmillan, New York

Goldberg M (2010) Death and injury rates of US Military Personnel in Iraq. Mil Med 175(4):220

Guria J (2009) Required expenditure. Road safety improvements in low and middle-income countries. Report to the World Bank Global Road Safety Facility, Addendum. Revised estimates of fatalities and serious injuries and related costs. New Zealand Institute of Economic Research, Wellington

Haddon W Jr (1968) The changing approach to the epidemiology, prevention and amelioration of trauma: the transition to approaches etiologically rather than descriptively. Am J Public Health 58:1431–1438

McLuhan M (2003) Understanding media. The extensions of man, Critical edn. Gingko Press, Corte Madera, California

McNamara R (1992) The post-Cold War world: implications for military expenditure in the developing countries. In: Proceedings of the World Bank annual conference on development economics 1991. Supplement to the World Bank economic review and the World Bank research observer, Washington DC

Murray C, Lopez A (eds) (1996) The global burden of disease. A comprehensive assessment of mortality and disability from diseases, injuries and risk factors in 1990 and projected to 2020. Harvard University Press, Boston

Piketty T (2014) Capital in the twenty-first century. Belknap Harvard, Cambridge, Massachusetts

Rumar K (1999) Transport safety visions, targets and strategies: beyond 2000. 1st European transport safety lecture. European Transport Safety Council, Brussels

Stern N, Dethier J-J, Halsey Rogers F (2005) Growth and empowerment. Making development happen. Munich lectures in economics. The MIT Press, Cambridge, Massachusetts; London, England

Tingvall C (1995) The zero vision. In: van Holst H, Nygren A, Thord R (eds) Transportation, traffic safety and health: the new mobility. In: Proceedings of the 1st international conference. Springer, Gothenburg, Sweden, Berlin

United Nations (2015) Resolution adopted by the General Assembly, 70/1 transforming our world: the 2030 agenda for sustainable development, New York

Wegman F, Aarts L (2006) Advancing sustainable safety: national road safety outlook for 2005–2020. SWOV, Dutch Institute for Road Safety Research, Leidschendam

Wegman F, Elsenaar P (1997) Sustainable solutions to improve road safety in the Netherlands. SWOV, Dutch Institute for Road Safety Research, Leidschendam

Chapter 4
Legislation, Enforcement and Education for Traffic Safety: A Brief Review of the Current State of Knowledge

Rune Elvik

4.1 Introduction

This paper is intended to give a brief overview of the current state of the art with respect to the effects of legislation, enforcement and education as road safety policy instruments. Key questions addressed in the review include:

What types of road user behaviour are regulated by means of legislation? What types of behaviour remain unregulated?
What can we learn from the history of road safety legislation with respect to changing road user behaviour and reducing the number of accidents and injuries?
How do legislation, enforcement and sanctions interact in bringing about changes in road user behaviour and in traffic injury?
Why are laws never one hundred percent effective in solving a problem and what can be done to make them one hundred percent effective?
What are the effects of educating road users? Why have most educational measures had limited success and how can they be made more effective in improving road safety?

The review starts by discussing legislation and continues by discussing enforcement. Education is discussed at the end of the review.

R. Elvik (✉)
Institute of Transport Economics, Gaustadalléen 21, NO-0349 Oslo, Norway
e-mail: re@toi.no

G. Tiwari and D. Mohan (eds.), *Transport and Safety*, Springer Transactions in Civil and Environmental Engineering,
https://doi.org/10.1007/978-981-16-1115-5_4

4.2 Legislation—Some Examples and General Lessons

Current legislation designed to promote road safety has developed during the last 40–45 years that is mostly after the peak in the number of traffic fatalities was reached in most highly motorised countries. Before this peak was reached, many forms of behaviour were not regulated by laws the way they are today.

Norway has always had strict road safety legislation. Speed limits on all public roads were first introduced in 1912 and have remained in force since then. For a long period, however, speed limits were raised as cars and road improved. The last time general speed limits were raised in Norway was in 1965. The general speed limit in urban areas was then raised from 40 to 50 km/h. In rural areas, the general speed limit was raised from 70 to 80 km/h. These speed limits remain in force today and are comparatively low by international standards. However, exceptions from the general speed limits are increasingly made, in particular on motorways. Motorways got a speed limit of 90 km/h in the late nineteen-sixties (there were very few motorways in Norway at that time). In 2001, this was raised to 100 km/h and in 2014 to 110 km/h on the best motorways. This still remains low by international standards.

Until the energy crisis in 1973, quite a few European countries did not have speed limits at all, at least not in rural areas. Today, few people believe that this could really be the case, but it was. There was a free choice of speed. In many countries, until about 1970, there was also a free choice about whether to drink and drive. Well, strictly speaking, drinking and driving was not allowed, but the legislation made effective enforcement almost impossible. In the first place, the police could normally only stop a driver if they suspected him of drinking and driving. If there was nothing peculiar about driving behaviour, there was no legitimate reason for suspicion. In the second place, the police had to prove that the driver was impaired. This involved various assessments of behaviour that were often contestable in court. The police might say, for example, that the driver could not walk a straight line. The driver would protest, and unless there was other evidence, it was word against word and a driver with a good attorney might get acquitted. Today, this sounds like a tale from the past, as indeed it is. Nowadays, the freedom of road users to make their own choices has been circumscribed, and many motorised countries have legislation that lays down:

- Permitted driving speeds. There are speed limits everywhere, except on German motorways.
- Use of protective devices. Wearing seat belts is required in very many countries. Using motorcycle helmets is still voluntary in some places (see below).
- Drinking and driving. All highly motorised countries have per se laws, i.e. laws stating that a driver is regarded as drunk if blood-alcohol concentration exceeds a certain value (0.05% is the most common).
- Use of daytime running lights. Some countries require motor vehicles to use headlights at all times.

Over time, legislation has expanded and included more and more aspects of road user behaviour. Since the expansion of legislation can be regarded as curtailing freedom, it is important to determine if legislation does accomplish its stated objective of improving road safety. Below, a few examples are reviewed in order to answer this question.

Figure 4.1, taken from Ross (1982), captures the history of legislation regarding the use of motorcycle helmets in the USA. Following the creation of the National Highway Traffic Safety Administration in 1966, many safety initiatives were launched and states were encouraged to make the use of crash helmets mandatory for motorcyclists. Most states complied. As can be seen from Fig. 4.1, there was a dramatic decline in the number of fatalities per registered motorcycle. After a few years, opponents of these laws gained the upper hand and several states started to repeal the laws. Helmet wearing declined and motorcycle fatalities increased. Most of the safety benefits of mandatory helmet wearing were lost.

There are two main lessons to be learnt from this example:

1. Introducing a law can lead to a large improvement in road safety.
2. If the law is repealed, a large part of the safety improvement may be lost.

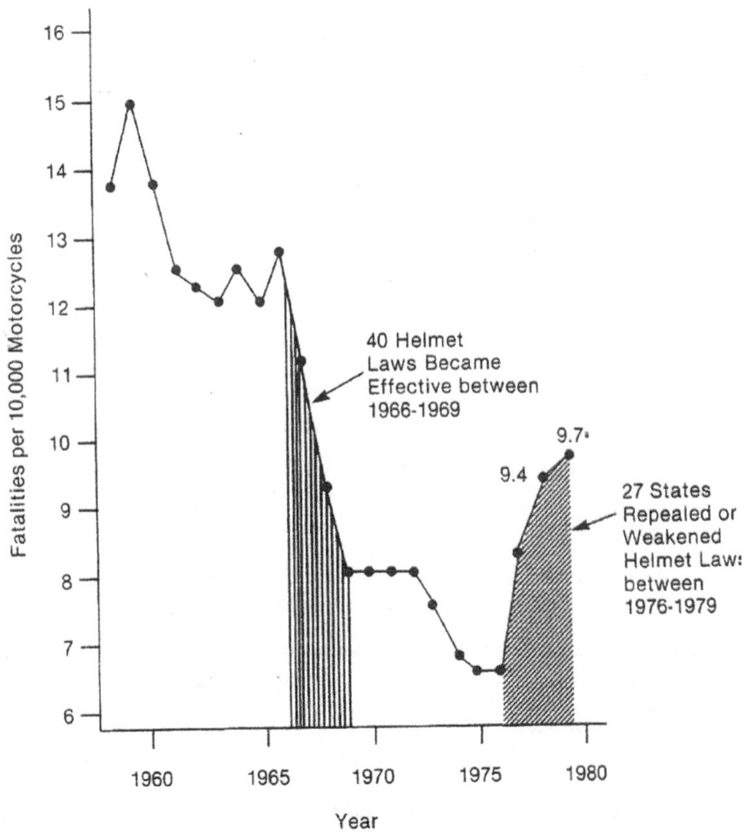

Fig. 4.1 Laws on helmet wearing by motorcyclists in the USA. *Source* Ross (1982)

Laws do, therefore, not necessarily lead to learning or to changes in behaviour that are sustained permanently. As soon as motorcyclists were allowed to take off their helmets, many did so, although both scientific and anecdotal evidence should have convinced motorcyclists that wearing helmets improve their chances of surviving an accident or getting less seriously injured.

The next example is given in Fig. 4.2, also taken from Ross (1982). It shows changes in the number of fatal and serious injuries during weekend nights (between 10 P.M. and 4 A.M.) before and after the per se law on drinking and driving was passed in Great Britain in 1967. This law made it illegal to drive with a blood alcohol concentration above 0.08%.

It is seen that the number of fatalities and serious injuries dropped sharply when the law took effect. It is also seen that there was a rather quick rebound. The number of fatalities and injuries started to increase again almost immediately after the sharp drop. The increase did not quite eliminate the initial safety effect. It did, however, reduce it considerably. As Ross (1982) shows later in his book, the most likely explanation for the rebound is that there was too little enforcement. Drivers soon discovered that the risk of being detected was low. The effect of the law then eroded. The lesson from this, which is the third lesson from using legislation to improve road safety is

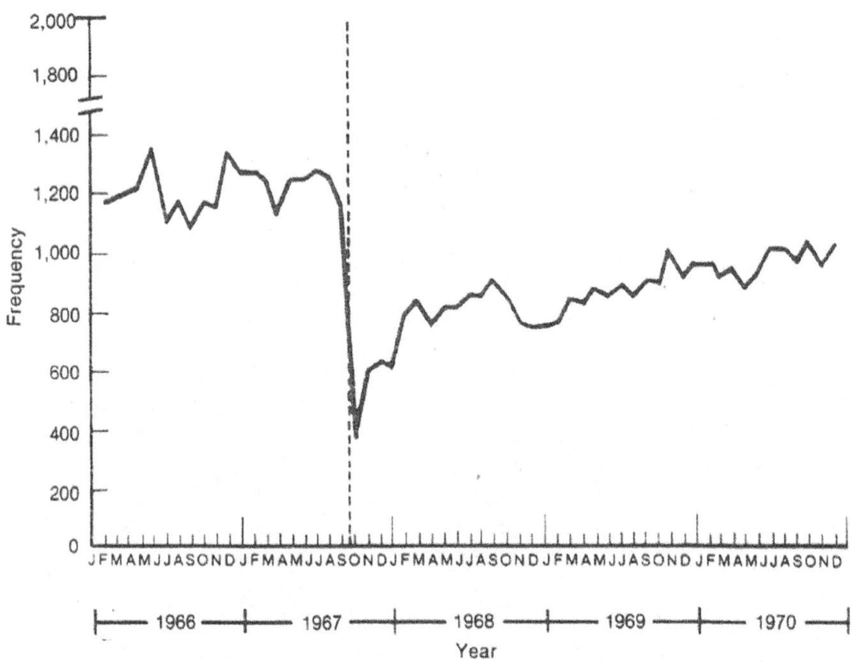

Fig. 4.2 Number of fatalities and serious injuries during weekend nights (between 10 P.M. and 4 A.M. Friday/Saturday, Saturday/Sunday) before and after the per se law for drinking and driving took effect in Great Britain. *Source* Ross (1982)

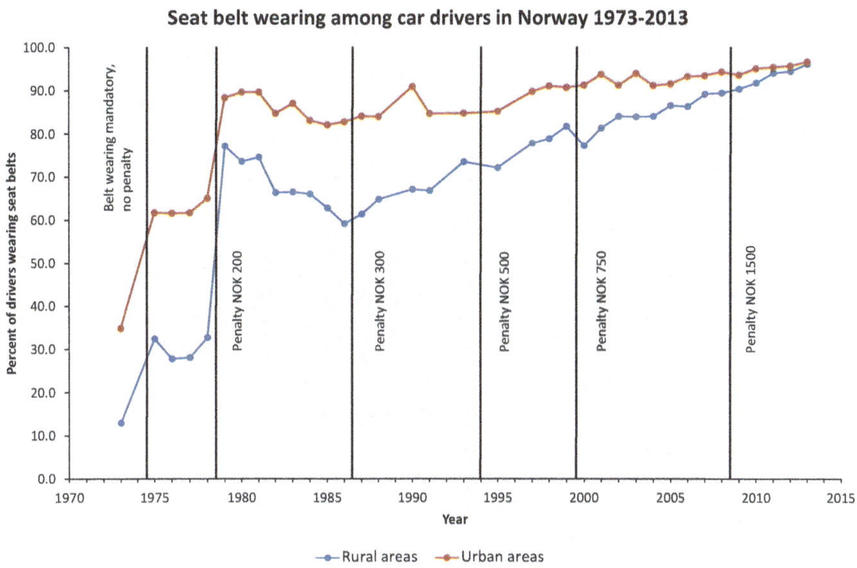

Fig. 4.3 Seat belt wearing among car drivers in Norway 1973–2013

3. To sustain the effects of a law in the long term, it must be effectively enforced.

Enforcement is, however, not always possible or at least not very effective. How can that be? It sometimes happens that laws are passed that cannot be effectively enforced, either because there are no sanctions for violating the law or because only secondary enforcement is possible. An example of the first type of law is the law requiring seat belts to be worn in Norway. The law was passed in 1975. However, until October 1979, those who violated the law were not punished in any way. All the police could do was to give drivers and passengers a friendly reminder of the law. Figure 4.3 shows seat belt wearing among car drivers in Norway from 1973 to 2013.

When the law passed, there was an increase in seat belt wearing. But then, in the next year, seat belt wearing remained unchanged at about 60% in rural areas and about 30% in urban areas. In 1979, a fixed penalty of NOK 200 (1 NOK = 0.12 US Dollars in July 2015) for not wearing seat belts was introduced. Wearing then soared to almost 90% in rural areas and around 70% in urban areas.

The fixed penalty was not changed until 1987. In the meantime, as shown in Fig. 4.3, there was a downward trend in seat belt wearing, in particular in urban areas. This trend was turned around when the fixed penalty increased from NOK 200 to NOK 300. Subsequent increases in the fixed penalty have also been associated with an increase in seat belt wearing. An evaluation by Elvik and Christensen (2007) found that increasing the fixed penalties for not wearing seat belts was associated with increased seat belt wearing.

A similar story can be told about legislation regarding seat belt wearing in the USA (Shults et al. 2004). Many states were reluctant to make seat belt wearing mandatory.

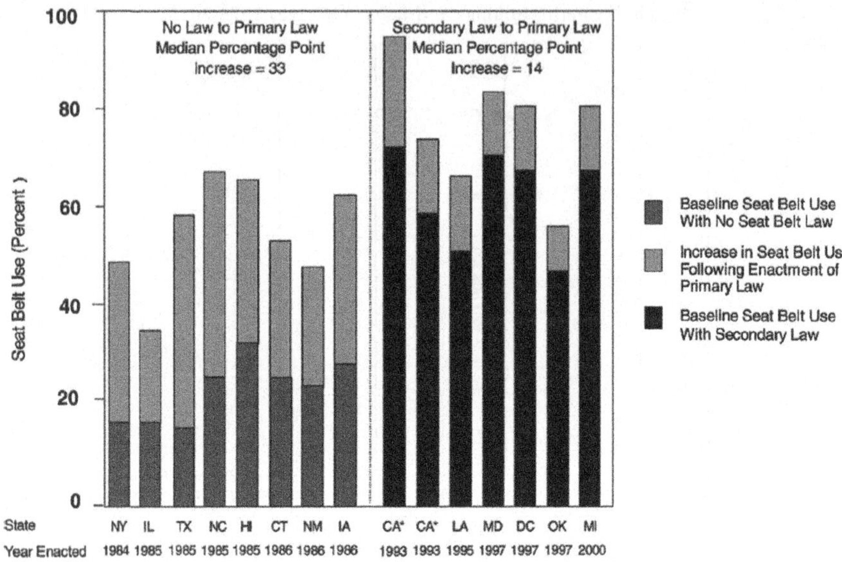

Fig. 4.4 Effects of primary and secondary enforcement on seat belt wearing. *Source* Shults et al. (2004)

A compromise chosen by many states was to pass a law, but only allowing secondary enforcement of it. Secondary enforcement means that a driver could only be given a traffic ticket for not wearing a seat belt if another offence had been committed. The police were not allowed to cite drivers if their only violation was not wearing seat belts. Secondary enforcement was, predictably, not very effective. Figure 4.4, taken from Shults et al. (2004) shows that primary enforcement is more effective than secondary enforcement. The following lessons can be learnt (the numbering of lessons continues consecutively):

4. For a law to be effective, there must be a sanction for violating the law. Without a sanction, the law cannot be enforced.
5. For a sanction of violations of a law to be effective, it must be possible to apply the sanction whenever the law is violated. Sanctioning should not depend on having violated a different law.

As shown in Fig. 4.3, seat belt wearing among car drivers in Norway is now more than 95%. It is tempting to conclude that virtually all of the safety benefits from using seat belts have now been harvested and that there little point in trying to bring seat belt wearing closer to 100%. This is wrong. Although seat belt wearing in general traffic has exceeded 95%, it remains low among car occupants who are killed in road accidents (Haldorsen 2015). Figure 4.5 illustrates the relationship between seat belt wearing in traffic and among killed car occupants in Norway during the period from 2005 to 2013.

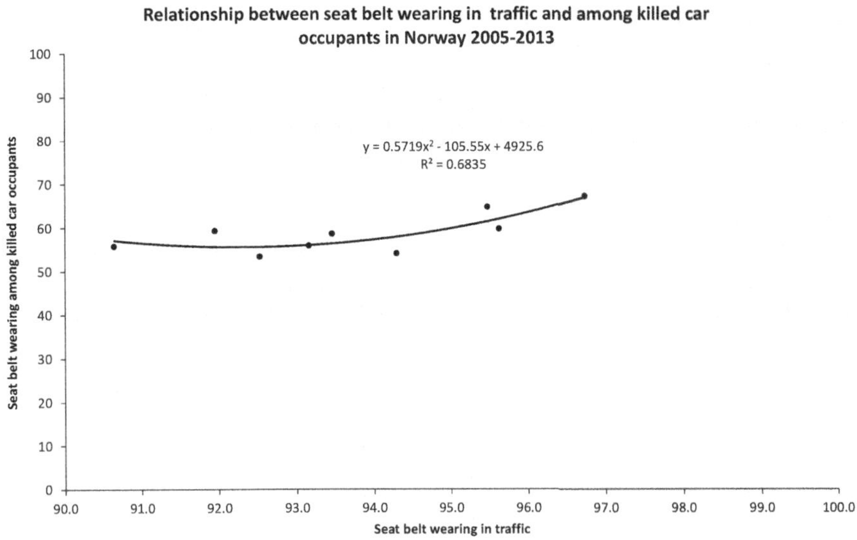

Fig. 4.5 Seat belt wearing in traffic and among killed car occupants in Norway 2005–2013. Based on Haldorsen (2015)

During this period, seat belt wearing in traffic increased from about 90.5% to close to 97%. In other words, the non-wearing of seat belts was greatly reduced, from close to 10% to a little more than 3%. During the whole period, however, seat belt wearing among killed car occupants remained at about 60%. Only a weak tendency can be seen for seat belt wearing among killed car occupants to increase as seat belt wearing in traffic increased. There is therefore still a quite large potential for reducing traffic fatalities in Norway by increasing seat belt wearing. A very similar, even more extreme, pattern is found in Sweden (Trafikverket 2015). In Sweden, 98% of car drivers in traffic wore seat belts in 2014, i.e. only 2% did not wear seat belts. Among killed car drivers in Sweden in 2014, 34% did not wear seat belts. How can such a dramatic difference be explained? Why does it arise?

A full explanation cannot be given, but a study by the Norwegian traffic police (Pasnin et al. 2009) shows that a high percentage of drivers involved in fatal accidents have a criminal record. The criminal record included not just traffic offences, but other types of crime, like burglary, violence, rape, drug offences and so on. While not all drivers involved in fatal accidents had a criminal record, it is clear that they were hugely over-represented. This leads to the sixth and final lesson about road safety legislation

6. No law is ever 100% effective. Every society has a group of hard-core criminals who will not comply with laws. This group is likely to be over-involved in accidents.

4.3 Enforcement—Key Lessons

As noted above, laws do not enforce themselves. Indeed, were that the case, the laws would not be needed, as road users would then adopt safe behaviour without being forced to do so by means of law. Enforcement is therefore very important for maintaining the effects of laws. What do we know about the effects of enforcement on road safety?

One lesson that was learnt many years ago is illustrated in Fig. 4.6. It is taken from the book by Ross (1982) and shows effects on accidents of enforcing the per se law on drinking and driving in New Zealand. A publicity campaign was first conducted. This was associated with a reduction in the number of accidents. Then there was

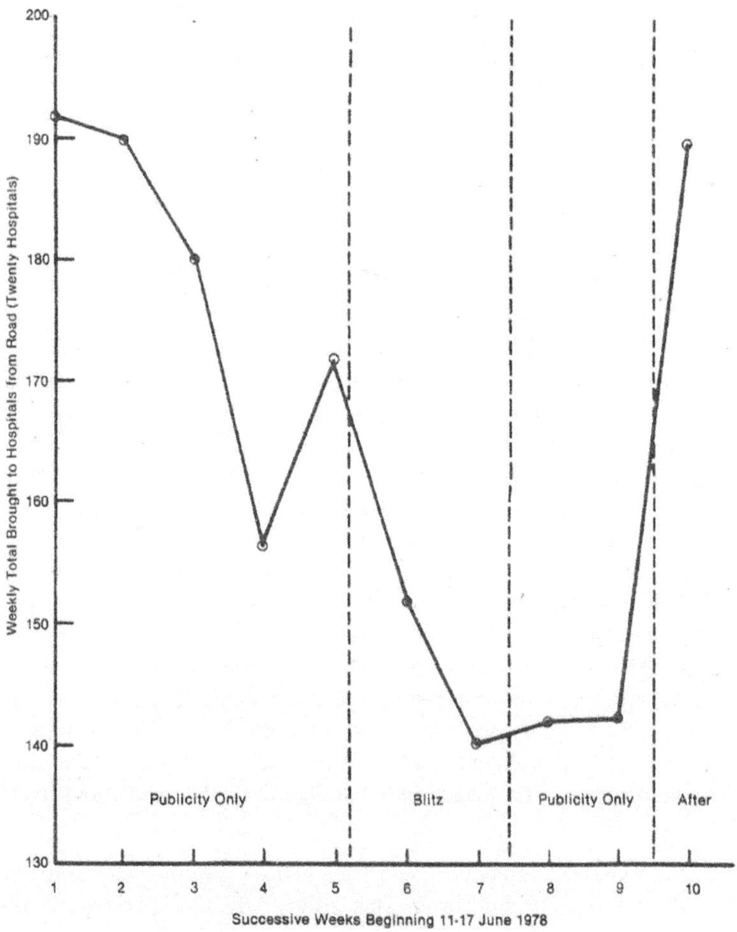

Fig. 4.6 Effects of enforcing the per se law on drinking and driving in New Zealand. Based on Ross (1982)

an enforcement blitz. This was associated with a large reduction in the number of accidents. The reduction lasted a couple of weeks after the blitz. Then accidents reverted to their original level. Several similar figures are given in the book by Ross. Lesson number one about enforcement is therefore

1. The effects of enforcement are transient. When enforcement ceases, its effects disappear quickly.

Road users adapt their behaviour to even minor changes in enforcement. This is shown in data collected in a recent Norwegian study (Elvik and Amundsen 2014). The study collected data for ten years on speed enforcement. During these ten years, examples could be found both of increases in enforcement and reductions of it. Using annual change in the rate of speeding (kilometres driven while speeding as a percentage of all kilometres driven) as dependent variable, 54 data points were identified by combining years (9 annual changes), urban or rural area (2 values) and three levels of speeding (3 values). These data points are shown in Fig. 4.7. Although the data are somewhat noisy, there is still a tendency for speeding to go up as enforcement is reduced and go down as enforcement is increased. Figure 4.8 shows the relationship more clearly. It is based on a review of a large number of studies of speed enforcement for the purpose of developing accident modification functions describing the effects of enforcement (Elvik 2015a).

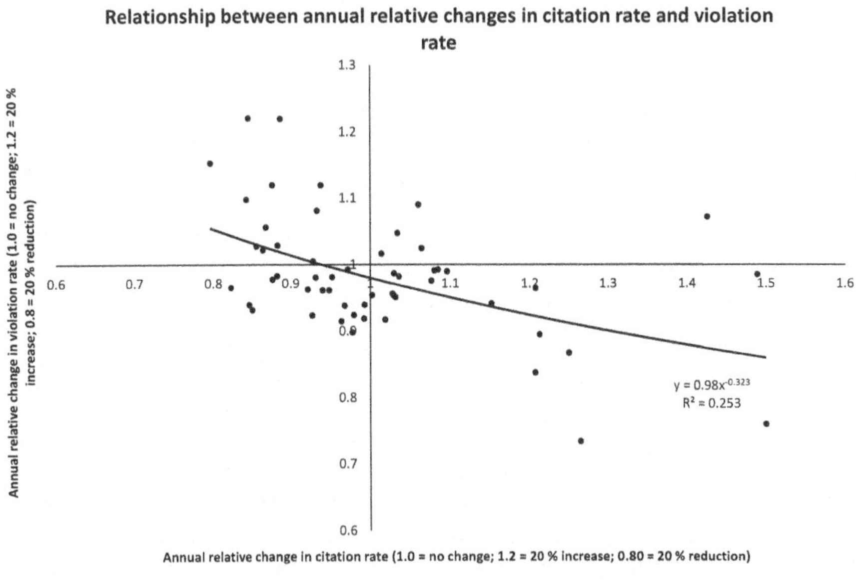

Fig. 4.7 Relationship between changes in speed enforcement and changes in speeding in Norway. Based on Elvik and Amundsen (2014)

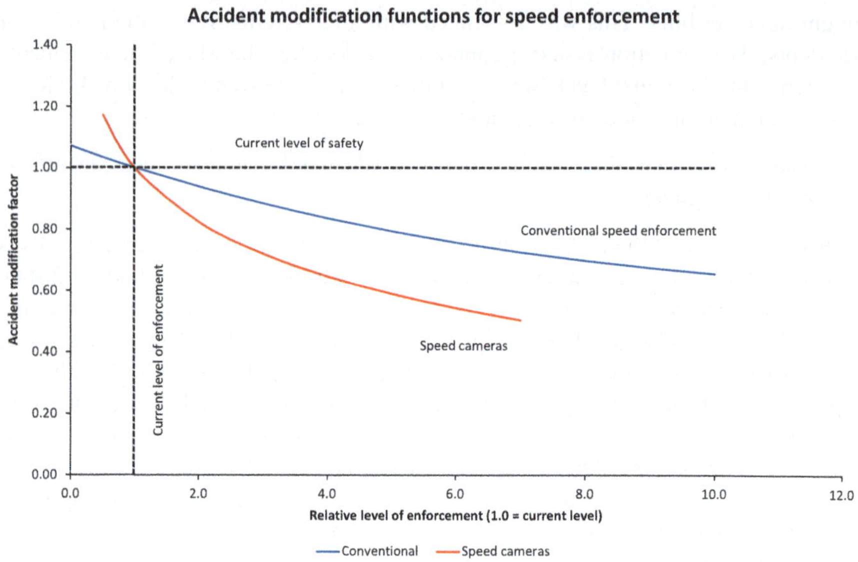

Fig. 4.8 Accident modification functions for speed enforcement. Based on Elvik (2015a)

Based on Figs. 4.7 and 4.8 and the studies underlying them, the following lessons can be learnt:

1. The current level of enforcement maintains the current level of safety.
2. When enforcement is reduced, violations and the number of accidents increase.
3. When enforcement is increased, violations and the number of accidents are reduced.
4. Road users react to even small changes in the level of enforcement.
5. There are greater changes in behaviour and accidents when the risk of apprehension is high than when it is low. This explains why speed cameras are more effective than traditional police enforcement (steeper curve in Fig. 4.8).

In Norway, the risk of apprehension when speeding is very low, close to 10 per million kilometres driven while speeding. An average driver in Norway drives about 13,000 kms per year. On average, therefore, a driver could be speeding continuously on every trip, all the time, for almost 8 years before, statistically speaking, he would be caught speeding. It stands to reason that even after a huge increase in speed enforcement in Norway, the risk of getting caught would remain low.

At speed cameras, the risk of getting caught is very much higher. Although not all cameras are operated all the time, at least not in Norway, they are operated often enough that most drivers do not gamble on the camera being turned off when they approach one. As can be seen in Fig. 4.8, the curve for speed cameras is steeper than the curve for traditional speed enforcement.

The advantage of speed cameras is that they can be operated continuously. The disadvantage is that the effects are very local. It is possible to connect several speed

Table 4.1 Game-theoretic model of the interaction between drivers and the police in determining the level of enforcement

		The police			
		Enforce		Not enforce	
Drivers	Violate speed limit		– 10,000		– 20,000
		– 300		50	
	Not violate speed limit		– 10,000		0
		– 50		– 50	

cameras with one another, in the form of section control. The cameras would then calculate the mean speed of driving over a certain distance of road. Section control may thus enlarge the area where speed cameras have an effect. However, section control is not feasible or practical everywhere. If the road has many junctions, there will be many cars entering and leaving the road and thus passing only one of the speed cameras. Installing and operating cameras are expensive. Section control would therefore be most effective on roads with a high traffic volume.

It is therefore difficult to believe that section control by means of speed cameras could ever be used on more than a small fraction of all roads in a country. Until vehicle technology monitoring and possibly regulating speed (ISA-systems; Intelligent Speed Adaptation) becomes more widely used, traditional enforcement performed by police officers will be the most important form of enforcement. This leads on to the final lessons regarding enforcement. These lessons concern the interaction between road users and the police in determining the level of enforcement.

A game-theoretic model of this interaction was proposed many years ago by Bjørnskau and Elvik (1992). The model can be explained by reference to Table 4.1.

In Table 4.1, driver has a choice between violating the speed limit and complying with it. The police have a choice between enforcing or not enforcing. The choices made by drivers will depend on what the police do and vice versa. The numbers given in the table are intended to illustrate the consequences of the various choice; in game theory, these are often referred to as "payoffs". The payoffs to drivers are shown in the lower left corner of each cell of the table, and the payoffs to the police are shown in the upper right corner of each cell of the table.

If we start in the upper left cell, it can be seen that drivers can improve their payoff (from −300 to −50) by complying with the speed limit. This will result in a move to the lower left cell of the table. However, once drivers comply with speed limits, it is seen that the police can improve their payoff (from −10,000 to 0) by not enforcing. This results in a move to the lower right cell of the table. From that cell, it is seen that drivers can improve their payoff (from −50 to 50) by speeding. This results in a move to the upper right cell of the table. However, when drivers are speeding, the police can improve their payoff (from −20,000 to −10,000) by enforcing. This brings the game back to the upper left cell where it started, and the circle can go on forever. The game, in other words, has no solution in pure strategies.

It does have a solution in mixed strategies. A mixed strategy is to choose between the pure strategies with certain probabilities. Thus, with the payoffs used as example in Table 4.1, the police should enforce with a probability of 0.2857 and not enforce with a probability of 0.7143. Drivers should speed with a probability of 0.50 and not speed with a probability of 0.50. See the paper by Bjørnskau and Elvik (1992) for details regarding how the mixed-strategy solution was obtained.

Recently, an attempt was made to test the model empirically (Elvik 2015b). It was supported, although the results regarding how the police adapt to changes in violation rate were not statistically significant. Still, the lessons are clear, both from this and other studies

1. Enforcement will never be one hundred percent effective, in the sense of eliminating violations.
2. The police always apply a certain tolerance margin when doing enforcement. This means that a certain level of minor violations is tolerated.
3. If violations go down, the police will tend to reduce enforcement.

4.4 Education—Key Lessons

Education is a very broad concept and includes many measures taken to enhance the knowledge and skills of road users. It is therefore not possible in this paper to review all studies that have evaluated the role and effects of education in improving road safety.

One of the road safety problems that has proven difficult to solve in all highly motorised countries is the high accident rate of young drivers, in particular male drivers. Elvik (2010) discusses why this problem has proven difficult to solve and offers the following remarks on it:

> It has long been the hope of educators that novice drivers can learn not just the skills needed for safe driving, but also acquire an understanding that these skills develop slowly and have not been fully learnt by the time a driver is licensed. However, teaching young people not simply to acquire certain skills, but also to correctly assess the limits of their skills is an almost impossible task. It is, so to speak, impossible to teach people that they do not know anything, or that what they know is only a very small part of what they need to know. Gregersen (1996) reported a very interesting experiment that shows this. He compared two groups of novice drivers. One group had been given skills training to make the driver as skilled as possible in braking and performing an evasive manoeuvre. The other group had been instructed that this task was very difficult and that they could not necessarily be expected to perform it successfully. The two groups then performed an evasive manoeuvre on a test track. Actually, the group that had been taught to master the skill and who erroneously believed that they did in fact master the skill, did a little worse on the task than the group who had been taught to have more modest expectations about their own performance.
>
> Novice drivers regularly overestimate their competence. A recent study shows that 30–40% of young drivers in Finland and the Netherlands rated their competence as better than driving licence examiners did (Mynttinen et al. 2009). Those who overestimated their competence (rated their competence higher than the rating given by the licence examiner) failed the

driving test more often than those who slightly underestimated their competence. Overestimating driving skills is hazardous, as it can make the driver accept a higher level of risk than he or she would if skills were understood more correctly.

As far as deliberate risk taking is concerned, Evans (2006) proposes the hypothesis that it can be caused by hormonal factors, in particular the high level of testosterone in young males. While introducing biological explanations of social phenomena is controversial and not always taken seriously, the data presented by Evans at the very least indicate that his hypothesis is plausible, although these data do not confirm it. Support for the hypothesis that testosterone levels influence risk taking comes from a study of financial risk taking and career choice, although in that study the largest effect of testosterone was found among women (Sapienza et al. 2009; both sexes produce both male and female sexual hormones, but in different mixtures).

The Handbook of Road Safety Measures (Elvik et al. 2009) summarises many studies that have evaluated the safety effects of basic driver training. Although the results of these studies vary, on the average no effect on safety has been found. It would obviously be unscientific and overly pessimistic to suggest that a successful way of teaching young people to drive more safely can never be found. It does, however, seem to be a challenge.

A closer look at different types of training shows a pattern in effects that may guide future efforts in a direction that gives more grounds for optimism. Several studies, summarised in the Handbook of Road Safety Measures, show that training advanced skills can be counterproductive. Possibly the best known example of this is skid training, i.e. training on driving on slippery road surfaces and learning how to control the car when it skids. There has been extensive programmes of skid training in the Nordic countries, where slippery roads in winter makes driving more difficult. Skid training normally takes place on a driving range, in low speed and under controlled conditions. In their original form, the skid training courses involved first teaching drivers how to produce a skid, then how to regain control of the car after it started skidding. After a few attempts, most driver mastered this skill. And it was fun. Producing a skid and then controlling it gives you a wonderful feeling of control.

The dangers of this kind of training should have been obvious. It is one thing to be able to do something at low speed and under controlled conditions, something entirely different to do it at high speed when the skid is not something you are prepared for, but something that just suddenly and surprisingly happens. Besides, it was a skill a driver very rarely needed to practice. It is therefore prone to be forgotten by the time you need to practice it. What many young drivers did not forget, however, was the joy of feeling that they were in control. Some drivers may erroneously have thought that after a few hours on a driving range, they really knew how to drive fast on slippery roads, while still being in full control.

The results were predictable. Skid training increased in the number of accidents by making drivers overconfident.

A different approach has been taken in the USA. There, graduated driver licensing has become the norm. Rather than teaching drivers advanced skills, the philosophy underlying graduated driver licensing is that novice drivers will simply not master advanced skills and should therefore not be driving in conditions when such skills are called for. In a graduated licensing system, a novice driver will therefore first be given a restricted license. Common restrictions include a prohibition on night-time driving, a ban on carrying same-age passengers and a zero blood-alcohol limit. If a driver complies with the restrictions, a full license can be acquired after a period of one to two years. There have been very many evaluations of graduated driver licensing. Vaa et al. (2015) summarise the results of these studies. There are many versions of graduated driver licensing programmes and not all are equally effective. The best programmes are associated with a reduction of accidents.

Another promising initiative is lowering the age when a driver can start training on the road. This was done in Sweden (Gregersen et al. 2000). The minimum age for starting driver training was lowered from 17½ to 16 years. The idea was to give young drivers more time to practice driving before taking the license test. Youngsters were encouraged to drive with parents, older siblings or others who could guide them. It was found that the number of kilometres driven during driver training increased considerably. This was associated with a lower accident rate after the driving license test had been passed.

Based on these studies, the following lessons are proposed with respect to driver training:

1. Training novice drivers advanced skills that are rarely practised in traffic may generate overconfidence that increases the number of accidents.
2. Imposing restrictions on novice drivers, which prevent them from driving in difficult conditions is associated with a reduction of the number of accidents.
3. Lowering the age for driver training, permitting novice drivers to drive more kilometres before getting licensed may reduce the number of accidents.

In general, learning is very influenced by motivation. If something does not interest you, or if you think, correctly or not, that you already know it, you will not make the effort to learn more. An important part of the task of any teacher is therefore to make learning fun.

Another motivating factor is that you clearly see the benefits to yourself or someone you love from learning something. You may dislike going to the gym at first, but once you see that you can run faster, lift heavier and perhaps become more good-looking, you will feel it was worth the investment and start to enjoy it. One very clear example of how education can be effective when the motivation to learn is strong is teaching parents to correctly install child restraints in cars. One may assume that most parents care about protecting their children and therefore want them to ride as safely as possible in the car. Two recent studies (Tessier 2010; Brown et al. 2011) found that educating parents about how to correctly install child restraints in cars was associated with large increases in the share of parents who correctly installed the child restraints. The lesson is

1. If there is motivation, education can be highly effective in producing safe behaviour.
2. The source of motivation may vary, but would often be that a person experiences an immediate personal benefit in learning the safe behaviour.

4.5 Concluding Reflections

It is clear that legislation, enforcement and education can all contribute importantly to improving road safety. Yet, it is also clear that these policy instruments may not always be as effective as their proponents are hoping for.

Laws can be repealed if they do not have popular support. Laws requiring motorcyclists to wear helmets survived only for a few years in many states in the USA. The national maximum speed limit of 55 miles per hour, introduced as an energy conserving measure in 1974, survived until 1987. States were then allowed to raise speed limits to 65 miles per hour. In 1995, the national speed limit was abolished entirely, and states could set any speed limit they wanted, including no speed limit at all. There is, therefore, no guarantee that even highly effective legislation can be sustained in the long term.

It is important that laws regulate important risk factors, not risk factors that make small contributions to accidents or injuries. Current legislation is not entirely consistent with such a principle.

Legislation is never one hundred percent effective. In all societies, there are individuals who are more or less delinquent, who live their lives more or less as criminals and who are difficult to influence.

Legislation will not be very effective without enforcement. Enforcement is effective, but only when and where it takes place. As soon as enforcement ceases, violations tend to increase. Current enforcement technology, in particular speed cameras, is highly effective but cannot entirely replace enforcement performed by police officers.

Educating road users to behave more safely is difficult, a lot more difficult than regulating their behaviour by means of laws and their enforcement. Laws and enforcement are often viewed as repressive policy instruments. They are negative. They tell people what they should or should not do. They command rather than encourage.

Proponents of education have therefore long argued that laws and enforcement reflect an impoverished view of human nature, that these policy instruments do not recognise the positive capabilities of humans to learn, help each other, love each other and excel in sports and academic performance. Would it not be better to educate road users, to use positive policy instruments that foster human development, proponents of education ask.

Modern road traffic is an advanced system in which you cannot travel safely without being, in a wide sense of that term, highly educated. However, this does not mean that road traffic offers great opportunities for education in the form of formal instructions or lectures delivered in classroom style. In the first place, the system is remarkably self-instructing and forgiving. It normally allows a margin for error and thereby for learning all by yourself. To put it elliptically, safe driving can certainly be learnt, but it cannot be taught. It is all about learning by doing, learning by trial and error, learning by imitation, learning from mistakes, etc.—all of which a driver can do entirely on his or her own. The technical skills needed to drive a car are easy to learn. Any teenager can learn it in a few hours. The danger is that when the technical operation of a car has been learnt, the driver may erroneously think that he or she can drive.

In the second place, road users are remarkably reliable in behaving in ways that prevent accidents. Given the very many opportunities there are for accidents to happen, extremely few of these opportunities result in an accident. There is, in other words, little room for improvement. There is little to learn. Most road users know everything they need to know to travel safely.

This does not mean that human factors do not contribute importantly to accidents. They obviously do. The driver is the problem. The solution is therefore to abolish the driver, not try to reform him, which is in most cases a hopeless project. Self-driving cars are rapidly being developed and hold great promise for improving road safety.

References

Bjørnskau T, Elvik R (1992) Can road traffic law enforcement permanently reduce the number of accidents? Accid Anal Prev 24:507–520

Brown J, Finch CE, Hatfield J, Bilston LE (2011) Child Restraint Fitting Stations reduce incorrect restraint use among child occupants. Accid Anal Prev 43:1128–1133

Elvik R (2010) Why some road safety problems are more difficult to solve than others. Accid Anal Prev 42:1089–1096

Elvik R (2015a) Methodological guidelines for developing accident modification functions. Accid Anal Prev 80:26–36

Elvik R (2015b) Speed enforcement in Norway: testing a game-theoretic model of the interaction between drivers and the police. Manuscript submitted to Accident Analysis and Prevention

Elvik R, Amundsen A (2014) Utvikling i oppdagelsesrisiko for trafikkforseelser. En oppdatering. Rapport 1361. Oslo, Transportøkonomisk institutt

Elvik R, Christensen P (2007) The deterrent effect of increasing fixed penalties for traffic offences: The Norwegian experience. J Safety Res 38:689–695

Elvik R, Høye A, Vaa T, Sørensen M (2009) The handbook of road safety measures. Bingley, Emerald

Evans L (2006) Innate sex differences supported by untypical traffic fatalities. Chance 19(1):10–15

Gregersen NP (1996) Young drivers' overestimation of their own skill—an experiment on the relation between training strategy and skill. Accid Anal Prev 28:243–250

Gregersen NP, Berg H-Y, Engström I, Nolén S, Nyberg A, Rimmö P-A (2000) Sixteen years age limit for learner drivers in Sweden—an evaluation of safety effects. Accid Anal Prev 32:25–35

Haldorsen I (2015) Dybdeanalyser av dødsulykker i vegtrafikken 2014. Rapport 396. Oslo, Statens vegvesen, Vegdirektoratet

Mynttinen S, Sundström A, Vissers J, Koivukoski M, Hakuli K, Keskinen E (2009) Self-assessed driver competence among novice drivers—comparison of driving test candidate assessments and examiner assessments in a Dutch and Finnish sample. J Safety Res 40:301–309

Pasnin LT, Løsnæs S, Smedstad G, Moe D, Nordtømme ME (2009) Hvem fortjener politiets oppmerksomhet? En studie av dødsulykkene i trafikken i 2004 og 2005. Utrykningspolitiets temahefte 2/2009. Utrykningspolitiet, Stavern

Ross HL (1982) Deterring the drinking driver. Legal policy and social control. Lexington Books, Lexington, MA

Sapienza P, Zingales L, Maestripieri D (2009) Gender differences in financial risk aversion and career choices are affected by testosterone. Proc Natl Acad Sci 106(36):15268–15273

Shults RA, Elder RW, Sleet DA, Thompson RS, Nichols JL (2004) Primary enforcement seat belt laws are effective even in the face of rising belt use rates. Accid Anal Prev 36:491–493

Tessier K (2010) Effectiveness of hands-on education for correct child restraint use by parents. Accid Anal Prev 42:1041–1047

Trafikverket (2015) Analys av trafiksäkerhetsutvecklingen 2014. Publikation 2015:073. Borlänge, Trafikverket

Vaa T, Høye A, Almqvist R (2015) Graduated driver licensing: searching for the best composition of components. Latin Am J Manage Sustain Dev (forthcoming)

Chapter 5
What and How of Effective Police Enforcement

Dinesh Mohan and Rahul Goel

5.1 Introduction

Road traffic injury (RTI) reduction depends on interventions in institutional arrangements, road and environment design, vehicle safety features, post-crash care and ensuring safer road user behaviour by better policing systems. Regulation of traffic by police enforcement can be an effective strategy to reduce the public health burden resulting from traffic injuries (Peden et al. 2004; Elvik and Vaa 2004; Blais and Dupont 2005). As with many traffic safety interventions, the outcomes are not always as expected, and a weak theoretical foundation in traffic safety research makes it difficult to predict the effectiveness of different enforcement measures. For example, an increase in fixed penalties for speeding or jail terms for drinking and driving offences have not been found to be very effective deterrent measures in some studies (Elvik and Christensen 2007; SWOV 2013; Briscoe 2004; Criminal Justice Policy 2000; Li et al. 2006, Wagenaar et al. 2007). Given the large variation in road designs and types of traffic mix, a given intervention is likely to have varying effects across different settings. Traffic enforcement measures can be costly, lead to additional workload for enforcement agencies and may involve additional costs in publicising these measures through various platforms. It is therefore important to assess whether a given enforcement measure, though seemingly beneficial in its intent, actually results in any reduction of delinquent behaviour of drivers and number of crashes.

D. Mohan (✉)
Indian Institute of Technology Delhi, New Delhi, India

R. Goel
MRC Epidemiology Unit, University of Cambridge, Cambridge, UK

G. Tiwari and D. Mohan (eds.), *Transport and Safety*, Springer Transactions in Civil and Environmental Engineering,
https://doi.org/10.1007/978-981-16-1115-5_5

5.1.1 Theoretical Framework for Enforcement Measures

Elvik (2004) discusses a simple theoretical model that was developed by Evans (1991) which can be used to understand the finding of a road safety evaluation study (Fig. 5.1). The basic understanding according to this model is that there are two causal chains which connect a road safety measure to its final outcome—engineering and behavioural. In the context of traffic enforcement, we are concerned with the causal chain through the behavioural effect. There are, therefore, two main theoretical strands based on which we can explain the effectiveness of traffic enforcement. First is the theory which explains why drivers correct their behaviour when an enforcement measure is implemented. Second is the theoretical basis which explains why that particular change in behaviour would lead to higher safety. For example, an enforcement measure targeting over-speeding would likely result in reducing the proportion of drivers driving above a certain speed limit. This is the behavioural effect of the enforcement. The final outcome, i.e. number of crashes and accidents would then be dependent on the relationship between speed distribution and crashes.

The underlying theory which explains the effectiveness of different enforcement measures is called the 'deterrence theory', where deterrence is 'the omissions or curtailment of a crime from the fear of legal punishment' (Gibbs 1975). According to this theory, the fear of punishment encourages potential offenders to comply with the law. The enforcement measure works not only by apprehending the offenders, which is often a very small proportion of all road users and in fact a small proportion of all offenders, but also by discouraging 'potential' offenders because of the perceived certainty getting caught (Bjornskau and Elvik 1992; Ross 1982, 1992; SWOV 2013; Briscoe 2004).

> The principal opportunity for criminal law to be effective in reducing drunk driving is paradoxically, not by affecting the apprehended law violators, who stand within its power. Rather, it lies in affecting unapprehended individuals who are sensitive to the threat that, should they behave illegally, they will be punished. (Ross 1992)

There are two types of deterrence, specific and general. Specific deterrence primarily focuses on punishing apprehended offenders and assumes that they will be deterred from repeating their offence in the future to avoid punishment. On the other hand, general deterrence focuses on the population in general and assumes that the threat of punishment will deter people from violating the law in the first place. The

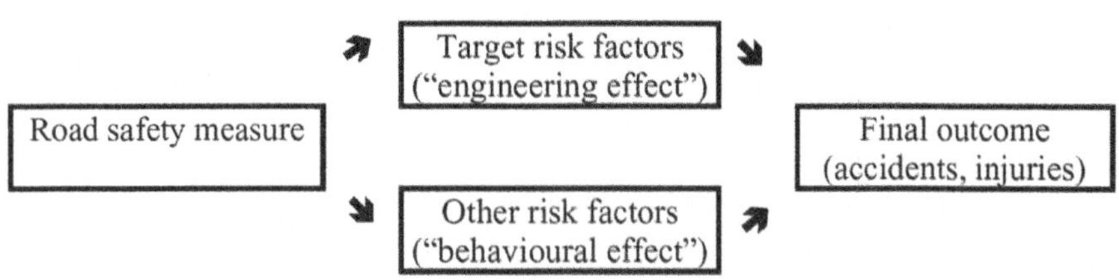

Fig. 5.1 Theoretical model of road safety. Adapted from Elvik (2004)

greater the perception of risk of punishment, the greater the likelihood that general deterrence will be effective. For an enforcement policy to be effective, it needs to ensure both types of deterrence are at work, so that a sanction not only impacts the individual who is being punished but also others who do not directly experience the sanction. The understanding of theoretical aspects that explain the effectiveness of enforcement measures is important to develop hypotheses for future application of these measures in different settings.

5.2 Objectives

In this paper, we assess the evidence base of effectiveness of on-road enforcement measures by conducting a review of systematic reviews on this topic. In this review, we focused only on the objective police programmes or strategies and excluded the reviews which assessed the effectiveness of a traffic enforcement law. This is because in different settings across the world a law may translate to actual implementation on the road by varying degrees in terms of how soon it is implemented as well as its spatial coverage. In some countries, while a law may exist, but its implementation may be limited because police may think of it as less of a priority or because there is lack of capacity to implement it (Shults et al. 2004b; Blais and Dupont 2005). We will use this review to answer the following questions:

(1) What are the different road safety enforcement measures for which evidence is available in systematic reviews and how current is this evidence?
(2) What are the different limitations or drawbacks of different studies as reported by the systematic reviews and what are their implications on results?
(3) What are the different factors which limit the generalisations of available evidence across different settings or across different types of modes?
(4) What is the theoretical basis of different enforcement measures?

5.3 Summary of Systematic Reviews

To find relevant studies, we used three main sources with a database of systematic reviews of road traffic injuries. These are The Handbook of Road Safety Measures (Elvik et al. 2009), Cochrane Injuries Review Group (https://injuries.cochrane.org/our-evidence) and Community Preventive Services Task Force (https://www.thecommunityguide.org/content/task-force-findings-motor-vehicle-injury). We also searched for the systematic reviews using the ancestry approach. We have not included any reviews published before 1990. Among the traffic enforcement measures, we found reviews covering four offences: speeding, red-light running, alcohol-impaired driving and seat belt use. In some cases, we found multiple reviews for the same enforcement measure. For example, the review of red-light cameras by Aeron-Thomas and Hess (2005) has been updated by Perkins et al. (2017). The speed camera review by Wilson et al. (2006) published in 2006 (Cochrane collaboration) was updated by Wilson et al. (2010) in 2010, and a study to update and expand the Cochrane systematic review, to provide a comprehensive account of the range of automatic speed enforcement strategies employed worldwide has been initiated by Steinbach et al. (2016). For effectiveness of speed cameras, we have also included a later contemporary review by Høye (2014) as it added value in terms of discussing some other facets that are missing from Steinbach's review.

5.3.1 Speed Control

One of the earliest reviews of the effectiveness of speed cameras was done by Pilkington and Kinra (2005) in 2005. The authors did not conduct a meta-analysis given the differing nature of the studies included. The review found that all the studies reported reduction in various crash outcomes (collisions, injuries and deaths). However, the authors noted that the level of evidence was relatively poor, and most studies lacked adequate comparison groups. The other two meta-analysis of speed enforcement were conducted more recently and includes all types of speed camera measures.

The most common speed enforcement methods are point-based where vehicle speeds are detected at fixed locations on the road. With point-based speed enforcement methods, the drivers get familiar with locations of cameras and modify their behaviour only in the immediate vicinity of speed enforcement. Hence, innovative approaches were needed to make speed enforcement more effective. Average speed enforcement method was developed as an alternative to point-based method. This is also referred to as 'average speed section control', 'point to point', 'time over distance' cameras or section control or trajectory control (21, 24). This type of enforcement involves the installation of a series of cameras at multiple locations along a road section. The average speed of a vehicle over a section of a road is calculated by capturing its licence plate number at more than one camera locations. In case this speed exceeds the posted speed limit, the vehicle information is communicated to a central unit. Almost all current installations throughout the world involve some degree of human verification to assess the validity of detected infringements. In such a system, there are stopping sites for manual enforcement.

A meta-analysis of speed cameras and average enforcement method was conducted by Høye (2014). The study reported the following:

- Speed cameras resulted in reduction of all crashes (20%) and larger reduction of fatal crashes (51%), though the latter may be affected by regression-to-the-mean.
- Section control resulted in larger reduction in all crashes (30%) compared to speed cameras, and reduction in KSI crashes (56%) was even greater than reduction of all crashes by section control and reduction of fatal crashes by speed cameras.
- The effect of speed cameras reduce as the distance from the camera increases.

- The authors also compared the reduction in the number of crashes as estimated by the two of the reviewed studies to the estimate from the power model of speed as reported by Elvik et al. (2009). Both the studies found crash reduction to be greater than what would be expected from the reduction in speed alone as predicted by the power model.
- The implementation of speed cameras may be accompanied by crash migration when drivers tend to slow down close to the cameras and then driver faster than they would have otherwise away from the cameras. The review found no evidence that this phenomenon, known as kangaroo driving, resulted in adverse safety effects.

Another review of speed cameras was done by the Cochrane group (Steinbach et al. 2016). This review includes mobile and fixed cameras, including the average enforcement methods. Unlike fixed cameras, mobile cameras are operated from parked motor vehicles, and therefore can be moved from one place to another. The study concluded the following:

- There was no difference of effect between the covert and overt cameras or between the urban and rural areas.
- There is a strong evidence suggesting that the implementation of speed cameras is associated with reduction in speed and crash outcomes.
- There was a reduction in percentage of vehicles exceeding the speed limit (50–64%).
- The effects do not account for the differences in the posted speed limit though percentage reduction is likely to be a function of the speed limit.
- There is evidence of some halo effect, i.e. the greater reduction in speed and crash outcomes in the vicinity of the cameras.
- No study provided empirical information on the effects of camera programmes on speeding and crash outcomes in the wider areas within which speed cameras are implemented, in order to assess whether general deterrence theory might be supported.
- No studies reported on the sizes of fines or penalties issued to offenders. By linking the size of fine with the specific road and camera where the driver had offended, it would be possible to assess whether larger fines and penalties are more effective. It is possible that 'persuasive' letters to offenders once caught speeding are equally effective a deterrent as being caught and brought to justice.

A review of the effectiveness of average speed enforcement methods was reported by Soole et al. (2013) in 2003. The review concluded the following:

- In general, drivers show higher level of acceptance of average speed enforcement. The traditional camera-based measures using instantaneous speed are criticised on the grounds that drivers need to speed at certain points due to unforeseen reasons.
- The limited evidence suggests that average speed enforcement method may be more effective than instantaneous speed enforcement methods.
- Studies have found the implementation of this method is associated with the reduction in average and 85th percentile speeds, the proportion of speeding vehicles

and speed variability. The approach has been specifically effective in reducing excessive speeding behaviour.

- In addition to reduction in speed, studies have also found considerable reduction in fatal and serious injury crash rates.
- There is a lack of distance 'halo' effect resulting from average speed enforcement implementation. This means that reduction in speed and crash rates have not been found outside the area of enforcement. Therefore, this enforcement method should be used as complementary to the existing fixed and mobile speed enforcement methods.
- Studies suffered from multiple drawbacks because of which the evidence needs to be carefully interpreted. None of the studies used the control/comparison site. Other drawbacks include lack of driving exposure data and studies not accounting for regression-to-the-mean effect.

There is a strong theoretical understanding based on which effectiveness of average speed enforcement method can be explained. Reduction in excessive speeding behaviour has considerable implications for road safety given the exponential relationship between vehicle speed and crash risk (Doecke et al. 2018; Elvik 2014; Koornstra 2007).

5.3.2 Red-Light Cameras

Red-light running results mostly in side-collision crashes which are more severe than other types of intersection crashes. In case there is a dedicated signal for the left-turning vehicles (in right-hand traffic), red-light running also results in head-on collisions. The implementation of red-light cameras (RLCs) is also associated with an increase in rear-end crashes resulting from drivers' tendency to apply break abruptly in order to avoid the fine. Since both the head-on and right-angle crashes have higher severity than rear-end crashes, even if the number of crashes is cancelled out, the severity level of crashes is still likely to reduce with the implementation of RLCs. A review by Høye (2013) summarises the empirical evidence of the effects of RLCs on intersection crashes.

- The present study found a non-significant decrease of all injury crashes by 13% and a non-significant increase of all crashes by 6%.
- Right-angle collisions were found to decrease by 13% (not statistically significant) and rear-end collisions were found to increase by 39% (statistically significant).
- For right-angle injury collisions, a far larger decrease was found (−33%, statistically significant), and for rear-end injury collisions, a smaller increase was found (+19%, statistically significant).
- The results seem to be affected to some degree by publication bias and the effects may, therefore, be less favourable than indicated. The direction of the effects does, however, not change when controlled for publication bias.

- The effects for crashes with unspecified severity are likely to be still more favourable when RLC-warning signs are not set up at each RLC-intersection, possibly because of drivers getting a habit of respecting red lights and expecting other drivers to braking. If this assumption is correct, one may also expect RLC to become more favourable over time.

A systematic review of the effectiveness of red-light cameras by Perkins et al. (2017) concluded the following:

- RLCs can be effective in reducing red-light violations and some types of traffic crashes, particularly, right-angle crashes, right-angle injury crashes and total injury crashes.
- RLCs also appear to be linked to an increase in rear-end crashes which is likely a result of drivers abruptly breaking to prevent the offence.
- The presence or absence of warning signs did not appear to have an impact on RLC effectiveness.
- While a number of studies reported that spillover (or diffusion of benefits) occurred, the magnitude of this effect is not established.
- Studies are limited to four countries: USA, Canada, Singapore and Australia. The authors caution the use of this evidence in the UK since the intersections in the USA and Australia are much larger in size than the UK; hence, drivers may have greater feeling of openness and more likely to jump the light. Further, the speed limits across the settings are different which may also influence the likelihood of red-light running.
- This review did not include studies which evaluated the effectiveness of red-light cameras used both for red-light running as well as enforcing speed limit during the green.
- Due to the rarity of death or severe injury events, most studies use a combined measure of crashes and do not differentiate between the severity levels of crashes.

In some cases, additional time is given to yellow times and successful RLC programmes may include many on-site modifications such as red-light visibility, addition of warning signs and amelioration of intersections geometry. This is clearly a case where engineering and enforcement measures are highly interrelated or at least the relationship between the two can be established (McGee and Eccles 2003).

5.3.3 Police Patrol for Alcohol-Impaired Driving

Control of drivers under the influence of alcohol has a strong empirical justification. A meta-analysis demonstrates that there is no evidence of a threshold effect for alcohol. Alcohol gradually affects driving skills. There is no sudden transition from unimpaired to impaired occurring at a particular BAC level. A review from the USA (Compton and Berning 2015) indicates that crash risk grows exponentially with increasing blood alcohol concentration (BrAC). The study shows that at low levels of alcohol (e.g. 0.03 BrAC) the risk of crashing is increased by 20%, at moderate alcohol levels (0.05 BrAC), risk increases to double that of sober drivers, and at a higher level (0.10 BrAC), the risk increases to five and a half times. At a BrAC of 0.15, the risk is 12 times, and by BrACs of 0.20+, the risk is over 23 times higher. Another meta-analysis concludes that 'most skills which are relevant for the safe operation of a vehicle are clearly impaired by BACs of 0.05%, with motor functions being more affected than cognitive functions and complex tasks more than simple tasks. Generally, the results provided no evidence of a threshold effect for alcohol. There was no driving-related performance category for which a sudden transition from unimpaired to impaired occurred at a particular BAC level' (Schnabel 2011).

A systematic review by Goss et al. (2008) of effectiveness of increased police patrols for preventing alcohol-impaired driving (including studies evaluating increased police patrols, either alone or combined with other interventions) targeting alcohol-impaired motor vehicle drivers concludes that:

- The 32 eligible studies included one randomised controlled trial, eight controlled before-after studies, 14 controlled interrupted time series (ITS) studies, six ITS studies and three studies with both ITS and controlled before-after analyses. Most interventions targeted only alcohol-impaired driving (69%) and included additional interventions such as media campaigns or special training for police officers (91%).
- Only two studies reported sufficient information to assess study quality completely. Two-thirds of studies were scored 'not adequate' on at least one feature. Five of six studies evaluating traffic fatalities reported reductions with the intervention, but differences were statistically significant in only one study. Effects of intervention on traffic injuries were inconsistent in the six studies evaluating this outcome, and no results were statistically significant.
- All four controlled studies evaluating fatal crashes reported reductions with the intervention, which were statistically significant in one study. All 12 controlled studies assessing injury crashes reported greater reductions with the intervention, though effects were minimal or not significant in several studies. ITS studies showed less consistent effects on fatal crashes (three studies) and injury crashes (four studies), and effect estimates were typically imprecise. Thirteen of 20 studies showed reductions in total crashes and about two-thirds of these were statistically significant.
- Therefore, the available evidence does not firmly establish that increased police patrols reduce the adverse consequences of alcohol-impaired driving. Good quality controlled studies with adequate sample size are needed to evaluate increased patrols. Also needed are studies assessing the cost-effectiveness of this intervention.

Evidence shows that an increase in the perceived risk of arrest appears to deter alcohol-impaired driving more effectively than increasing the severity of penalty after arrest and police patrol intervention increase the presence of police and the perception of being caught (Goss et al. 2008).

5.3.4 *Driving Under the Influence (DUI) Checkpoints*

These checkpoints refer to police operations where one or more police cars are standing beside the road and where police officers pull out drivers in order to check whether or not he or she has an illegal blood alcohol level (BAC). At these checkpoints, also known as sobriety checkpoints, drivers can be stopped even if they do not give any indication of driving under the influence of alcohol, and therefore, by correcting their driving behaviour close to these checkpoints does not necessarily prevent the drivers from being stopped. Erke et al. (2009) conducted a meta-analysis of the effectiveness of DUI-checkpoints. The review concludes:

- Crashes involving alcohol (or proxy measures of such crashes) are reduced at least by 17% and all types of crashes are reduced by 10–15%. Proxy measures of alcohol-related crashes include night-time or weekend night crashes.
- The largest reductions were found during the first 6 months of the DUI-checkpoint implementation, which may be confounded because the intensity of implementation may be much higher for short-term programmes.
- DUI-checkpoints in Australia result in the highest reduction in crashes indicating the Australian methods of booze buses and intensive publicity are highly effective. A similar approach when implemented in New Zealand also found large reductions, thus, strengthening the evidence of their effectiveness.

A practical implication from this meta-analysis is that highly visible checkpoints where many drivers are pulled out and tested, following the Australian example, are likely to be most effective.

5.3.5 *Seat Belts*

Dinh-Zarr et al. (2001) conducted a systematic review systematic review of the effectiveness of primary seat belt laws in the USA which included five evaluations of the effect of primary laws on observed seat belt use. These studies examined belt use in 12 states and the District of Columbia that enacted primary laws during the 14-year period from 1984 to 1997 and a couple of years later Shults et al. (2004a) re-examined the studies included in the systematic review to explore whether the benefits of a primary law differ based on: (1) the baseline seat belt use rate or (2) whether or not the primary law replaces a secondary law. This review includes studies from 1980 to 2000 and is restricted to the studies from USA. This review also estimates the effect of seat belt enforcement where the law is graduated from secondary to primary. A primary seat belt law implies that a driver can be stopped by enforcement officers solely for not wearing a belt. On the other hand, within a secondary seat belt law, the driver can be fined for seat belt only after the driver has been stopped for another offence. The authors hypothesised that a primary law has a greater effect on drivers' perceived risk of detection and punishment, and public in general may also perceive seat belt law as important. These factors may result in making a primary law more effective than a secondary law. The study concluded the following:

- All the studies evaluating primary versus secondary law found primary seat belt law to be more effective than secondary law. The studies which reported fatalities as outcome, found median decrease of 8% higher among primary law states than secondary law states, though statistical significance of this estimate was not reported.
- Enforcement enhancement programmes are associated with an increase in seat belt use (median 16 percentage points) and decrease in injuries.
- Based on the studies which carried out a follow-up of the enforcement enhancement programmes after they had concluded, there is evidence that the seat belt use somewhat declined after the programmes are ended.

Elvik et al. (2009) have reported meta-analysis of seat belt enforcement with no restriction to country and conclude the following:

- The results show the enforcement increases seat belt use by 21% during the enforcement period and by 15% afterwards.
- The covertness of the enforcement improves the effectiveness of seat belt use. Greater effects have been found when checkpoints are not announced compared to when they are. This may be possible if the drivers think that they will fasten the seatbelts close to a checkpoint, and therefore, general compliance may be lower.
- The change in seat belt usage rate is higher when the baseline rate is lower. A scatterplot of increase in usage rate versus the baseline usage rate shows a negative relationship between the two.

5.4 Effect of Intensity of Enforcement and Penalties on Deterrence

Though a great amount of research has been done on the mechanisms and processes of deterrence over the past four decades, the exact situations under which sanctions (or the threat of sanctions) are likely to influence or change a person's behaviour are still not known in certainty. The difficulty associated with determining casual relationships arises partly from the problem of eliminating competing explanations. Some of these include effect environmental design changes on-road user behaviour, changes in modal shares on the road and secular changes in people's behaviour over time. Another problem is that police enforcement levels and intensities can change over short periods of time due to economic and political changes, and so it is difficult to do long-term studies in many locations. Because there are no systematic reviews of the effect of penalties on deterrence in this section, we discuss the results of studies that are available.

5.4.1 Intensity of Enforcement

In 2002, Koornstra et al. (2002) published a report where they attempt to find a relationship between intensity of police enforcement and level of traffic law violation as an approach to get more insight about which enforcement level is needed in order to change road user behaviour and fatality risks. The results are shown in Fig. 5.2 illustrated by belt wearing and drunk driving data on enforcement and violation levels in Sweden, the UK, and the Netherlands at that time. The authors cautioned that this curve needs to be validated with research results because of the complexity of that research when it comes to differentiating police enforcement efforts (combined with publicity) and the complexity of data-collection. To the best of our knowledge, no serious efforts have been made to determine such curves for speed control, seat belt use, helmet use, DUI control and other violations for different modal shares in different countries of the world. What the curve does show is that per cent law violation decreases as enforcement intensity increases and that enforcement levels have to be different for different types of violations. For example, the curve shows that in Sweden the enforcement levels needed for control of DWI and for enforcing seat belt use so that violations were limited to about 12%, there had to be 250 checks per 1,000 driver licence holders for DWI and 8 for seat belt use.

Table 5.1 shows information regarding the alcohol-related BAC limits, road fatalities and enforcement measures in selected European countries (ECORYS 2014). This report commissioned by the DG for Mobility and Transport, European Commission, concluded that 20–28% of all road fatalities in the EU in 2012 could be attributed to drink-driving. This is a significant decrease from the 1980s when many countries

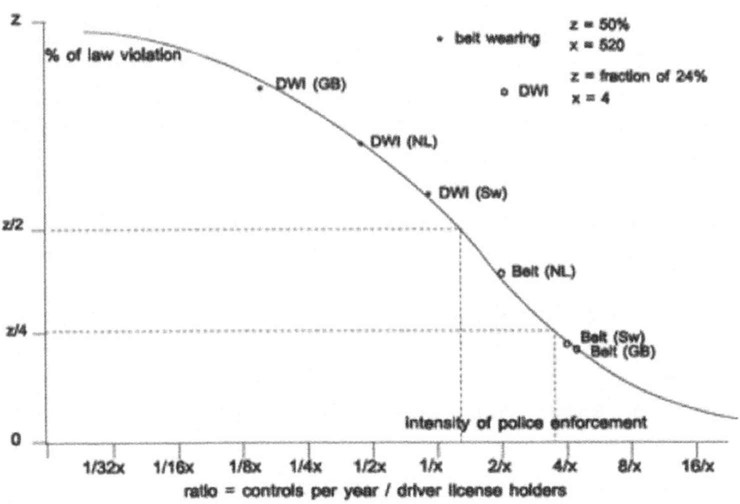

Fig. 5.2 Relationship between enforcement intensity and law violation levels (*source* Koornstra et al. 2002). GB—UK, SW-Sweden, NL–Netherlands, DWI—driving while intoxicated

Table 5.1 Overview of all information regarding the alcohol-related BAC limits, road toll and enforcement measures in selected European countries, data 2007–2010

Country	Legal blood alcohol limit (g/L)	Police tests per 1000 inhabitants	Share of alcohol-related road fatalities		Share alcohol offenders (above legal)	Share respondents who had at least once a week 5 or more drinks
		Number	Expert estimates	Official statistics	%	%
Poland	0.2	47	13	7	9.5	19
Portugal	0.5	63	35	6	5.9	28
Austria	0.5	87	18	6	5.8	36
Spain	0.5	112	NA	31	1.8	34
Hungary	0	130	8	31	3.1	24
France	0.5	190	29	31	3.3	20
Sweden	0.2	287	25	16	0.8	13
Finland	0.5	385	24	29	1.3	22

Adapted from ECORYS (2014)

reported share of alcohol-related fatalities to be in the range 30–45% (Sweedler et al. 2004). The data also show that in some countries expert estimates of the share of alcohol-related alcohol fatalities can be higher than the official statistics. Moreover, the definition of 'impaired' is different for each country. It ranges from 0.2 g/l in Sweden to 0.5 g/l in many countries and so a comparison of countries based on numbers of deaths from drink-driving crashes is not really possible. There is general agreement that there was a significant reduction in the period 1980–2010 which can be attributed to stronger laws, vigorous enforcement, and changes in social norms which all contributed to the progress that has been made though not much change has been observed over the last decade.

However, what is not clear is the level of enforcement that ensures a significant reduction in fatalities attributed to drinking and driving. Table 5.1 shows that the enforcement levels range from 100 to over 300 tests per 1,000 inhabitants per year. This means that in megacities like Delhi, Shanghai, Beijing or Mexico (populations in excess of 15 million persons), about 5,000–15,000 drivers would have to checked every day for effective control of drinking and driving in these cities. At present, we do not have reliable studies available to inform us about the minimum level of enforcement that needs to be put in place in a cost-effective manner in low and middle-income countries.

5.4.2 Effect of Penalties

In a review of deterrence-based measures on-road user behaviour, Davey and Freeman (2011) state:

- In order for the 'fear of punishment' to be effective, individuals must believe that the likelihood of apprehension for breaking the law is relatively high.
- A considerable body of early research demonstrated a weak negative relationship between perceived severity of sanctions and a range of illegal behaviours. That is, as perceptual severity increases, the likelihood of an individual committing that offence decreases; *however, an opposing body of research demonstrates that perceptions regarding the severity of penalties do not have the salient deterrent impact that was once assumed* (emphasis added). In fact, some researchers have reported a counter-intuitive relationship, with crime rates actually increasing with increases in the severity of the penalty. Nevertheless, it may be suggested that the greatest deterrent impact in regards to severity of sanctions will be found among those who have never committed an offence, rather than habitual offenders.
- It is recognised that for road safety, the swiftness of impending penalties is an important aspect for achieving deterrence. However, despite the link between the speed of the response and learned behaviour, the effects of the celerity of legal sanctions are by far the least studied of the three major deterrent mechanisms.
- In regards to general deterrence, a considerable body of evidence suggests that the threat of apprehension and subsequent legal sanctions, especially when supported by well-publicised media campaigns, can produce a deterrent effect, even if short, on offending behaviour.
- In order to create and maintain a deterrent effect, policing operations should be highly visible, sustained and widespread. This ensures that all motorists, whether newly licenced or experienced, perceive a constant high risk of apprehension. If drivers do not regularly observe policing operations, they may become undeterred which may be then reinforced by successfully engaging in offending behaviours that remain undetected, e.g. punishment avoidance.
- Any deterrence-based method employed in isolation does not offer a panacea for the problem of road accidents and fatalities.
- Our current understanding of the mechanisms of deterrence is based heavily on studies that have focused on younger populations. In fact, the bulk of published deterrence-based studies are from a small number of highly industrialised countries (e.g. USA, Canada, Australia, etc.), and thus deterrent forces are likely to fluctuate with the surrounding environment.

The above summary highlights the fact that we do not have enough systematic reviews that assess the effectiveness of general deterrence policies on-road safety that may have universal applicability. For the present, we have to rely on the studies that seem to point in a similar direction. A brief summary from different countries is given below.

5.4.2.1 Fixed Penalties—Fines

Norway

"For speeding in general, no effect of increasing fixed penalties can be found. For speeding close to speed camera sites, there is a weak tendency for the violation rate to go down. This tendency is not statistically significant at conventional levels. For seat belt wearing, wearing rates are found to increase as fixed penalties have increased. In recent years, however, enforcement of the seat belt law has stepped up, making it impossible to separate the effect of enforcement from that of fixed penalties" (Elvik and Christensen 2007).

The Netherlands

"Many studies have demonstrated that the combination of enforcement and penalties prevent the violation of traffic regulations and increase road safety. However, the most common type of penalty at the present time, a fine, has been found to have little effect... When road users consider the subjective probability of detection to be sufficiently likely, they will avoid violating a regulation... The combination of enforcement and penalty is generally preventative when road users avoid traffic violations on the basis of the expected negative consequences. In other words, road users adapt their behaviour without having already been punished. In particular, frequently conducted and very visible traffic checks, which are unpredictable in terms of time and place and are combined with public information campaigns, bring about the general prevention of traffic violations. Many studies have demonstrated that combining enforcement and penalties prevents violations and increases road safety. Of course, the penalty must match the seriousness of the violation and must be substantial enough to influence behaviour, but particularly the frequency, visibility and unpredictability of inspections are responsible for the general prevention of traffic violations. Making penalties heavier, as an isolated measure, has been found to have little extra effect. Research into the specific preventative effect of penalties shows that the effect of the currently most common type of penalty, a fine, is negligible when expressed in time. The effects are also negligible in terms of recidivism." (SWOV 2013).

Australia

"What we do know from the available evidence, however, is that the certainty of detection, apprehension and conviction does matter and in fact may matter more than punishment severity in deterring potential offenders. Informal sanctions from family, peers and colleagues who learn about the offence, and the resulting feelings of shame and embarrassment, are also anticipated costs associated with apprehension and conviction for an offence. Policies that can successfully increase the perceived certainty of detection and prosecution for drink-driving offences are therefore likely to have a greater impact on offending and, subsequently, road accident rates than those advocating harsher penalties." (Briscoe 2004).

"It is suggested that substantial increases in fines and licence disqualifications would have limited potential in deterring recidivist offenders. The present analysis, failed to find any evidence for a significant relationship between fine amount and the likelihood that an offender will return to court for a new driving offence. Nor was there any evidence from our analyses to suggest that longer licence disqualification periods reduced the likelihood of an offender reappearing before the courts." (Moffatt and Poynton 2007).

USA

"Speeding citations and their legal consequences are the most common enforcement tools to identify and control speeders, yet little is known about the effectiveness of a speeding citation. There was no significant effect of receiving legal consequences on the risk of receiving a subsequent speeding citation (adjusted RR 0.98, 95% CI 0.83–1.16)... Increasing drivers' perception that they are at risk of being caught speeding and awareness of the consequences from receiving points may improve the effectiveness of speeding law enforcement." (Li et al. 2006).

"We examined effects of state statutory changes in DUI fine or jail penalties for first time offenders from 1976 to 2002. Results: Twenty-six states implemented mandatory minimum fine policies and 18 states implemented mandatory minimum jail penalties. Estimated effects varied widely from state to state. Using variance weighted meta-analysis methods to aggregate results across states, mandatory fine policies are associated with an average reduction in fatal crash involvement by drivers with BAC $\geq$ 0.08 g/dl of 8% (averaging 13 per state per year). Mandatory minimum jail policies are associated with a decline in single-vehicle night-time fatal crash involvement of 6% (averaging 5 per state per year), and a decline in low-BAC cases of 9% (averaging 3 per state per year). No significant effects were observed for the other outcome measures. Conclusions: The overall pattern of results suggests a possible effect of mandatory fine policies in some states, but little effect of mandatory jail policies." (Wagenaar et al. 2007).

"Driving under the influence (DUI) is a significant public health problem... The results showed support for the swiftness and certainty of punishment, there was no support for the severity of punishment. That is, the relationship between the amount of the fine and DUI relapse was not significant. However, deterrence theory would expect certainty and severity of punishment to show a multiplicative relationship, meaning that severity would have its strongest effects when certainty of punishment was high. This interaction was not tested in either study; therefore, firm conclusions regarding the influence of fines cannot be drawn at this time." (Nochajski and Stasiewicz 2006).

New Zealand

"The question arises whether it is fair and appropriate to have flat-rate penalties (irrespective of prior records) for more and more offences, particularly in the cases of first offenders who can receive no concession and those who continue to re-offend and incur no additional penalty...There must be principled means for adjusting the amount of a fine to take account of both the offender's culpability and his or her resources and there must be efficient and reliable systems of collection and enforcement to ensure that most fines that are imposed will be paid in full and on time… Large fines are often difficult to collect and prove costly to enforce…as with infringements (although to a lesser extent) they may get to be perceived as a method of raising additional public revenue rather than as appropriate penalties for offences." (Criminal Justice Policy 2000).

5.4.3 Summary of the Evidence on Deterrence

- Legislation and enforcement are effective when violations are visible and easy to detect.
- Stricter punishment not as effective as subjective perception of being caught.
- Severe punishment and laws sometimes reduce enforcement by police officials and conviction rates in courts
- There is little evidence that severe penalties reduce violations in traffic, including jail sentences given in isolation.
- Announcement of severe punishments can have a deterrent effect over a short period and the beneficial effect disappears over time.
- All violations that are not considered serious in terms of threat to life or wilful negligent acts endangering the community (serious injury or death), and those that do not require court judgement should have fixed penalties. Penalties for such offences should be in proportion to the ability of the defaulter to pay.
- There is an absence of studies that could provide guidelines on police enforcement for low and middle-income countries on the following issues:
  - Influence of road and infrastructure design on traffic violations and the need of enforcement or effectiveness of enforcement.
  - Critical/minimum levels of enforcement necessary for different traffic violations.

5.5 Conclusions

There is a need to translate the results from car-based studies to settings where motorcycles and cyclists share the road space with cars. In such a context, what car-based studies refer to as property-damage only crashes may translate to higher severity crashes if the parties involved are cars/buses/trucks and vulnerable road users. This is the same for intersection crashes resulting from red-light running. The side crashes are often lead to high-severity crashes in case of cars. These will result in even higher severity injury crashes if between a four-wheeled vehicles hitting a motorcycle. It is possible that some of the enforcement measures which proved to be successful in car-based societies may lead to higher reduction in severity of crashes if not the number of crashes in contexts where vehicular mix consists of cars and a high proportion of vulnerable road users.

The reviews included focused on answering multiple questions. The outcomes include both the compliance rate for the law that is being enforced as well as the crash rates. The first outcome indicates how effective enforcement measure has been to reduce the delinquent behaviour of the drivers that was being targets. The second outcome which includes various metrics of crashes indicates whether enforcement measure translates to reducing the crashes which is not always a given. For instance, red-light camera enforcement results in overall increase in the number of crashes because increase in rear-end crashes may offset the decrease in side and head-on crashes resulting from red-light running.

The reviews have not discussed the injuries classified by the road user types. This means that there is a potential for a revised review of the same studies to understand the effect of the enforcement measures on-road users outside the cars such as pedestrians, cyclists and motorcycle riders.

In summary:

- Legislation and enforcement are effective when violations are visible and easy to detect.
- Stricter punishment not as effective as subjective perception of being caught.
- Severe punishment and laws sometimes reduce enforcement by police officials and conviction rates in courts
- There is little evidence that severe penalties reduce violations in traffic, including jail sentences given in isolation.
- Announcement of severe punishments can have a deterrent effect over a short period and the beneficial effect disappears over time.
- All violations that are not considered serious in terms of threat to life or wilful negligent acts endangering the community (serious injury or death), and those that do not require court judgement should have fixed penalties. Penalties for such offences should be in proportion to the ability of the defaulter to pay.
- There is an absence of studies that could provide guidelines on police enforcement for low and middle-income countries on the following issues:
  - Influence of road and infrastructure design on traffic violations and the difficulties of enforcement when designs are not adequate for the kind and volume of road users present.
  - Critical/minimum levels of enforcement necessary for different traffic violations.
  - Enforcement methods that would be cost effective in situations with high proportion of motorcycles and other vulnerable road users.

References

Aeron-Thomas A, Hess S (2005) Red-light cameras for the prevention of road traffic crashes. Cochrane Database Syst Rev (2). https://doi.org/10.1002/14651858.CD003862.pub2
Bjornskau T, Elvik R (1992) Can road traffic law enforcement permanently reduce the number of accidents? Accid Anal Prev 24(5):507–520
Blais E, Dupont B (2005) Assessing the capability of intensive police programmes to prevent severe road accidents: a systematic review. Br J Criminol 45(6):914–937
Briscoe S (2004) Raising the bar: can increased statutory penalties deter drink-drivers? Accid Anal Prev 36(5):919–929
Compton RP, Berning A (2015) Drug and alcohol crash risk. J Drug Addict Educ Eradication 11(1):29
Criminal Justice Policy, Group (2000) Review of monetary penalties in New Zealand. Ministry of Justice, Auckland
Davey JD, Freeman JE (2011) Improving road safety through deterrence-based initiatives: a review of research. Sultan Qaboos Univ Med J 11(1):29–37
Dinh-Zarr TB, Sleet DA, Shults RA, Zaza S, Elder RW, Nichols JL, Thompson RS, Sosin DM, Se TFCP (2001) Reviews of evidence regarding interventions to increase the use of safety belts. Am J Prev Med 21(4):48–65. https://doi.org/10.1016/S0749-3797(01)00378-6
Doecke SD, Kloeden CN, Dutschke JK, Baldock MRJ (2018) Safe speed limits for a safe system: The relationship between speed limit and fatal crash rate for different crash types. Traffic Inj Prev 19(4):404–408. https://doi.org/10.1080/15389588.2017.1422601
ECORYS (2014) Study on the prevention of drink driving by the use of alcohol interlock devices. European Commission: DG for Mobility and Transport, Rotterdam
Elvik R (2014) Speed and road safety—new models. Institute of Transport Economics, Oslo
Elvik R, Vaa T, Hoye A, Sorensen M (2009) The handbook of road safety measures, 2nd edn. Emerald Publishing Limited, Bingley, UK
Elvik R, Vaa T (2004) The handbook of road safety measures. Elsevier, Amsterdam
Elvik R (2004) To what extent can theory account for the findings of road safety evaluation studies? Accid Anal Prev 36(5):841–849
Elvik R, Christensen P (2007) The deterrent effect of increasing fixed penalties for traffic offences: the Norwegian experience. J Safety Res 38(6):689–695
Erke A, Goldenbeld C, Vaa T (2009) The effects of drink-driving checkpoints on crashes—a meta-analysis. Accid Anal Prev 41(5):914–923. https://doi.org/10.1016/j.aap.2009.05.005
Evans L (1991) Traffic safety and the driver. Van Nostrand Reinhold, New York
Gibbs JP (1975) Crime, punishment, and deterrence. Elsevier, New York
Goss CW, Van Bramer LD, Gliner JA, Porter TR, Roberts IG, DiGuiseppi C (2008) Increased police patrols for preventing alcohol-impaired driving. Cochrane Database Syst. Rev. (4). https://doi.org/10.1002/14651858.CD005242.pub2
Høye A (2013) Still red light for red light cameras? An update. Accid Anal Prev 55:77–89. https://doi.org/10.1016/j.aap.2013.02.017
Høye A (2014) Speed cameras, section control, and kangaroo jumps—a meta-analysis. Accid Anal Prev 73:200–208. https://doi.org/10.1016/j.aap.2014.09.001
Koornstra M (2007) Prediction of traffic fatalities and prospects for mobility becoming sustainable-safe. Sadhna—Acad Proc Eng Sci 32(4):365–396
Koornstra M, Lynam D, Nilsson G, Noordzij P, Pettersson H-E, Wegman F, Wouters P (2002) SUNflower: a comparative study of the development of road safety in Sweden, the United Kingdom, and the Netherlands. SWOV, Leidschendam
Li J, Lawpoolsri S, Braver ER (2006)Speeding tickets: effective deterrents for future violations or not? In: TRB 85th Annual Meeting Compendium of Papers CD-ROM 29. Transportation Research Board, Washington DC, pp 1–29
McGee HW, Eccles KA (2003) Impact of red light camera enforcement on crash experience, vol 310. Transportation Research Board, Washington DC

Moffatt S, Poynton S (2007) The deterrent effect of higher fines on recidivism: driving offences. Attorney General's Department, NSW Bureau of Crime Statistics and Research, Sydney

Nochajski TH, Stasiewicz PR (2006) Relapse to driving under the influence (DUI): a review. Clin Psychol Rev 26(2):179–195

Peden M, Scurfield R, Sleet D, Mohan D, Hyder AA, Jarawan E, Mathers C (2004) World report on road traffic injury prevention. Geneva World Health Organization

Perkins C, Steinbach R, Edwards P, Beecher D, Hess S, Aeron-Thomas A, Cohn E, Kakar S (2017) Red light enforcement cameras to reduce traffic violations and road traffic injuries. In: What works: crime reduction systematic review series. London

Pilkington P, Kinra S (2005) Effectiveness of speed cameras in preventing road traffic collisions and related casualties: systematic review. BMJ 330(7487):331–334

Ross HL (1992) Are DWI sanctions effective? Alcohol, Drugs and Driving 8 (HS-041 390)

Ross HL (1982) Deterring the drinking driver. Lexington Books, Lanham, MD

Schnabel E (2011) Alcohol and driving-related performance—a comprehensive meta-analysis focusing the significance of the non-significant (Alkohol und fahrrelevante Leistungen - Eine umfassende Metaanalyse). Ph.D., Fakultät für Humanwissenschaften, Universität Würzburg

Shults RA, Elder RW, Sleet DA, Thompson RS, Nichols JL (2004) Primary enforcement seat belt laws are effective even in the face of rising belt use rates. Accid Anal Prev 36(3):491–493

Shults RA, Nichols JL, Dinh-Zarr TB, Sleet DA, Elder RW (2004) Effectiveness of primary enforcement safety belt laws and enhanced enforcement of safety belt laws: a summary of the Guide to Community Preventive Services systematic reviews. J Safety Res 35(2):189–196

Soole DW, Watson BC, Fleiter JJ (2013) Effects of average speed enforcement on speed compliance and crashes: a review of the literature. Accid Anal Prev 54:46–56. https://doi.org/10.1016/j.aap.2013.01.018

Steinbach R, Perkins C, Edwards P, Beecher D, Roberts I (2016) Speed cameras to reduce speeding and road traffic injuries: protocol for a systematic review. London

Sweedler B, Biecheler M, Laurell H, Kroj G, Lerner M, Mathijssen M, Mayhew D, Tunbridge R (2004) Worldwide trends in alcohol and drug impaired driving. Traffic Inj Prev 5(3):175–184

SWOV (2013) Penalties in traffic. In: SWOV Fact sheet. SWOV, Leidschendam, NL

Wagenaar AC, Maldonado-Molina MM, Erickson DJ, Ma L, Tobler AL, Komro KA (2007) General deterrence effects of U.S. statutory DUI fine and jail penalties: long-term follow-up in 32 states. Accid Anal Prev 39(5):982–994

Wilson C, Willis C, Hendrikz JK, Bellamy N (2006) Speed enforcement detection devices for preventing road traffic injuries. Cochrane Database Syst Rev (2). https://doi.org/10.1002/14651858.CD004607.pub2

Wilson C, Willis C, Hendrikz JK, Le Brocque R, Bellamy N (2010) Speed cameras for the prevention of road traffic injuries and deaths. Cochrane Database Syst Rev (10). https://doi.org/10.1002/14651858.CD004607.pub3

Chapter 6
Principles for Development of Safer Rural Highway Systems for Conditions Prevailing in Low and Middle-Income Countries

Geetam Tiwari

6.1 Introduction

Road traffic safety has been recognised as a global health problem by all stakeholders in the new millennium. Since 2004 several international organisations have committed themselves to work towards reducing the burden of road traffic injuries (RTIs) such as the World Health Organisation (WHO), the World Bank and the United Nations. Following statements have been reiterated in several reports:

- A disproportionately high burden of road traffic deaths and injuries occur in low and middle-income countries (LMICs) (90%), and the burden is expected to increase due to rapid urbanisation and motorisation in LMICs.
- Globally, the number of fatalities per 100,000 population (mortality rate) ranges from less than 3 to almost 40. The rate is less than 9 in high-income countries (HICs) but averages around 20 in LMICs, with the African region demonstrating the highest rate (26.6) (WHO 2015).
- Road traffic injuries are predictable and preventable. Despite the growing burden of RTIs globally, multiple intervention strategies and projects have contributed to a significant reduction of the burden of road traffic injuries in many high-income countries (Peden et al. 2004).
- Empirical evidence for effective interventions is extensive, including enforcement of legislation on speed control and alcohol consumption, promotion of seatbelt and helmet utilisation, and safer design and use of roads and vehicles (Peden et al. 2004).

A recent study (Staton et al. 2016) published a systematic review and a meta summary of effectiveness of road traffic injury measures in LMICs. The authors referred to the Global Plan for the Decade of Action for Road Safety 2011–2020

G. Tiwari (✉)
Indian Institute of Technology Delhi, New Delhi, India

G. Tiwari and D. Mohan (eds.), *Transport and Safety*, Springer Transactions in Civil and Environmental Engineering,
https://doi.org/10.1007/978-981-16-1115-5_6

prepared by the WHO (WHO 2011). The plan has been developed to guide efforts at national and local levels to reduce the forecasted level of road traffic fatalities around the world. One of the important objectives of the plan is to strengthen institutional capacity on road safety management and improve the health system for post-crash response. Researchers have pointed out that most of the analysis to date has been conducted in high-income settings and has focused on vehicle occupants rather than the vulnerable road users in LMICs (Ameratunga et al. 2006; Staton et al. 2016).

The study included a search of 8560 articles and out of that 18 met all the inclusion criteria. Table 6.1 gives the list of included studies and type of intervention identified. Nine studies out of 18 deal with the effect of legislation, followed by education in 5 studies, enforcement in 2 studies and speed control and road improvement in one study each. Important conclusions from the study include the following:

- Legislative interventions had the strongest evidence for reduction in crashes and injuries.
- Enforcement of legislations had an impact on the success of intervention.
- Educational interventions had limited effect.
- Road improvements (paved vs. unpaved road) lead to increase in crashes.
- Two studies which included environmental changes by including speed humps resulted in lower crashes and injuries.
- Studies which included changing pedestrian behaviour to use pedestrian bridges and underpasses failed due to perceived risk and inconvenience and led pedestrians to increase their personal risk by creating their own path through the traffic.
- Further evaluations of road safety audits and implementation initiatives should be undertaken given the complex road environments found in LMICs.

This summary shows a focus on legislation and educational measures for improving road traffic safety problems in LMICs which have been evaluated. Only two studies evaluated impact of paved roads and speed humps.

More than 90% of studies selected by the authors to evaluate the effectiveness of legislation and educational strategies reflect the safety focus in these countries. This also implies that standards used for designing and constructing roads will ensure safety; therefore, only appropriate legislation regarding seat belt use, speed limits and alcohol control are required to improve traffic safety in LMICs. Often road standards (geometric design standards) in most LMICs have been based on either UK or USA design manuals. There are two important concerns in using or developing highway design standards mainly based on those in use in USA or UK. The traffic mix for which these standards have been developed is very different from the traffic existing in LMICs. Second concern is whether the design standards are based on traffic safety science even in the HICs (Hauer 1988).

In the next section, we show how the traffic and crash patterns are substantially different in LMICs. Therefore, highway design standards applicable to HICs highways may not have the same level of effectiveness in ensuring safe movement of traffic in LMICs.

Table 6.1 Road traffic injury prevention study characteristics (*source* Staton et al. 2016)

Authors[a]	Geographic Region	Prevention Type	Study Design	Targeted Population	Risk of Bias	Outcome Measures
Abegaz et al. (2014)	Ethiopia	Legislation	Interrupted time series	All road users	Low	RTC per 10,000 vehicles, RT deaths per 10,000 vehicles
Andreuccetti et al. (2011)	Brazil	Legislation	Interrupted time series	Vehicle drivers	Low	RT injuries, RT deaths
Bacchieri et al. (2010)	Brazil	Education	Non-randomised intervention-longitudinal with steeped wedge design	Cyclist	Low	Cyclist crashes, Cyclist "near miss"
Bishai et al. (2008)	Uganda	Enforcement	Interrupted time series/cost-effectiveness	All road users	Low	RT deaths, RT crashes
Chandran et al. (2014)	Mexico	Multifaceted (legislation and education)	Interrupted time series	All road users	Low	RT crashes, RT injuries, RT deaths
de Andrade et al. (2008)	Brazil	Legislation	Interrupted time series	All road users	Low	RT deaths
Espitia-Hardeman et al. (2008)	Colombia	Legislation	Interrupted time series	Motorcyclist	Low	RT deaths
Farange et al. (2002)	Brazil	Legislation	Cross-sectional	All road users	Moderate	TBI cases, RT crashes, RT deaths
Gomez-Garcia et al. (2014)	Mexico	Legislation	Interrupted time series	All road users	Low	Alcohol RT crashes, alcohol-related RT injuries
Guanche Garcell et al. (2008)	Cuba	Enforcement	Cross-sectional/time series	Vehicle drivers	High	RT crashes, RT deaths, RT injuries

(continued)

Table 6.1 (continued)

Authors[a]	Geographic Region	Prevention Type	Study Design	Targeted Population	Risk of Bias	Outcome Measures
Ichikawa et al	Thailand	Legislation	Interrupted time series	Motorcyclist	Low	Motorcycle crashes
Nadesan-Reddy and Knight (2013)	South Africa	speed control	Interrupted time series	All road users	High	Pedestrian/vehicle RT injuries
Passmore et al. (2010)	Vietnam	Legislation	Interrupted time series	All road users	Moderate	Risk for head injuries among RT injuries patient, RT deaths
Poli de Figueiredo et al. (2001)	Brazil	Legislation	Interrupted time series	All road users	Moderate	RT crashes, RT deaths
Rahimi-Movaghar (2010)	Iran	Community	Cross-sectional/community comparison	All road users	Moderate	RT injuries, RT deaths
Salvarani et al. (2008)	Brazil	Education	Interrupted time series	Vehicle drivers	Moderate	RT injuries, RT trauma severity, RT deaths
Swaddliwudhilpong et al. (1998)	Thailand	Education	Community RCT	Motorcyclist	Low	RT injuries
Zimmerrian et al. (2015)	Tanzania	Road improvement	Non-randomised intervention-longitudinal with steeped wedge design	All road users	Low	RT injuries rate

[a] All references from Staton et al. (2016)

6.2 Highway Traffic and Crash Patterns in LMICs

A substantial proportion of population in Asian and African LMICs continue to live in rural areas. Sixty-nine per cent of Indian population lives in rural areas at present and this will change to sixty per cent by 2030. Most African countries will also continue to have larger proportion of populations living in rural areas. Expansion of urban areas will continue along the highways in many LMICs as has been experienced in the past few decades in India. Therefore, the number of habitations along the highway will continue to grow. Presence of villages and small towns along the highways has resulted in a mixed traffic patterns on highways in most LMICs posing traffic safety challenges to road designers.

Table 6.2 shows the presence of motorised two-wheelers (MTW), bicycles and other slow moving vehicles on segments of national highways in India (Dey et al. 2006). Similar patterns were reported in an earlier study (Tiwari et al. 2000) covering different types of highways in India.

These highways experience high speed variations also as shown in Table 6.3 (Dey et al. 2006; Tiwari et al. 2000). A large portion of highways pass through small habitations in North India (Table 6.4) (MoRTH 2010; NHAI 2007, 2011, 2015, 2016; NHIDCL Sikkim 2016). Similar trends are observed in Bangladesh, Myanmar and Cambodia (Table 6.5) (ADB 2008, 2009; Ministry of External Affairs (India) 2013). Density of small towns and villages along the highway and presence of tractors, MTW and three-wheelers on the highway along with cars, buses, trucks and truck trailers presents a very different traffic mix as compared to North America and Western Europe where most of the highway standards have been developed.

Figure 6.1 shows traffic movement on a typical 2 lane undivided highway in India. Paved shoulders are used by bicyclists while fast moving heavy trucks pass on the adjacent lane. Figure 6.2 shows four-lane divided highway. Most of the National highways have this configuration in India. Left lane and shoulders are used by three-wheelers, two-wheelers and bicyclists (India observes left-hand traffic rules). All these vehicles are exposed to high speed motorised traffic moving on the right lane.

Figure 6.3 shows traffic moving on a four-lane divided highway in India. A tractor and other slow moving vehicles are on the left lane, and heavy vehicles use the right most lane. Other motorised traffic overtakes from the left lane (passenger side), because heavy traffic is on the right lane. Heavy vehicles find it difficult to use the left lane due to the presence of parked vehicles, tractors, bicyclists and motorcyclists. Figure 6.4 shows a four-lane highway passing through a small habitation. Traffic mix is very similar to what is often observed in an urban area. These highway segments have the presence of long-distance traffic as well as local traffic.

Traffic crash patterns in LMICs are also substantially different as compared to North America and Western Europe. Table 6.6 shows crash patterns observed in highways in India (Mohan et al. 2017). Pedestrian and motorcyclist involvement in fatal crashes on rural highways is greater than that of other road users. These highway crash patterns are similar to those observed in urban areas.

Table 6.2 Traffic mix in selected segments of rural highways in India (*source* Dey et al. 2006)

Highway No.	Traffic Vol	Composition (%)							Ratio of fast and slow moving vehicles	Sample size
		Fast moving vehicles				Slow moving vehicles				
		Car/jeep	Truck/bus	MTW	3-wheeler	Bi-cycle	Tractor	Others		
NH-58	244	31.9	40.2	9.00	–	6.90	8.7	3.3	81.1:18.9	405
NH-73	408	34.3	5.90	29.7	6.1	16.5	5.0	2.5	76:24	756
NH-31	412	22.3	26.9	23.8	4.9	17.7	2.9	1.5	77.9:22.1	780
NH-31	250	26.4	36.8	11.2	–	15.2	6.8	3.6	74.4:25.6	459
NH-73	412	16.5	28.2	22.8	1.7	19.7	4.8	6.3	69.2:30.8	752
SH-67	839	21.5	5.60	41.1	1.3	25.3	0.5	4.7	69.5:30.5	1407
NH-74	1,032	21.4	6.30	40.4	1.0	25.3	0.8	4.8	69.1:30.9	1672
NH-30	480	16.7	21.2	11.7	11	27.3	6.3	5.8	60.6:39.4	877
NH-58	580	27.0	32.4	12.6	1.6	18.6	4.5	3.3	73.6:26.4	1031
NH-7	1,211	36.5	19.0	16.5	6.8	9.50	10	1.7	78.8:21.2	1889
SH-5	228	42.1	17.1	24.6	2.2	6.10	6.6	1.3	86:14	393
SH-6	428	29.2	9.40	32.9	6.8	14.7	4.9	2.1	78.3:21.7	756
SH-14	623	21.5	7.50	41.6	1.8	17.2	9.0	1.4	72.4:27.6	1073
SH-59	600	22.5	13.0	35.0	–	20.0	8.0	1.5	70.5:29.5	1107
SH-47	425	17.0	26.6	31.5	3.3	8.90	10.1	2.6	78.4:21.6	735
SH-47	578	23.9	16.1	38.1	2.4	7.10	11.9	0.5	80.5:19.5	993
NH-73	293	16.0	27.7	30.4	3.4	15.0	6.1	1.4	77.5:22.5	547

Table 6.3 Speed variation observed on selected highway segments in India (*source* Dey et al. 2006)

Highway number	Traffic volume	V_{85}	V_{50}	V_{15}	$(V_{85}-V_{50})/(V_{50}-V_{15})$
NH-58	244	51.6	37.8	17.8	0.69
NH-73	408	63.6	48.6	15.4	0.452
NH-31	412	39.8	27.2	15.6	1.086
NH-31	250	48.0	35.6	13.4	0.559
NH-73	412	45.6	33.2	11.8	0.579
SH-67	839	37.4	27.6	10.6	0.576
NH-74	1032	43.2	31.0	11.4	0.622
NH-30	480	28.6	20.6	11.6	0.889
NH-58	580	45.0	35.4	20.2	0.632
NH-7	1211	54.8	44.8	18.2	0.376
SH-5	228	55.8	45.4	18.2	0.382
SH-6	428	49.2	35.2	24.8	1.346
SH-14	623	45.0	34.6	15.0	0.531
SH-59	600	45.0	34.8	17.4	0.586
SH-47	425	60.4	35.2	17.4	1.416
SH-47	578	59.0	34.8	20.8	1.729
NH-73	293	59.2	34.6	21.6	1.892

Table 6.4 Village settlement along selected rural highways in India per 100 km (*source* MoRTH 2010; NHAI 2007, 2011, 2015, 2016; NHIDCL Sikkim 2016)

NH No	State	Total length (km)	Section (From-To)	Percentage passing through developed area (%)	Number of villages	Frequency of settlements (km/village, km/town)
NH 510	Sikkim	33	Singtam to Rabangal	79	18	1.8
NH 34	West Bengal	84	Barasat to Krishnagar	68	66	1.3
NH 33	Jharkhand	56	Mahulia to Baharogora	NA	75	0.75
NH 6	Jharkhand	16	Baharagora to Chichra	NA	28	0.55
NH 74	Uttarakhand	99	Nagina to Kashipur	NA	48	2.9
NH 8E	Gujarat	210	Gadu to Dwarka	NA	29	7.2
NH 59	Madhya Pradesh	155	Indore to Jhabuwa	NA	80	1.93

Table 6.5 Villages in the influence zone of highways in LMICs

Country	Project name	Total length (km)	Section	Number of villages	Frequency of Settlement (km/village, km/town)
Myanmar[a]	Preparation of Detailed Engineering Project Report for Two-Lane Road from Paletwa to Zorinpui (India Myanmar Border) in Chin State of Myanmar, 2013	109.20	Paletwa to Zorinpui (Indo-Myanmar Border)	24 (Along the project road)	1.8
Bangladesh[b]	Road Network Improvement and Maintenance Project—II, 2008	241.47	6 road segments	9154 (In zone of influence)	0.02
Cambodia[c]	Primary Roads Restoration Project, 2009	577	3 major Roads (NR 5, NR6, NR7)	132 (Beneficiaries)	

[a]ADB (2008); [b]Ministry of External Affairs (India) (2013); [c]ADB (2009)

Fig. 6.1 Mixed traffic on a two-lane undivided highway in India

Fig. 6.2 Mixed traffic on a four-lane undivided highway in India

Fig. 6.3 Mixed traffic on a four-lane divided highway in India

In North America, 10% of the fatal crashes on highways involve pedestrians. The presence of motorcycles is negligible and long stretches of roads pass through wilderness. A large proportion of the highways are access controlled and designed for four-wheeled motorised traffic. Therefore, the road standards that have evolved to make an access controlled highways safe for motorised vehicles may not ensure safety to other road users present on LMICs highways. However, in India, standards similar to those in HICs have been adopted (IRC 2007; MoRTH 2010). In the past two decades, major investments have gone into expanding the national highway system in India. Yet the number of fatalities has continued to grow. Density of highways in a state and number of fatalities seem to have a strong correlation. There is a strong reason to question the safety aspects of current standards in use.

Fig. 6.4 Highway passing through habitation: mixing of long-distance traffic and local traffic

Table 6.6 Crash patterns observed on selected highway segments in India (*source* Mohan et al. 2017)

	Location	Fatalities by type of road users, per cent						
		Pedestrians	Bi-cycle	Motorised two-wheeler	Car	Bus	Truck	Unknown and others
Urban	Mumbai (2008–2012)	58	2	29	4	0	0	7
	Delhi (2013)	47	10	26	3	4	3	7
	Highways (1998)	32	11	24	15	3	14	1
Rural highways	2-lane NH 8 (2010–2014)	20	2	42	14	9	13	1
	4-lane NH24 (2010–2014)	27	5	44	8	7	4	4
	6-lane NH1 (2010–2014)	34	3	10	6	5	41	1

6.3 Road Standards and Safe Roads

Hauer (1988) has discussed extensively the role road standards can play in ensuring safety. Giving numerous examples from AASHTO and MUTCD, Hauer states that "What civil engineers do has a major effect on road safety. However, contrary to appearances, the level of safety built into roads is largely unpremeditated. Standards and practices have evolved without a foundation knowledge. At times the safety consequences of engineering decisions are not known, at others some knowledge exists but is not used". Hauer explains how "There are no safe highways .... However, it is correct to say that *highways can be built to be safer or less safe*. Road safety is a matter of degree". Perhaps, this statement is applicable to all highways regardless of region and geographical location. In LMICs, we have to investigate the safety of a road measured by the frequency and severity of crashes occurring on it.

Hauer further explains "Not all is known about the relationship between road design and safety. However, from research and experience, mostly from HICs, we have learned that building a wider median, placing obstacles further from the travelled lanes, providing more pavement friction, consistently designing curves with larger radii, providing full illumination, etc., all make for safer highways". However, the impact of these measures when the highway traffic consists of motorised two-wheelers and non-motorised vehicles (NMV) is not known. Impact of wide shoulders, paved shoulders, etc., are also unknown in case of mixed traffic on highways.

Highways in LMICs do not appear to have become safer. In India, the up-gradation of undivided highways into divided highways from 2 to 4 lanes have increased rate of traffic fatalities (Naqvi and Tiwari 2015). Pedestrian fatalities form a very small proportion of fatalities in HIC highways, whereas many LMICs continue to have 20–40% pedestrian fatalities on highways (Mohan et al. 2017).

The law of diminishing marginal returns to safety improvement measures is discussed by Hauer:

> Many highway design decisions are about dimensions for which there is no sharp border between safe and unsafe, only a gradual change diminishing in magnitude. Thus, e.g., increasing the width of the median from 50 to 60 m will save fewer crashes than an increase of the median width from 5 to 15 m. Eventually there is a width at which one says: "widening the median further cannot be justified because the improvement in safety is too small."

This kind of detailed information is not available for LMICs highways. Detailed studies have not been carried out in Africa or Asia where major expansion in highway network is expected in the coming decades. Wide medians and wide shoulders require more land acquisition resulting in higher costs or constrained by non-availability of land. In India, most highways pass through villages and towns with heavy density of pedestrians and bicycles and other NMVs. Wider medians and wider shoulders may be used by these vehicles, and the adjacent traffic lanes carry motorised vehicles moving at 80–100 km/h. The impact on traffic safety because of the presence of NMVs on wide shoulders and median is unknown. The relationship between highway features like lane width, shoulder width, median design and safety is not known with the kind of precision that is possible in other engineering disciplines that allow experimentation.

Hauer (1988) also questions the current practice of conducting safety audits where the focus is on mere compliance with the current standards. Hauer (1988) states:

> No road in use is entirely crash-free, and therefore, in the interest of honest human communication no road can be called safe. The safety of a highway does not change abruptly when some highway dimension changes slightly. It follows that meeting or not meeting a dimension standard does not correspond to a road being 'safe' or 'unsafe'. Also, highway design standards evolve with time. We used to build lanes 3.6 m (12 feet) wide, now the standard calls for 3.75 m wide lanes. This does not mean that the entire stock of old highways with 3.6 m lanes is unsafe. It means only that the information, the judgements, and the economic considerations that go into the formulation of design standards change in time. In short,

highway design standards are not the demarcation line between what is safe and unsafe. They are a reflection of what a committee of professionals of that time considers to be overall good practice.

Hauer (1988) has provided useful insights into the relationship between safety and highway standards; however, it is true that a large number of road safety measures have been found to be effective in HICs, and major improvement in road safety has been observed in many countries since 1970s.

6.4 An Overview of Factors Affecting Road Traffic Crashes

Factors that can influence traffic crashes are broadly related to traffic patterns, road geometry, driver and vehicle characteristics, and pattern of surrounding environment. Traffic characteristics (such as traffic flow and speed) and road characteristics (such as road geometry and the quality of infrastructure) might affect road accidents. Wang et al. (2013) have discussed some important factors in an exhaustive review published in 2013.

6.4.1 Speed

Vehicle speed has a very strong correlation with severity and occurrence of traffic crashes (Elvik et al. 2004a, b). Higher speeds are associated with higher severity of crashes if other factors (e.g. environment and vehicle design) remain the same. This has been shown by a large number of researchers (Hauer 2009; Aarts and Schagen 2006; Kockelman and Kweon 2002; O'donnell and Connor 1996; Shankar et al. 1996). Nilsson (2004) employed before-after studies in Sweden using the Power Model to investigate the impact of speed on traffic safety. It was found that changes in the number of accidents (or accident rate) can be associated with the changes in speed according to a power function. Similarly, Elvik et al. (2004a, b) undertook an extensive evaluation on the effect of speed on accidents again using the Power Model. They concluded that there is a causal relationship between changes in speed and changes in road accidents. Some studies have discussed the role of speed variance to increase in accident frequency (Lave 1985). However, a later study (Davis 2002) argued that such a claim of "variance kills" may be subject to ecological fallacy. Therefore, there remains the question of the role of speed variance in road safety.

6.4.2 Traffic Flow

Traffic flow is a measure of a number of vehicles present on the road. It is reasonable to assume that more vehicles present on the road will lead to more conflicts and hence more traffic crashes. Therefore, a large number of researchers have examined the relationship between traffic flow and accidents (Belmont and Forbes 1953; Ceder 1982; Ceder and Livneh 1982; Gwynn 1967; Turner and Thomas 1986). The results have been mixed. Earlier studies found that the accident rate increases linearly with the hourly traffic flow for two-lane road (Belmont and Forbes 1953). However, it was found later that for different types of accidents, the relationships between accident rates and hourly traffic flow are different. Ceder and Livneh (1982) found, for example, hourly traffic flow was found to be inversely related with accident rates for single-vehicle accidents in all cases; while in some cases, hourly traffic flow was found to be positively related with accident rates for multi-vehicle accidents. Under different flow conditions, the relationship between the total accident rate and hourly flow was found to follow a U-shaped curve under free-flow conditions while for the case of "congested" flow data the accident rate increased more sharply. Clearly, the impact of traffic flow on accident rate was found to be different under different traffic flow conditions. Martin (2002) found that accident rates are highest in light traffic compared to heavy traffic, especially on three-lane French motorways. The study also showed that hourly accidents were much worse in a night-time and light-traffic situation. Therefore, the author concluded that light traffic (low traffic flow) is a safety problem both in terms of accident rate and severity. As many things could affect road safety during night-time, however, such as lighting, this is an area requiring further study.

Similar work by Golob and Recker (2003) demonstrated that accident severity generally tracks the inverse of traffic volume. Overall, research studies suggest that the total number of accidents increases as traffic flow increases. In terms of accident rates, it seems to have a U-shaped relationship with hourly traffic flow. As the number of vehicles increases, the number of crashes will most probably increase, however, exactly how the number of crashes increases is important. The change in number of crashes in proportion to the traffic flow is defined as risk. Several researchers have found that the risk decreases with increase in traffic flow (Hauer and Bamfo; Ardekani Hauer).

6.4.3 Traffic Density

Traffic density expresses the combined effect of traffic flow and speed. The relationship between traffic density and crashes has been investigated often using volume by capacity ratio (V/C) (Ivan et al. 2000; Shefer 1994). The hourly accident rates (per million vehicle kilometres) and the V/C ratio on a US interstate highway were examined in a study (Zhou and Sisiopiku 1997). The results showed that the relationship follows a U-shaped pattern and accidents involving injury and fatalities tended to decrease while the V/C ratio increases. Another study (Ivan et al. 2000) found a negative exponential relationship with density (volume/capacity ratio), controlling for land use, environment and lighting. The results showed that the accident rate is the highest at low V/C ratio.

A freeway segment study (Lord et al. 2005) found an overall inverse relationship with the number of accidents (per year per km). Accident-density and accident-V/C relationships were also examined according to different accident categories such as total, single-vehicle and multi-vehicle accidents. The result showed an U-shaped relationship for total and single-vehicle accidents but a positive relationship for multi-vehicle accidents.

Generally, mixed relationships have been found between density and safety in the literature, depending on the measurements of density and types of accidents. Clearly, this is an area that has been studied less and as such further research is required.

6.4.4 Road Characteristics

Engineering theory suggests that road designs—lane width, shoulder presence, number of lanes, median design—influence driving behaviour (operating speeds, lane changes, etc.); therefore, it is reasonable to expect that the road geometry can influence traffic safety. Improvement in road geometry is expected to improve traffic safety. The suggestion of different operating speed different road terrain and geometry is based on the notion of road safety in highway design (Lamm et al. 1999). Findings from several researchers support this hypothesis. Shankar et al. (1995) explored the effects of various roadway geometrics (e.g. horizontal and vertical alignments) on road accident frequency and found that the increased number of horizontal curves per kilometre on rural freeways increase the possibility of an accident resulting in "possible injury" relative to "property damage only". Another study (Milton and Mannering 1998) found that short sections are less likely to experience accidents than longer sections; narrow lanes (less than 3.5 m) and sharp horizontal curves tend to decrease accident frequency in Eastern Washington. Noland and Oh (2004) analysed the county-level highway data from the State of Illinois in the USA and found that an increase in the number of lanes and lane widths was associated with increased fatalities; and an increase in the outside shoulder width was found to be associated with reduced accidents. Positive relationship between the increased number of lanes and accident numbers was reported by Kononov et al. (2008) also.

Four important studies have reported relationship between road curvature and its association with traffic accidents (Haynes et al. 2007, 2008; Milton and Mannering 1998; Wang et al. 2009). All four studies found that road curvature is a protective factor meaning that more curved roads in an area result in less road accidents. Contrary to these results, Abdel-Aty and Radwan (2000) found that the degree of curve increases the number of accidents on a road segment. Wang et al. (2009) have given an explanation for this:

> This may be because different curvature measurements were used, such as minimum radius, number of horizontal curves per mile, mean horizontal deflection angle, degree of horizontal curve per 100 m arc, bend density (Abdel-Aty and Radwan 2000; Haynes et al. 2007; Noland and Oh 2004; Shankar et al. 1995). In addition, these studies were conducted at different scales, which may also be subject to the modifiable areal unit problem (MAUP) (Openshaw 1984). Curvature may be risky considering its engineering effect; however, from the behavioural aspect, drivers may drive more slowly and cautiously on curved roads. On the other hand, on straight roads as mentioned above, drivers are more likely to fall asleep or feel bored (physiological theory). Therefore the overall safety effect of road curvature (compared to straight roads) is likely to be mixed.

These findings that road infrastructure designs do affect road safety. In the context of LMICs, if the presence of curvature on the road reduces speed, a number of crashes are likely to reduce. However, this has to be established with data from LMICs.

Road infrastructure improvements (e.g. road upgrading and pavement) and roundabout design are also found to be beneficial for safety. In the case of HICs, not only does better vehicle design, but also improvements in road safety engineering reduce the severity of whiplash injuries when accidents occur, and this could be done by enhanced signal visibility or through complex intersection geometric upgrades (Navin et al. 2000; Perez 2006).

In the case of LMICs, the safety benefit of roundabouts is clear, however, up-gradation involving improved pavement surface, wider lanes, wider shoulders may lead to higher speeds and increase opportunities for lane changing and conflicts. Pedestrians and slow vehicles on the curb side lane or shoulders will be exposed to motorised vehicles moving at much higher speeds. Safety benefit of road up-gradation using present standards is unclear for LMICs.

The *Handbook of Road Safety Measures* gives a comprehensive overview and illustration of how various interventions impact road safety (Elvik et al. 2009). Almost all the research in the handbook comes from studies done in HICs and very little for LMICs. Can this knowledge be simply transferred to LMICs?

6.5 Safety Science and Safety Vision

Study of international literature of how countries have improved their safety performance over the years shows a multitude of potential explanations. Researchers have developed benchmarking methodologies to learn from international comparisons (Koornstra et al. 2002; Wegman and Hagenzieker 2010). This is a difficult task due to the presence of confounding factors. Wegman (2017) states "It is, to the best of our knowledge, fair to say that no country has a full explanation of the progress made. However, it is also fair to say that our knowledge and understanding of why countries made progress has increased significantly over the last few decades".

In the last few decades, traffic safety measures have been focused on the following:

- Improving human behaviour (speed, alcohol, seat belts and helmets) through legislation, enforcement and campaigns.
- Safer infrastructure through planning and design.
- Safer vehicles through better crashworthiness, active vehicle safety and vehicle inspections.

We examine these interventions in the context of theories of safety science to understand what role different interventions may have on safety consequences.

It has been a tradition in road safety to analyse road safety data for understanding why crashes occur, which factors influence risks, and what determines crash severity, and based on this understanding, to arrive at reliable conclusions on how to prevent them most effectively and efficiently. We call this a data-driven approach. In this approach, we derive priorities by using crash data, background data, exposure data and data of safety performance indicators. This is what the researchers call a scientific method and evidence-based interventions. However, there has to be a fundamental vision or theory which drives what data should be collected and what should be evaluated. For example, if the theoretical understanding is that driver error causes traffic crashes and driver training can reduce traffic crashes the data collection process focuses on collecting driver-related data and modelling impact of driver characteristics and knowledge about driving rules. If the theoretical understanding is that driving behaviour is influenced by the road and traffic characteristics, then road geometry and traffic characteristics (operating speeds, traffic volume, type of vehicles) are modelled for controlling traffic crashes.

Stoop et al. (2017) provide a detailed discourse on safety science in general and its application to transportation safety. They highlight the development of three basic notions as the cornerstones for safety science as a scientific discipline, no matter what domain it is related to: interdisciplinarity, problem-solving orientation and systems approach.

Interdisciplinarity is a first necessary condition to deal with complex phenomena that exist in reality: such phenomena cannot be reduced to paradigmatic notions within one scientific domain. However, a decomposition of their complexity is a prerequisite for unravelling their control laws, properties, relations, variables and performance indicators.

Problem orientation: Achieving consensus on a common problem definition is considered a second prerequisite for a scientific approach of safety. Interdisciplinary discourses may result in controversies and rivalries, up to the level of schools of thinking. It also may lead to individual antagonism, defining minorities as dissenting voices in a homogeneous scientific community.

Systems approach: While an interdisciplinary approach may provide coverage of broad issues on which a variety of actors, disciplines and stakeholders may have achieved consensus, resolution of the problem and enhancement of the actual safety level of performance to the required level requires a third necessary condition: the application of a systems theory. Systems theory facilitates in structuring a complex reality. A decomposition in elements, components, aspects and relations provides oversight and coherence across levels and entities that interact with each other. Putting events in the context of systems in which they operate requires a distinction between event and system, similar to a patient and the health system or a convict and the judicial system. While accidents should be prevented for the sake of their unacceptable consequences, the object of research for safety interventions is at the systems level (ESReDA 2009; Stoop 2015).

Perrow (1984) has discussed some fundamental properties of complex systems in terms of tight and loose coupling between different elements and nonlinear interactions between elements, and therefore, the need for building systems which do not depend on user or operator alone to ensure safety. Traffic crashes present an excellent example of complex systems, uncertainties and nonlinear interactions between human beings, vehicles and the road environment. This makes a strong case for moving away from focusing on the errors that road users make to concentrating on road and vehicle designs that can reduce the propensity and severity of crashes. Safety science has had a major influence on the traffic safety theories in the last fifty years.

6.5.1 Traffic Safety Theories

Elvik et al. (2004a, b) have discussed two important road safety theories that are related to engineering and human behavioural effects. Road safety measures could affect road safety by influencing relevant factors through engineering effect and behavioural adaptation. This suggests that engineering and human behaviour related factors are two important sources of risks. For example, road lighting improves visibility (engineering effect) but road users tend to be less alert (behavioural adaptation). Most factors can be related to either engineering or human behavioural effects. Vehicle-related factors can also be explained through engineering effects. For instance, compared to cars large trucks have unique characteristics, most notably high gross weight, long vehicle length and poor stopping distance, which can be associated with different levels of risk (Chang and Mannering 1999).

Many other safety theories can be explained based on the engineering and behavioural theories. For instance, drivers can modify their behaviour based on what they see on the road ahead of them (e.g. increasing speed or reducing attention), especially when the lower risk is brought about by a road design countermeasure (Assum et al. 1999; Hauer 2017). Physiological theory may be related to both engineering and behavioural theory to some extent. For instance, it was suggested that drivers are more likely to fall asleep or feel bored on straight, monotonous, dual carriageway roads with little traffic (Sagberg 1999). In this case, drivers changed their behaviour on certain types of road (e.g. straight and monotonous roads); and on the other hand, road engineers could alter the road environment in order to reduce driver boredom. However, in some cases, fatigue or boredom are linked more to the characteristics of the person themselves rather than engineering or behavioural adaption. For instance, it was found that individuals with a higher level of anxiety may be more likely to feel fatigue (Jiang et al. 2003). In addition, some groups of people (e.g. older people) are inherently more vulnerable than others, thus, more likely to be involved in an accident or to be more seriously injured if an accident occurred (Bedard et al. 2002).

Safety science has influenced the traffic safety interventions in HICs, primarily leading to the emergence of safe systems approach in The Netherlands and Vision Zero in Sweden.

Vision Zero accepts, as a basic starting point, that human beings make conscious and subconscious mistakes. That is why accidents occur, and the safety work must in the first instance be directed at those factors which can prevent accidents leading to death and serious injury. Accidents in themselves can be accepted, but not their serious consequences.

According to Vision Zero, the principal cause as to why people die and are seriously injured is that the energy to which people are exposed in a traffic accident is excessive in relation to the energy that the human frame can withstand. Vision Zero is, among other things, based on the research that the famous American road safety expert William Haddon conducted in the 1960s (Haddon Jr 1968, 1970, 1973, 1980). Knowledge of energy and tolerance has to a great extent served as a basis for the development we have seen of the passive safety characteristics of vehicles and for the development of different protection systems such as child safety seats, helmets and seat belts. One important consequence of Vision Zero as a general policy for safety work is that the view of knowledge which has served as a basis for the development of a sub-component in the road transport system, namely the vehicle, also has become a general principle for the entire road transport system (Belin 2016).

According to Vision Zero, it is not the individual road-user who has the ultimate responsibility but rather the so-called system designers. The responsibility for safety is thus split between the motorists and the system designers (i.e. infrastructure builders and administrators, the vehicle industry, the haulage sector, taxi companies and all the organisations that use the road transport system professionally), on the basis of the principles that:

- The system designers have ultimate responsibility for the design, upkeep and use of the road transport system and are thus responsible for the safety level of the entire system.
- As before, the road users are still responsible for showing consideration, judgement and responsibility in traffic and for following the traffic regulations.
- If the road users do not take their share of the responsibility, for example, due to a lack of knowledge or competence, or if personal injuries occur or for other reasons that lead to risk, the system designers must take further measures to prevent people from being killed or seriously injured.

In Vision Zero, the responsibility for safety is a chain of responsibility that both begins and ends with the system designers (Belin 2016).

Sustainable safety approach of The Netherlands is based on similar principles. Wegman (2017) has noted that:

There are two good reasons why the traditional approach (working on reducing "spikes in distributions") will become less effective and efficient in countries with mature road safety policies. The first reason lies in the fact that serious road crashes will occur as long as we leave the inherent unsafe conditions in road traffic untouched: the inherent risks come from a combination of the physical vulnerability of the human body and the levels of kinetic energy in crashes (a combination of speed and mass). These inherent risks also stem from the fact that the road transport system cannot be designed from the perspective of the human being as long as it fails to defend against human errors and offenses that can result in crashes. Because of this, we are almost fully dependent on how well drivers, riders, and pedestrians perform their tasks. It is remarkable that, while the road transport system puts its faith in individual driving skills, the rail system and the aviation system are designed from a safety perspective—and even well-trained professionals like train drivers and airplane pilots are only allowed to operate under rather strict conditions. A second good reason lies in the fact that our traditional policies have become less effective and efficient. Traditional interventions dealing with reducing relatively high risks are in the process of coming to the ends of their life cycle, suggesting that they may be subject to the law of diminishing returns. In the Netherlands, these two underlying reasons have triggered a paradigm shift and resulted in the development of Sustainable Safety, the Dutch version of the Safe System approach.

6.6 Principles for Safe Highways in LMICs

Given the understanding from traffic safety theories of the last fifty years, the systems approach and Vision Zero, we can propose some basic principles which can form the corner stones for developing safe highways in LMICs.

Safe systems approach has three key principles (Chen and Meuleners 2011a, b; Transport Research Centre 2008):

- Principle 1—Recognition of human frailty
- Principle 2—Acceptance of human error
- Principle 3—Creation of a forgiving environment and appropriate crash energy management.

Current highway standards for geometric design of highways can be reviewed in the context of these three basic principles. Principle 1 and 2 must recognise that highways in LMICs will have the presence of NMVs and pedestrians along with motorised traffic. Principle 3 becomes the operational principle for setting appropriate speed limits for ensuring a forgiving environment for all road users. Pedestrians will make mistakes in judging the possible risk in the system, whereas drivers can make mistakes in adopting an appropriate speed.

Design speed and design vehicle are the two most important elements which have been used to set highway standards in the past. Stopping distance of a modern car is very different from a tempo (three-wheeler) or 2-axle trucks present on LMICs highways. Therefore, the selection of a design vehicle itself becomes important for setting the minimal standards for stopping distance, sight distance and overtaking distances.

Design speed governs the design of horizontal curve, vertical curve and the safe stopping distance. Conventional practice of keeping design speed higher than operational speed has been questioned by several researchers. Therefore, the design speed must be in line with the requirement of principle 3—"creation of a forgiving environment and appropriate crash energy management". This implies that for setting appropriate design speed, presence of NMVs, presence of activities along the highway, and density of built up area along the highway, frequency of towns and villages through which the highway passes must be taken into consideration. Design speed may vary from 30 to 90 km/h with a road cross section designed for appropriate crash energy management depending on the surrounding land use present along the highway.

Speed compliance by design: We started this paper quoting the success of legislation and enforcement, however, taking lessons from number of studies in HICs, most effective measure for speed compliance in LMICs will be by design: active speed control measures. LMICs have weak institutional capacity, weak enforcement of legislation, therefore, speed control by texture change, audible markers, rumble strips, change in geometric standards, median designs, lowering speeds at intersections by introducing roundabouts, raised stop lines and speed humps on minor roads are expected to be more successful in speed compliance by all road users-good drivers, bad drivers, young drivers, knowledgeable drivers, drivers with poor driving education, etc., ensuring compliance with the principle 2.

Many of the current standards for highway cross section require revisions (Chen and Meuleners 2011a, b; Mohan et al. 2017) to comply with the principle 3. Appropriate design of service roads, width of shoulders, and design of medians, has to be reviewed to ensure safe designs for NMVs and different kinds of vehicles on the road.

Experience from HICs is that standards alone cannot ensure safe roads for all unless the safety performance is evaluated. Vision Zero accepts that it is possible to design a transport system that will not have any deaths and serious injuries. Therefore, the realisation of Vision Zero also requires generation of new knowledge and establishing a process which enables generation of new knowledge to ensure safe highways in LMICs. Given the complexity of traffic safety science and its implementation in field, continuous experimentation is required in LMICs to develop safe highways based on the principles of safe systems approach.

References

Aarts L, van Schagen I (2006) Driving speed and the risk of road crashes: a review. Accid Anal Prev 38(2):215–224

Abdel-Aty MA, Radwan AE (2000) Modeling traffic accident occurrence and involvement. Accid Anal Prev 32(5):633–642

ADB (2008) Road network improvement and maintenance Project II: Final resettlement action plan (No. 34415-013). Asian Development Bank, Manila

ADB (2009) Project performance evaluation report of the Primary Roads Restoration Project. Asian Development Bank, Manila
Ameratunga S, Hijar M, Norton R (2006) Road-traffic injuries: confronting disparities to address a global-health problem. The Lancet 367(9521):1533–1540
Assum T, Bjørnskau T, Fosser S, Sagberg F (1999) Risk compensation—the case of road lighting. Accid Anal Prev 31(5):545–553
Bedard M, Guyatt GH, Stones MJ, Hirdes JP (2002) The independent contribution of driver, crash, and vehicle characteristics to driver fatalities. Accid Anal Prev 34(6):717–727
Belin M-A (2016) Vision zero as a new way of thinking. J Austr Coll Road Saf 27(3):60
Belmont DM, Forbes TW (1953) Effect of average speed and volume on motor-vehicle accidents on two-lane tangents. Highway Res Board Proc 32:383–395
Ceder A (1982) Relationships between road accidents and hourly traffic flow—II: probabilistic approach. Accid Anal Prev 14(1):35–44
Ceder A, Livneh M (1982) Relationships between road accidents and hourly traffic flow—I: analyses and interpretation. Accid Anal Prev 14(1):19–34
Chang L-Y, Mannering F (1999) Analysis of injury severity and vehicle occupancy in truck-and non-truck-involved accidents. Accid Anal Prev 31(5):579–592
Chen H, Meuleners L (2011a) A literature review of road safety strategies and the safe system approach
Chen HY, Meuleners L (2011b) A literature review of road safety strategies and the safe system approach. Curtin-Monash Accident Research Centre (C-MARC), Curtin University, Perth
Davis GA (2002) Is the claim that 'variance kills' an ecological fallacy? Accid Anal Prev 34(3):343–346
Dey PP, Chandra S, Gangopadhaya S (2006) Speed distribution curves under mixed traffic conditions. J Transp Eng 132(6):475–481
Elvik R, Christensen P, Amundsen A (2004a) Speed and road accidents. An evaluation of the Power Model. TØI report, 740, 2004
Elvik R, Christensen P, Amundsen A (2004b) Speed and road accidents: an evaluation of the Power Model. Institute of Transport Economics, Oslo
Elvik R, Vaa T, Hoye A, Sorensen M (2009) The handbook of road safety measures, 2nd edn. Emerald Publishing Limited, Bingley, UK
ESReDA (2009) Guidelines for safety investigations of accidents. European Safety Reliability and Data Association, Kaunas, Lithuania
Golob TF, Recker WW (2003) Relationships among urban freeway accidents, traffic flow, weather, and lighting conditions. J Transp Eng 129(4):342–353
Gwynn DW (1967) Relationship of accident rates and accident involvements with hourly volumes. Traffic Q 21(3)
Haddon W Jr (1968) The changing approach to the epidemiology, prevention, and amelioration of trauma: the transition to approaches etiologically rather than descriptively based. Am J Public Health Nations Health 58(8):1431–1438
Haddon W Jr (1970) On the escape of tigers: an ecologic note. Am J Public Health Nations Health 60(12):2229–2234
Haddon W Jr (1973) Energy damage and the ten countermeasure strategies. Hum Factors 15(4):355–366
Haddon W Jr (1980) Advances in the epidemiology of injuries as a basis for public policy. Public Health Rep 95(5):411
Hauer E (1988) A case for science-based road safety design and management. Paper presented at the Highway safety: at the crossroads: proceedings of a conference
Hauer E (2009) Speed and safety. Transp Res Rec J Transp Res Board 2103:10–17
Hauer E (2017) Road safety research and practice: problems of coexistence. University of Toronto
Haynes R, Jones A, Kennedy V, Harvey I, Jewell T (2007) District variations in road curvature in England and Wales and their association with road-traffic crashes. Environ Plan A 39(5):1222–1237

Haynes R, Lake IR, Kingham S, Sabel CE, Pearce J, Barnett R (2008) The influence of road curvature on fatal crashes in New Zealand. Accid Anal Prev 40(3):843–850
IRC (2007) Manual of specifications & standards. Indian Roads Congress, New Delhi
Ivan JN, Wang C, Bernardo NR (2000) Explaining two-lane highway crash rates using land use and hourly exposure. Accid Anal Prev 32(6):787–795
Jiang N, Sato T, Hara T, Takedomi Y, Ozaki I, Yamada S (2003) Correlations between trait anxiety, personality and fatigue: study based on the Temperament and Character Inventory. J Psychosom Res 55(6):493–500
Kockelman KM, Kweon Y-J (2002) Driver injury severity: an application of ordered probit models. Accid Anal Prev 34(3):313–321
Kononov J, Bailey B, Allery B (2008) Relationships between safety and both congestion and number of lanes on urban freeways. Transp Res Rec J Transp Res Board 2083:26–39
Koornstra M, Lynam D, Nilsson G, Noordzij P, Pettersson H-E, Wegman F et al (2002) SUNflower: a comparative study of the development of road safety in Sweden, the United Kingdom, and the Netherlands. SWOV, Leidschendam
Lamm R, Psarianos B, Mailaender T (1999) Highway design and traffic safety engineering handbook. McGraw-Hill, New York
Lave CA (1985) Speeding, coordination, and the 55 mph limit. Am Econ Rev 75(5):1159–1164
Lord D, Washington SP, Ivan JN (2005) Poisson, Poisson-gamma and zero-inflated regression models of motor vehicle crashes: balancing statistical fit and theory. Accid Anal Prev 37(1):35–46
Martin J-L (2002) Relationship between crash rate and hourly traffic flow on interurban motorways. Accid Anal Prev 34(5):619–629
Milton J, Mannering F (1998) The relationship among highway geometrics, traffic-related elements and motor-vehicle accident frequencies. Transportation 25(4):395–413
Ministry of External Affairs (India) (2013) Preparation of detailed engineering project report for two-lane road from Paletwa to Zorinpui (India Myanmar border) in Chin State of Myanmar, Detailed Project Report, Volume IV: environmental impact assessment report. Ircon Infrastructure & Services Limited, New Delhi
Mohan D, Tiwari G, Bhalla K (2017) Road safety in India: Status report 2016. Transportation Research & Injury Prevention Programme, Indian Institute of Technology Delhi, New Delhi
MoRTH (2010) Manual for specifications and standards for four laning of highways. Ministry of Road Transport & Highways, Government of India, New Delhi
Naqvi HM, Tiwari G (2015) Assessing risk factors for pedestrian fatal crashes on National Highways in India. Paper presented at the EASTS, Cebu, Philippines
Navin F, Zein S, Felipe E (2000) Road safety engineering: an effective tool in the fight against whiplash injuries. Accid Anal Prev 32(2):271–275
NHAI (2007) Consultancy Services for preparation of Feasibility cum Preliminary Design for 4/6 Laning of Indore to Gujarat/MP border section of NH-59 (NHAI/BOT II/ DPR/04) to be executed as BOT (Toll) project on DBFO Pattern under NHDP Phase III. Wilbur Smith Associates Pvt. Ltd., New Delhi
NHAI (2011) Feasibility and detailed project report for rehabilitation and upgrading to 4/6 lane divided carriageway of Barasat - Baharampore (km 31 to km 193) section of NH 34 in the state of West Bengal, Package 1. The Louis Berger Group Inc., New Delhi
NHAI (2015) Final detailed project report Volume I (Main Report) Mahulia-Baharagora-JH/WB border at Chichra (NH-33 &NH-6). SMEC and Rodic Consultant Pvt Ltd., Gurugram, Haryana
NHAI (2016) Consultancy services for preparation of feasibility study for 4/2 laning with paved shoulder of Gadu-Dwarka section ofNH-8 from km 263.00 (design chainage km 287.430) to km 473.00 (design chainage km 497.320) in the state of Gujarat through public private partnership (PPP) on design, build finance, operate and transfer (DBFO) toll basis. STUP Consultants Pvt. Ltd., Mumbai
NHIDCL Sikkim (2016) Detailed project report for widening to 2-Lane of NH 510 (Singtam-Tarku-Rabongla-Legship-Gyalshing) in the state of Sikkim. CM Engineering & Solution, Gurugaon, Haryana

Nilsson G (2004) Traffic safety dimensions and the power model to describe the effect of speed on safety. Bulletin-Lunds Tekniska Högskola, Inst för Teknik och Samhälle, Lunds Universitet, 221
Noland RB, Oh L (2004) The effect of infrastructure and demographic change on traffic-related fatalities and crashes: a case study of Illinois county-level data. Accid Anal Prev 36(4):525–532
O'donnell C, Connor D (1996) Predicting the severity of motor vehicle accident injuries using models of ordered multiple choice. Accid Anal Prev 28(6):739–753
Openshaw S (1984) Concepts and techniques in modern geography number 38: the modifiable areal unit problem. Geo Books, Norwick
Peden M, Scurfield R, Sleet D, Mohan D, Hyder AA, Jarawan E et al (2004) World report on road traffic injury prevention. World Health Organization, Geneva
Perez I (2006) Safety impact of engineering treatments on undivided rural roads. Accid Anal Prev 38(1):192–200
Perrow C (1984) Normal accidents: living with high risk systems. Basic Books, New York
Sagberg F (1999) Road accidents caused by drivers falling asleep. Accid Anal Prev 31(6):639–649
Shankar V, Mannering F, Barfield W (1995) Effect of roadway geometrics and environmental factors on rural freeway accident frequencies. Accid Anal Prev 27(3):371–389
Shankar V, Mannering F, Barfield W (1996) Statistical analysis of accident severity on rural freeways. Accid Anal Prev 28(3):391–401
Shefer D (1994) Congestion, air pollution, and road fatalities in urban areas. Accid Anal Prev 26(4):501–509
Staton C, Vissoci J, Gong E, Toomey N, Wafula R, Abdelgadir J et al (2016) Road traffic injury prevention initiatives: a systematic review and metasummary of effectiveness in low and middle income countries. PLoS ONE 11(1):e0144971
Stoop J, de Kroes J, Hale A (2017) Safety science, a founding fathers' retrospection. Saf Sci 94:103–115
Stoop JA (2015) Challenges to the investigations of occurrences: concepts and confusion, metaphors, models and methods. The European Safety, Reliability and Data Association, Baden
Taylor MC, Baruya A, Kennedy JV (2002) The relationship between speed and accidents on rural single-carriageway roads: TRL
Tiwari G, Fazio J, Pavitravas S (2000) Passenger car units for heterogeneous traffic using a modified density method. Paper presented at the Proceedings of fourth international symposium on highway capacity.
Transport Research Centre (2008) Towards zero: ambitious road safety targets and the safe system approach. International Transport Forum, OECD, Paris
Turner D, Thomas R (1986) Motorway accidents: an examination of accident totals, rates and severity and their relationship with traffic flow. Traffic Eng Control 27(7–8):377–383
Wang C, Quddus M, Ison S (2009) The effects of area-wide road speed and curvature on traffic casualties in England. J Transp Geogr 17(5):385–395
Wang C, Quddus MA, Ison SG (2013) The effect of traffic and road characteristics on road safety: a review and future research direction. Saf Sci 57:264–275
Wegman F (2017) The future of road safety: a worldwide perspective. IATSS Res 40(2):66–71
Wegman F, Hagenzieker M (2010) Editorial safety science special issue road safety management. Saf Sci 48:1081–1084
WHO (2011) Saving millions of lives: decade of action for road safety 2011–2020. World Health Organisation, Geneva
WHO (2015) Global status report on road safety 2015. World Health Organization, Geneva
Zhou M, Sisiopiku V (1997) Relationship between volume-to-capacity ratios and accident rates. Transp Res Rec J Transp Res Board 1581:47–52

Nelson G (2004) Traffic safety dimensions and the power model to describe the effect of speed on safety. [illegible] Lund Institute of Technology, Lund University. Bulletin 221
Noland RB, Oh L (2004) The effect of infrastructure and demographic change on traffic-related fatalities and crashes: a case study of Illinois county-level data. Accid Anal Prev [illegible]
O'Cinneide C, [illegible] D (1998) Enhancing the safety of minor rural roads [illegible]
[illegible]

Chapter 7
Traffic Calming: The Way Ahead in Mixed Traffic

Christer Hydén

7.1 Introduction

Motorised traffic is a strong contributor to low quality of life for many people in our cities. Volumes are big, and speeds are often very high. Even if safety of prime concern in every country I dare to say that the issue today in developed countries has shifted from being a question about survival of the pedestrian to survival of the city; while in developing countries, the question is one of pedestrian survival. One main ingredient in survival of the city is to produce liveable conditions for pedestrians, which of course includes the elimination of the physical threat. So, even though the perspective is different the main question today, as it "always" has been, is how to reach decent living conditions for pedestrians so that they can be safe and also feeling safe, being comfortable enough, not having to live with noise, emissions, etc. "Business-as-usual is no longer an option.

Current approaches to road safety in the world's poorest countries are both indefensible and unsustainable. They are indefensible because they will result in millions of deaths and injuries that could be prevented through affordable investments. And they are unsustainable because no country can afford the economic and social costs associated with current approaches. The fact that these costs are hidden from view does not detract from their devastating effects".

(https://www.makeroadssafe.org/publications/Documents/road-traffic-injuries-kevin-watkins.pdf).

The key to a new era has its origin in "Traffic Calming". The phrase gives a certain flavour of softness and emphasises the need for a motorised traffic that is less dominant both in numbers, speeds and priority. Let me immediately say that the concept as such is nice, but we are still very far away from being able to demonstrate

---

C. Hydén (✉)
Department of Technology and Society, Lund University, Lund, Sweden
e-mail: Christer.hyden@tft.lth.se

G. Tiwari and D. Mohan (eds.), *Transport and Safety*, Springer Transactions in Civil and Environmental Engineering,
https://doi.org/10.1007/978-981-16-1115-5_7

on a large scale or in a sustainable way the success of this concept. There is still a very long way to go before "life in cities" is as attractive and sustainable as traffic calming promises to be.

7.2 Pedestrian Safety

7.2.1 History

Pedestrians' life qualities is the best indicator for defining the "maturity" of the whole system. My interpretation is that pedestrians have always been the poor citizens in many senses, and still are. They have no voting power, and they are given low priority in questions about allocating space and priority in traffic. (Just think about pedestrians about to cross a road. They have priority but only when they use a zebra crossing, often forcing them to take long detours.) They are exposed to high risks. This was true as long ago as in the nineteenth century. Between 1840 and 1900, the mortality rate due to road accidents in the UK was on average about 50 pMa (per million persons living per annum) in the UK, probably mainly pedestrians (Ishaque and Noland 2006). Today—more than 100 years later—the mortality rate due to road accidents is around 40 pMa in countries like UK, Sweden, Norway and the Netherlands. In the USA, it is as high as 140 pMa, approx 7 pMa killed pedestrians. Globally, the figure is a bit over 200 pMa, out of which pedestrians account for roughly half, i.e. 100 pMa! Mortality rates have gone down in highly motorised countries, but that is of course primarily due to large decreases of walking illustrated by the situation in a Copenhagen main street in the 1920; see Fig. 7.1.

7.2.2 Developing Countries

In developing countries, mortality rates are still increasing in most countries, and the vulnerable road users are most hit by road accidents as can be seen in the following example from India (Table 7.1).

The most common striking vehicles are trucks, followed by buses and cars; see Fig. 7.2. Motorised two-wheelers are the striking vehicle only in 4% to 8%. Regarding their share of vehicles, 38% in Mumbai, 57% in Delhi and 71% nationally, motorcycles do not seem to be overrepresented in fatalities, rather the opposite.

Undoubtedly pedestrians are by far the most vulnerable group in India in general.

The fatality figures are quite high as in other low- and middle-income countries (LMIC). For instance, Sayer and Palmer reported that "more than 40%" of all road fatalities in African countries was with pedestrians, and in Middle Eastern countries, the figure was 50% (Hydén and Svensson 2009). But pedestrians are not the only

Fig. 7.1 Down town street in Copenhagen in the 1920s, high volumes of pedestrians

Table 7.1 Traffic fatalities by road user and type; Delhi (2001–2005), Mumbai (1996–1997) and Kota (2007) (Figure no. 24 in Mohan et al. 2009)

Type of road user	Mumbai	Delhi	Kota
Truck	2	3	6
Bus	0	3	1
Car	2	4	19
Three-wheeled scooter taxi	4	3	4
Motorised two-wheeler	7	26	33
Human and animal powered vehicle	0	3	1
Bicycle	7	10	5
Pedestrian	79	47	28
Other (Tractor, etc.)	0	1	4
Total	101	100	101

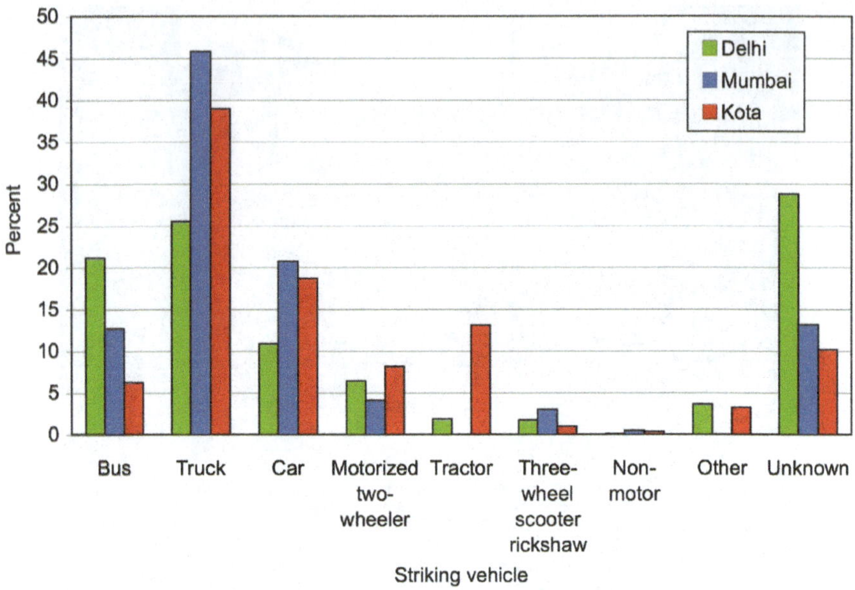

Fig. 7.2 Striking vehicles in the three cities in Table 7.1

vulnerable group. Altogether pedestrians, cyclists and motorcyclists make up almost half of those killed on the roads (WHO 2009).

7.2.3 Situation Today Regarding Vulnerable Road Users

The vulnerability of pedestrians, cyclists and motor cyclists is finally taken seriously by the "World Community". WHO in their Status Report (2009) reaffirmed the understanding of road traffic injuries as a global health and development problem and it: "… draws our attention to the needs of all road users—including these most vulnerable groups. They too must be considered and given equal priority when policy discussions on road safety, land use and urban planning are made". My interpretation of the situation is that decision-makers, and others all over the globe nowadays are putting road safety quite high on the agenda. This is of course encouraging. However, it is one thing to put it high on the agenda. A completely different thing is to put relevant and efficient action on the agenda. I think that is what we still are lacking. Road safety seems to be a sensitive area for decision-makers. They think that the public do not want the necessary measures. They do not know, because there has almost never been an attempt to produce a holistic view on traffic. Blatter (1995) indicated that decision-makers were more conservative than people in general when it came to measures making the use of the car more difficult. Another example is that some Swedish traffic planners in a survey made at Lund University clearly demonstrated a hesitation regarding humps. They meant that it is something that not can be used on a large scale, and some also claimed that they got a lot of objections from people (not clear who) regarding humps (Risser 2010). Coming back to a lack of holistic solutions, I think this is a typical example. Compromise seems to be the keyword, and the solutions—according to my interpretation—are often conservative and not very much in favour of the living conditions for pedestrians.

7.2.3.1 Separation of Pedestrians

I will demonstrate my scepticism to today's planning for pedestrians by giving you the story about the prime strategy to safeguard pedestrians in cities. It started already in the nineteenth century and still goes on. Authorities in the UK were already discussing the issue in the nineteenth century. First attempt was to give separate space for the pedestrians, thus building foot paths (Ishaque and Noland 2006). However, that only solved parts of the problem. The other—more critical—question was how to safeguard pedestrians when crossing streets. The way to solve the problem was either to separate different road users by the help of traffic signals, or to separate pedestrians physically with tunnels/bridges. Both these solutions were tried in London in the 1870, but were turned down almost immediately and were not reconsidered until after several decades. The reason was that pedestrians were offered too long waiting times in the signal (30 s green once in 5 min) and because of those pedestrians would not accept them and they would therefore jay walk. In a similar way, tunnels/foot bridges were considered as useless because pedestrians would not use the tunnel/bridge because of the detours they had to make. A study in Delhi just emphasises the problem. A majority of pedestrian underpasses constructed recently in Delhi are not being used for their intended purpose. General observation and a pilot survey of some underpasses in Delhi indicate that pedestrians find an alternative surface route to cross the road. (Sinha AICP, Avinash Kumar https://pubsindex.trb.org/view.aspx?id=777751. 26/10/10). I can report on similar experiences from Sweden (see Fig. 7.3), i.e. I imagine that the problem is quite universal.

Even though it seemed impossible, a lot of research resources were spent in order to find out how and under what preconditions pedestrians were actually using the tunnel/bridge. The conclusion was that, unless the time used was considerably smaller using the tunnel/bridge pedestrians would not use it. This is of course almost impossible to achieve, so the conclusion was to force pedestrians to use the separation by installing guard rails and increase enforcement; see examples from Beijing in Fig. 7.4.

The examples from Beijing illustrate the dilemma. The safest solution in the short term is of course to see to it that pedestrians are not using the street to cross. However the comfort offered to pedestrians is very low. Besides car drivers were offered streets (almost) free from pedestrians, which in turn meant that pedestrians, who were trespassing anyway, were exposed to even higher risks, due to higher vehicle speeds. Besides the high costs for this kind of separation only allow these

Fig. 7.3 Separation of pedestrians and cyclists on a street with high speeds and low use of the separation

Fig. 7.4 Inconvenient foot bridge and guard rail and enforcement in Beijing, China

measures on an irregular basis, i.e. not producing any holistic view on "pedestrian welfare".

7.2.3.2 Separation with Traffic Signals

The "traffic signal story" is very similar to the one about tunnels/foot bridges for separation of vulnerable road users. Meta-analysis of the safety effects of traffic signals shows that introduction of pedestrian signals at intersections increase pedestrian risks if there are simultaneous green for cars and turning vehicles. However, there is a reduction of pedestrian risks when pedestrians have separate phases. The problem is that separate phases produce too much delay for all road users. The result is therefore that separate phases are not very common, at least not in Sweden. And, again, the costs for large-scale implementation are too high.

7.2.4 "The Zebra Crossing Story"

So neither separation nor signalling was considered feasible. New, more cost-effective solutions were to be found. This indicated the start of the efforts to find a compromise between driver and pedestrian interests. The basic idea was to mark crossings for pedestrians with special yielding rules for drivers and special rules of conduct for pedestrians. In UK, the first attempts to mark a crossing for pedestrians took place in the 1930. To make the following story a bit short—and a bit provocative—you can say that all the time up till today, there have been all sorts of attempts to find a solution where car drivers observed pedestrians and also managed to let pedestrians pass safely. The first zebra crossing came early with special rules for the conduct of both driver and pedestrians. Sweden has a very interesting history of the zebra crossing, which seems to be applicable here. In the 1980, we had 70,000 such crossings. It was found that even if zebra crossings had been present for more than 50 years, there was no evaluation made. So a colleague of mine made a thorough study and found out that a zebra crossing had more than two times higher risk for pedestrians than a crossing location without any facility (Ekman 1988), see Fig. 7.5.

These results were so shocking, primarily I guess for those who were responsible for these 70,000 crossings. This was one reason why nothing happened for more than 10 years. Once there was a discussion, the main conclusion that emerged was that the main problem was that car drivers did not yield to pedestrians at zebra crossing. So a new law was introduced in the beginning of the year 2000 making yielding for pedestrians mandatory when pedestrians are on or close to the zebra

Fig. 7.5 Pedestrian risk at different crossing types

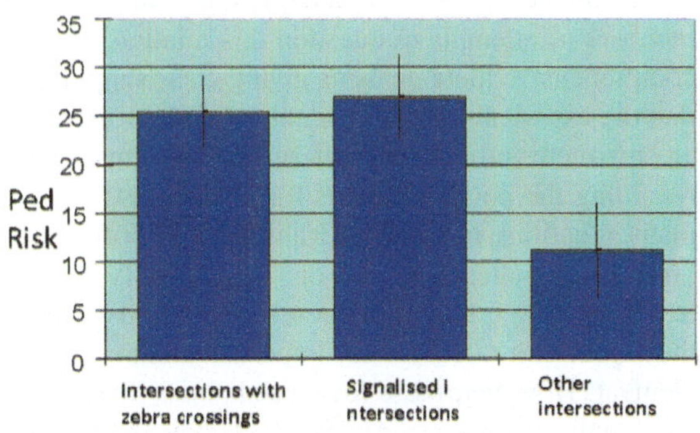

crossing. A follow-up study showed that the yielding rate went up from 5–20% to 40–50%. However, the number of injured pedestrians increased. The reason being risk compensation. Already in the Ekman study, the conclusion was that pedestrians felt safe on the zebra, and there was therefore an obvious risk that when pedestrians observed a much higher degree of driver compliance with the yielding rule, they would feel even safer. So, the result was that zebra crossings were officially classified not any more as a safety device, but a device to improve the mobility of pedestrians. UK has a similar story by taking away a lot of the crossings. I think that "the story of the zebra crossing" (or other type of crossing) will not end for many years still. A meta-analysis of the safety effects of zebra crossings by Elvik et al. (2010a) showed that the introduction of a zebra crossing produces an increase of injury accidents for pedestrians by 44% (−6; +121). My main conclusion is that we still have not implemented any solutions balancing safety for pedestrians with keeping up the mobility for motor vehicles in a large scale. Motorised traffic is still given the highest priority.

It is striking that so little is done to really produce a sustainable change of pedestrians' conditions; particularly in view of all the knowhow we have today on how to improve pedestrian safety.

7.2.4.1 Speed

Speed reducing measures that actually lower speeds is definitely the most effective solution. Meta results from Elvik et al. (2010b) clearly show this. The introduction of speed reduction at a zebra crossing reduces injury accidents for pedestrians by 42% (−70; +11). The effect is dependent on the actual speed reduction. Nilsson (2004) presented the so-called power model saying that the effects of changes in speed on the number of accidents and the severity of injuries can be estimated by means of a set of power functions. For fatal accidents, the relation is:

$$(\text{Fatal accidents after})/(\text{Fatal accidents before}) = (\text{Speed after}/\text{Speed before})^4$$

If speed is reduced by 10%, fatal accidents are estimated to be reduced by 0.94 = 0.66, which corresponds to a reduction of fatal accidents by 34%. There are endless numbers of attempts to question this relation. However, Elvik et al (2004) made a comprehensive meta-analysis giving clear support to the power model. They found that the exponents were not identical with those proposed by Nilsson, but "they are close to them and exhibit a pattern that conforms to the Power Model". The scepticism regarding the power model is for me a reflexion of the tendency I have seen in so many situations, namely to devaluate the importance of speed as a regulator of safety. "It is not a matter of speed" or "it is not only" speed that counts is something you hear quite often. In an interview study with Swedish planners and decision-makers, they agree that safety is very important and that low vehicle speeds are an important element. However, these low speeds must "be reasonable". "Humps are good, but they cannot be used everywhere; we have received a lot of complaints". Etcetera.

As one Swedish planner said at the interviews "we get a lot of complaints when we introduce humps" (Risser 2010). That can be compared with the experience in the Norwegian city of Bergen where humps are introduced on almost all streets in a city (more than 2000 in a city of 230,000 inhabitants). Their experience is that once humps are introduced on a street neighbours contact the city asking for humps on their streets as well. Bergen has a more than 20 years experience of systematic introduction of humps, which has resulted in humps (or raised crossings) even on streets with heavy bus traffic. This is still something impossible in Sweden. Another example, the Swedish Vision Zero, adopted in principle in many countries, claims that the long-term goal is that there should be no seriously injured or killed in road traffic (https://www.visionzeroinitiative.com/). One of the main principles taken is that the speed of vehicles should be no higher than 30 km/h where motorised and non-motorised traffic interact. Still 13 years later, there is a long way to go to reach this target.

My interpretation of the present situation is that this is a way of avoiding the critical question "How to introduce efficient speed reducing measures", with the anticipation that this will be very unpopular. One more strong indication of the avoidance behaviour to a new strategy in Sweden is to introduce new speed limits—in built-up areas 40 and 60 km/h in addition to the former ones, 30 and 50 km/h. One of the philosophies behind this is the idea that the speed limits should better reflect the expectations of drivers. So therefore the goal is both to reduce speeds and to improve the acceptance by car drivers. In this perspective, the results are not very encouraging. If the speed limit was reduced 10 km/h—from, e.g., 50 to 40 km/h—the actual mean speeds were only reduced by a bit more than 2 km/h (Hydén et al. 2008). This resulted in severe reductions of the compliance rate, in the case of 50 to 40 km/h from 80% compliance to 60%. So even though speeds were reduced a little the change clearly demonstrated that drivers did not take the change "seriously enough", as the compliance rate went down. In large, most local decision-makers are in favour of the new speed limits—especially 40 km/h.

7.2.5 The Promotion of Low Speeds

For me, the change of speed limits is yet another example on how the attempts to lower speeds is based on measures with minor effects; at the same time as they send the message to the public that this is the kind of measures used, resulting in very little annoyance. How can decision-makers be so hesitant in introducing efficient measures, in spite of the fact that the public is more in favour of these kinds of measures (efficient speed reduction with, e.g., humps) than the decision-makers? And how is it possible that one city—Bergen, Norway—can use "unpopular" humps on a very large scale, while most other cities consider it almost impossible? I think one of the most important reasons is the promotion of high speeds as something important and "enjoyable" for us as drivers. This kind of promotion can be seen every day in papers and magazines. The automobile industry spends billions on advertising in different media. Ads giving messages like "Zero to 100 in a whisper" or the car industry which also produces air planes stating that "we have jet in our genes", etc. We have been told that strong acceleration and speed performance are important from a safety point of view. However, we never see the counterargument that Zero to 100 km/h may not be very favourable with regard to pedestrian safety and well-being in our cities. I cannot claim that any of these stand points are "true". However, what I can claim is that "somebody" should take the responsibility for assessing the importance of the present characteristics of automobiles from a holistic point of view, i.e. not only safety for certain groups but safety for all road users, and—particularly—the attractively and sustainability of the city's traffic.

I have presented a lot of engineering efforts to improve safety and well-being for pedestrians. The red thread is speed. One major problem in this context is the speed and acceleration performance of cars. Every day you can read advertisements for cars with headings like "Zero to 100 km/h in a whisper", and messages saying that strong acceleration power is important from a safety point of view. Why do we go to such a great extent to neglect the importance of this kind of hiding the risks involved in such attitudes? How can we ask a driver to slow down in a residential area when he is sitting in a comfortable car with nice music, etc., and a capability of making 200–250 km/h and an acceleration that takes him to 100 km/h "in a whisper". How can authorities compete with messages like "it is important to adapt speed, 30 km/h is the speed limit, etc". It is no wonder that it is difficult to come through with this kind of perceived negative messages. The industry openly admits that they cannot sell cars with "speed management messages". However, they are now involved in many projects dealing with speed. One of the first—PROSPER—dealt with so-called ISA-applications. ISA stands for Intelligent Speed Adaptation, and includes different types of systems; voluntary systems with a beep signal when the desired speed is overridden, active gas pedal that can be overridable or not, etc. To make the story short, in Work Package 4 of PROSPER the following conclusions were presented (Carsten 2005).

- Clear additional benefits from microsimulation modelling for non-overridable system
- ISA has very substantial safety potential
- In macro-modelling, Authority-Driven Scenario delivers substantially greater benefits than Market-Driven Scenario
- Delay postpones safety benefits and reduces benefit-to-cost ratios.

The point is that the authorities have not driven the most efficient systems from a safety point of view—either in research or in implementation. My department carried out research on a non-over ridable active gas concept in urban areas in the 1990 with positive results. However, this research was ended in the beginning of 2000, and there is no research or concept demonstrating any really significant effects regarding "controlled speeds" and or substantial safety benefits to this day. Instead the industry is to solve the problem in other ways, night vision, collision warning and other similar systems that could solve problems without "direct" speed management (the aim is of course to give drivers relevant information (e.g. making a pedestrians visible in the dark) so that they can slow down in time). The most spectacular system today I think is the "automatic braking" function, e.g. in Mercedes-Benz PRE-SAFE® Brake and Volvo City Safety. The "simple idea" (not technically) is to detect obstacles before a collision occurs and thereby introduce an immediate braking, thus avoiding a collision or—at least—lowering the speed at a collision. It may very well be that such a system will decrease even pedestrian accidents which of course is quite positive in itself. The problem though is that again we will have a new system introduced where the benefit—and welfare of pedestrians is not assessed. How will this system change drivers' behavior, how will it be perceived by pedestrians and what general impact will it have on the attractivity and sustainability of a city? I think that the "pedestrian society" should demand some kind of relevant assessment before these systems are introduced! The problem is that they are already partly implemented. But there will be more systems coming. They will partly "compete" with new ways of organising traffic so as to make attractive and sustainable cities for all users. Ideally, it would be great if there was an interaction between the two of the most important actors on this scene, namely the road and planning on one side and the industry on the other.

7.2.6 *Traffic Calming*

I have now slowly approached what I mean by traffic calming. There is more than one definition but the concept is the same all over the world, i.e. measures to reduce vehicle speeds to below 50 km/h in order to reduce injury accidents, pollution and to make areas more liveable for people. Traffic calming measures include volume control measures for motorised traffic and accident reducing measures where speed reducing measures are the most efficient ones as they both reduce risks and make streets less attractive to car drivers. However, preconditions for traffic calming are different in different countries. Just to take one example, in India two-wheelers and three-wheelers stand for approx. 70% of all motorised vehicles while in Europe and North America the share is only a few percent (Hydén and Svensson 2009).

7.2.7 *Humps (Speed Breakers, Sleeping Policemen, etc.)*

I will start by describing some measures that have a well-documented effect on speeds, thereby are central in a traffic calming context. The number one with regard to speed is the speed hump. It was originally designed in the UK building on the very simple principle that the more uncomfortable car drivers perceive the passage to be the slower they will go. They definitely seem to be much more effective than for instance the presence of small children along the road. If the hump is constructed according to the original hump by Watts (1973), which is a round-top hump 3. 6 m long and 0.1 m high on top, and if humps are put on a longitudinal distance of maximum 50–75 m, then vehicle speeds will be around 20–25 km/h at the hump and 5–15 km/h higher

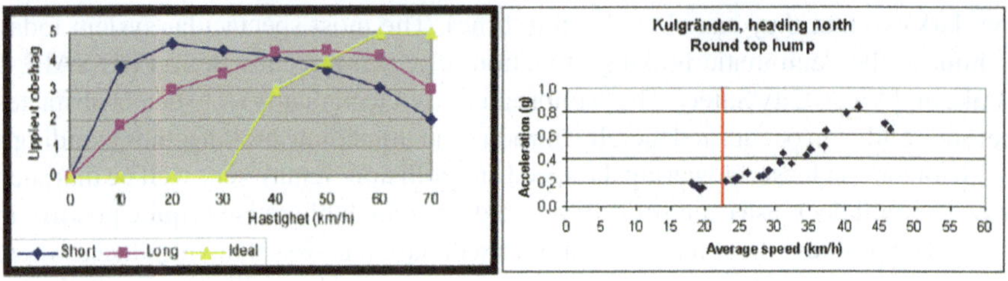

Fig. 7.6 Relation between experienced discomfort (the left y-axis) and speed (x-axis), theoretically (left) and empirically (right)

between the humps. The exact speeds will depend heavily on the hump design. I feel that I have to add a rather technical aspect here. Ideally a hump—or any other speed reducing measure should work in a way that the discomfort for drivers in principle is non-existing at speeds below the target speed, in the case of humps most often 30 km/h. In case, the target speed is exceeded the degree of discomfort should go up very quickly. The empirical part of Fig. 7.6 shows an example from one round-top hump in Lund, Sweden (Bjarnason 2004). It shows clearly that the discomfort starts at a rather high level, and then there is no distinct raise of the discomfort curve at 30 km/h, or elsewhere for that sake. That means that car drivers do not get any clear message, and they may therefore choose their passing speed partly on other criteria than the discomfort aspect, e.g. if they are in a hurry, etc. I claim that there is a strong need to try and do more theoretical and empirical work on the function of humps in order to optimise the design. It is important both from the point of view of making the humps more effective (i.e. more even speeds at the target speed) and also more attractive to the drivers (i.e. more comfortable to pass at speeds below the target speed).

As I wrote earlier the safety effect is quite good locally. However, even more interesting is that this has been proven on a large scale as well. In the city of Gothenburg in Sweden, effective speed-reducing measures have been introduced in an almost complete way on all streets except for on the largest arterials. The results are very encouraging; this traffic calming in Gothenburg has resulted in a large reduction in the numbers of deaths and serious injuries and a socio-economic benefit of more than 47 times the direct costs (Nilsson and Thulin 2004).

7.2.8 Small Roundabouts

One important tool in the traffic calming tool box is the small roundabout; see Fig. 7.7. Small is here defined as a roundabout with only one lane in approaches and exits as well as inside the roundabout. Optimally designed it can lower speeds to around 30 km/h at all approaches and gives all road users equal opportunity to interact with other road users, improves safety for all road users and reduces delay (even

Fig. 7.7 Small roundabouts, inexpensive and safe for all—and popular

for car users at least when the roundabout replaces a traffic signal) and reduces noise and emissions. Thanks to the design the roundabout can be used for enhancing aesthetics and create an attractive environment (SKL2008). Elvik et al. (2010c) report a reduction of injury accidents by 41% (−47; −34) when changing a four-armed, non-signalised intersection to a small roundabout. There are results from the USA indicating that the reduction of fatal and incapacitating injury crashes can be as high as 90% (Retting et al. 2001). A small roundabout can carry at least 23,000 incoming vehicles per day and can be made quite small which make them interesting as an element in creating new streets where the speed is reduced at every equipped intersection and also between the intersections (Hydén and Várhelyi 2000). The popularity of this concept is demonstrated by its rapid development. In Sweden, there were around 150 roundabouts in 1980, and in 2006, there were more than 1500 (Kolbenstvedt et al 2007).

It is important to understand that qualities of roundabouts can differ quite extensively depending on the design of the roundabout, which in turn will have large implications on the safety of the roundabout. This will be particularly true for LMIC where the experience so far is limited.

7.2.9 Speed and Interaction with Vulnerable Road Users

I mentioned equal opportunities for all road users interacting at roundabouts. Obviously, speed plays an important role in providing these opportunities. In a study on interaction between car drivers and bicyclists, it was shown that even if cyclists in most interactions were the ones who formally had to yield for cars; Figure 7.8 shows that the lower the car speeds were, the higher the yielding rate among car drivers was.

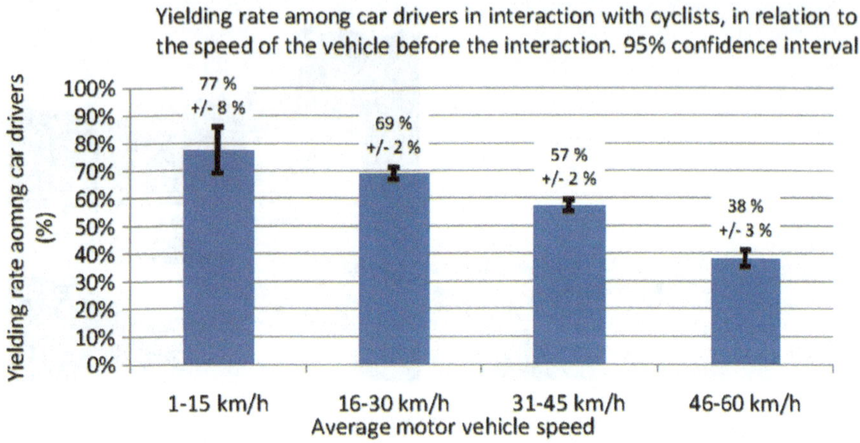

Fig. 7.8 Yielding rate versus speed for car drivers in interaction with bicyclists

These results are the most promising ones regarding future city traffic that I can think of. It shows that it is possible to reach desirable conditions without any strict rules, just by convincing (forcing) drivers to drive slowly. Road users in the study acted very "balanced"; cyclists did not force their way through and drivers who came with slow speed (below 30 km/h and lower) stopped and yielded in a very smooth way.

Regarding other qualities, I will only take it up shortly here; my main conclusion is that if speeds can be low—below 30 km/h—and with low variance both in a flow and along a road, then significant reductions of noise and emissions (incl. CO_2) can be reached (Hedström 2004). Speed reduction from 50 to 30 km/h typically reduces noise levels by 4–5 dB (Pharoah and Russell 1989), or more in certain circumstances (Take Back Your Streets, Conservation Law Foundation 1995).

7.2.9.1 Shared Space

There is one trend, primarily in Europe that I want to add. That is the so-called shared space concept (https://www.rudi.net/books/20024). The basic idea is to "deregulate" as much as possible and leave to all road users to interact themselves. No priority rules are in principle valid. A study was made in the city of Norrköping, Sweden. A signalised intersection with heavy traffic was rebuilt with the shared space principles; see Fig. 7.9). An evaluation showed that today there is a great mix of road users, very low vehicle speeds (around 15 km/h) and generally improved conditions for pedestrians and cyclists. However, visually impaired pedestrians who felt that they had no proper guidance through the place. And road users generally complained because rules were not clear which created insecurity and frustration (Jaredsson 2002). My conclusion regarding shared space is that it may work at individual locations with quite specific preconditions, primarily many bicyclists and pedestrians moving in

Fig. 7.9 Before and after introducing shared space in the Swedish city of Norrköping (*Photo* Sofia Jaredsson)

different directions. If shared space is working on larger scale is still not very clear, and my personal opinion is that it cannot work unless speed reducing measures are introduced.

7.2.9.2 Traffic Calming in Low- and Middle-Income Countries (LMIC)

There is one more important question that partly remains to be answered. That is whether traffic calming may work in LMIC. I will try and reflects on that (impossible) question by giving you a relevant example from India. My university in Sweden had the opportunity to carry out a project on Traffic Calming Guidelines in India thanks to the Swedish aid-organisation SIDA (Hydén and Svensson 2009). The main aims of the project were: (1) To understand pedestrian safety problem in a developing country and (2) to identify feasible traffic calming measures. International experiences were complemented by field studies at seven sites in the city of Jaipur. A wide range of field studies were made, out of which traffic conflict studies were the most comprehensive ones. The main conclusion of the studies is that there is an urgent need for traffic calming measures in Indian cities. Pedestrians are extremely exposed and vulnerable, and the speeds are too high to allow a safe and proper communication. One main problem is that road users are not offered any comfortable and safe crossing options. The existing pedestrian crossings are not suited for pedestrians. Behavioral studies showed that in 95% of interactions between pedestrians and motor vehicles, there was "no reaction" from the drivers' side. The conclusion from the studies is that there is "chaos" that has to be dealt with. The proposed measures are "standardised Traffic Calming measures", primarily by introducing humps at both entrances and exits to intersections and to introduce raised footpaths in all corners to see to it to make the approaches for pedestrians as comfortable and safe as possible by preventing cars from being able to use that space (see Fig. 7.10). Pedestrian crossings are located close to the intersections so that pedestrians do not have to make large detours to use them. The measures that are proposed are simple which is an important part of the scope. Without simplicity and low costs, there will never be any large-scale use.

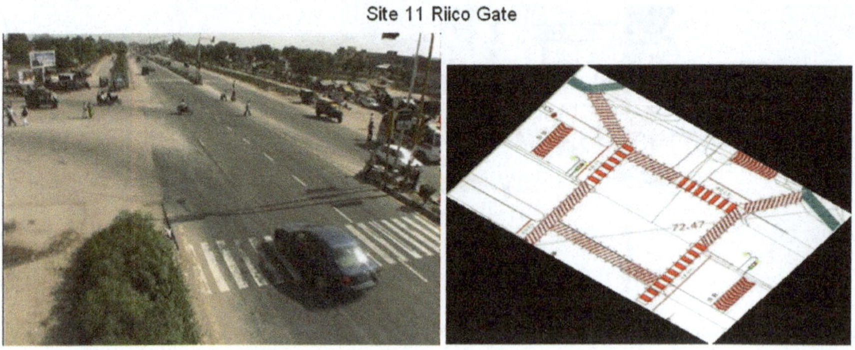

Fig. 7.10 One of the experimental intersections in Jaipur, India

Table 7.2 Speeds at sites with comparative humps in Jaipur, India, and Lund, Sweden (Hyden and Svensson 2009)

	Length (m)	Height (m)	Mean speed (km/h)	85-percentile speed (km/h)
Collectorate, Jaipur	3.8	0.10	21	24
Lalkothi, Jaipur	3.8	0.11	18	23
Average, Lund	3.6	0.10	18	21

Unfortunately, no measures are taken yet in Jaipur, which of course will be the next and very important step. When this is done, and after studies carried out, it would be time to decide on studies for studying the validity of the results in other parts of India. We sincerely hope that other Indian cities also take up this. We are ready to assist together with our colleagues at IIT.

As speed reducing is a central element in traffic calming we made some comparative studies of the speed reducing effect of similarly designed humps in Sweden and in Jaipur; see Table 7.2.

The table indicates that there are very small differences in car speeds at the studied humps in Jaipur compared with speeds at the humps in Lund. Speeds are at most two to three km/h higher in Jaipur. This may be due to the fact that humps in Jaipur are 0, 2 m longer. Even though no humps are exactly similar, the general conclusion is that the effects of humps in Jaipur and Lund seem to work in almost the same way with regard to speed reduction. This finding is very encouraging. Based on the positive experiences from traffic calming in other countries, there is all reason to initiate more comprehensive trials of the concept also in India.

7.3 Conclusions

It is quite clear that there is a great potential in introducing traffic calming measures. And it is also necessary. However, the important thing is to ensure the low speeds that one is targeting on. There are many attempts that more or less have failed. The zebra crossing is a typical example. It has become obvious that it is not sufficient "just" to introduce strict yielding rules or any other measure without ensuring low speeds at the same time. One big problem is that many measures are introduced without actually safeguarding the effects. Assessment of effects is lacking to a very high extent. The result is that theories on which measures are based often are vague or non-existent. The result is therefore that the evaluation of effects will be "arbitrary" in the sense that there is no guarantee that measures are targeting the right kind of behaviour.

I have indicated the lack of holistic solutions that can safeguard pedestrians and also make the life of pedestrians more attractive. As both parts are of importance for an attractive and sustainable city, there is a strong need to put more efforts in demonstrating holistic solutions on a large scale and to assess their effects. This is valid both for high-income countries and for low- and middle-income countries.

References

Bjarnason S (2004) Round top and flat top humps—the influence of design of the effects. Department of Technology and Society, LTH, Lund University. Thesis 125

Blatter J (1995) Möglichkeiten und Restriktionen für umweltorientierte Maßnahmen im Personennahverkehr - Analyse der Freiburger Verkehrspolitik unter besonderer Berücksichtigung der Parkraumpolitik und der regionalen Zusammenarbeit im ÖPNV. EURES discussion paper dp-47, Institut für Regionale Studien in Europa, Freiburg

Carsten O (2005) Presented at PROSPER Conference, Brussels 23th November 2005

Ekman L (1988) Fotgängares risker på markerat övergångsställe jämfört med andra korsningspunkter. (Pedestrian risk on marked crossing compared with other crossing locations). Department of Traffic Planning and Engineering, LTH, Bulletin 76. Lund 1988

Elvik R, Christensen P, Amundsen A (2004) Speed and road accidents. An evaluation of the Power Model. Transport Economics Institute. TOI report 740/2004. Oslo

Elvik R, Erke A, Vaa T (2010a) Trafikksikkerhetshåndboken, webversion https://tsh.toi.no/?22192#2219230 (Traffic Safety Handbook, Norwegian web-version, 2010–10–17)

Elvik R, Erke A, Vaa T (2010b) Trafikksikkerhetshåndboken, webversion https://tsh.toi.no/?22192#2219230 (Traffic Safety Handbook, Norwegian web-version, 2010–10–17)

Elvik R, Erke A, Vaa T (2010c) Trafikksikkerhetshåndboken, webversion https://tsh.toi.no/?21728#2172830 (Traffic Safety Handbook, Norwegian web-version, 2010–10–17)

Hedström R (2004) Miljöeffekter Av 30 Km/H I Tätort - Med Avseende På Avgasutsläpp Och Buller - En förstudie. VTI meddelande 869

Hydén C, Svensson Å (2009) Traffic calming in India. Report on the theory of Traffic Calming and empirical trials in the city of Jaipur. Department of Technology and Society, LTH, Lund University. Bulletin 252

Hydén C, Várhelyi A (2000)The effects on safety, time consumption and environment of large scale use of roundabouts in an urban area: a case study. Accid Anal Prev 32:11–23

Hydén C, Jonsson T, Linderholm L, Towliat M (2008) Nya hastighetsgränser i tätort - Resultat av försök i na°gra svenska kommuner. (Bulletin 240 A / 3000; Vol. Bulletin 240 A / 3000). Lund University Faculty of Engineering, Technology and Society, Transport and Roads, Lund, Sweden

Ishaque MM, Noland RB (2006) Making roads safe for pedestrians or keeping them out of the way?—an historical perspective on pedestrian policies in Britain. Centre for Transport Studies, Imperial College, London. From: https://pubsindex.trb.org/view.aspx?id=776669 (2010–10–17)

Jaredson S (2002) Utvärdering av Skvallertorget i Norrköping (Evaluation of Skvallertorget in Norrköping). Department of Science and Technology Institutionen för teknik och naturvetenskap. Linköpings Universitet. Linköping, Sweden

Kolbenstvedt M, Elvik R, Elvebakk KB, Hervik A, Braein L (2007) Effekter Av Den Svenske Trafikksikkerhetsforskningen 1971–2004 (Effects of Swedish Traffic Safety work 1971 – 2004). Vinnova Analys VA 2007:07. Stockholm, Sweden

Nilsson G (2004) Traffic safety dimensions and the Power Model to describe the effect of speed on safety. Bulletin 221. Lund Institute of Technology, Department of Technology and Society, Traffic Enginering, Lund

Nilsson G, Thulin H (2004) Trafiksäkerhetsutveckling i Göteborg (Traffic safety development in Gothenburg). VTI: R503. Linköping, Sweden

Mohan D, Tsimhoni O, Sivak M, Flannagan MJ, Road safety in India: challenges and opportunities, The University of Michigan. Transportation Research Institute, UMTRI-2009–1.

Pharoah T, Russell J (1989) Traffic calming: policy and evaluations in three European Countries, South Bank Polytechnic (London), February 1989. In: Litman T (ed) Benefits and costs, evaluating traffic calming benefits, cost and equity impacts. Victoria Transport Policy Institute, pp 1997–1999

Retting RA, Persaud B, Garder P, Lord D (2001) Crash and injury reduction following installation of roundabouts in the United States. Am J Public Health 91(4) (2001)

Risser R (2010) Mjuka" faktorer som indikatorer för hållbara transporter i staden ("Soft" factors as indicators of sustainable transport in cities). Department of Technology and Society, LTH (preliminar)

Take Back Your Streets, Conservation Law Foundation (Boston; www.clf.org), May 1995, p 27. In: Litman T (ed) Benefits and costs, evaluating traffic calming benefits, cost and equity impacts. Victoria Transport Policy Institute, pp 1997–1999

Trafikkontoret (2004) God effekt av trafiklugnande åtgärder i Göteborg (Good effect of Traffic Calming measures in Gothenburg). Internt Gatubolaget. Internal report. Projektnummer: 500 4110 80

Watts GR (1973) Road humps for the control of vehicle speed. Transport and road research laboratory, Laboratory report 597, Crowthorne 1973

World Health organization (2009) Global status report on road safety: time for action. World Health Organization, Geneva. www.who.int/violence_injury_prevention/road_safety_status/2009

Chapter 8
State of the Art of Roundabout Performance for Promoting of Urban Safety

Werner Brilon

8.1 Introduction

Town planning in the nineteenth century favoured large circular places as elements of agreeable city design. With increasing city traffic at the beginning of the twentieth century, these locations were the first where roundabouts (which just means one-way direction of traffic on the circle) were established like Columbus Circle in New York (1905) or Place Etoile in Paris (1907) (Todd 1988 and 1991). This happened in many countries around the world. In consequence, most of the large cities in the twentieth century had their monumental large traffic circles.

However, the traffic rules were quite different in various countries. In Germany, circulating traffic had priority; in other European countries, the entering traffic had the right-of-way with the consequence that under high traffic demand these intersections became gridlocked. In the USA, a variety of rules had been tested over the years. However, in 1966 the UK introduced the "off-site priority rule". This rule means: (a) the circular traffic has priority over the entering vehicles and (b) the vehicles on the inner lanes are privileged in a conflict over vehicles travelling further outside (GOV.UK). This rule is the background of the great success of roundabouts in the UK, and it is the reason for exceptionally high capacities at large roundabouts which can only be observed in the UK.

Meanwhile, outside the UK only part (a) of the "off-site priority rule" is valid in most countries, i.e. traffic on the circular roadway of the roundabout has priority over the approaching traffic. This rule, which has been valid in Germany since ever, has been adopted by the highway code in most countries of the Western Hemisphere during the last three decades.

W. Brilon (✉)
Ruhr-University Bochum, Bochum, Germany
e-mail: werner.brilon@rub.de

G. Tiwari and D. Mohan (eds.), *Transport and Safety*, Springer Transactions in Civil and Environmental Engineering,
https://doi.org/10.1007/978-981-16-1115-5_8

The acceptance and application of the valid traffic rules are the key for traffic safety of roundabouts. In the Western countries, the acceptance of these rules, usually, is quite good. Speaking about safety, thus, organization and acceptance of the traffic rules is a significant basic condition for all conclusions about traffic safety.

8.2 Classification of Roundabouts

Also the styles of roundabout design are specific to different countries. The traditional layout of roundabouts in the early twentieth century involved large multi-lane circles. These, however, were not successful regarding safety. Especially two-lane exits emerged as a major source of severe accidents. Thus, in the 1950s and the 1960s the larger circles were no longer favoured in most countries on the European continent or in the USA. Later in 1980, the big success of roundabouts in the UK incited planners and researchers in several European countries to study and experiment with roundabouts.

These studies unveiled unexpected gains in traffic performance and safety, however, only for the single-lane roundabouts. These compact intersections were found to be able to carry up to 25,000 veh/day combined with rather low delays for road users and with the highest potential to prevent accidents. They are still the most favoured type of roundabouts.

Later on, slightly larger—and also smaller—roundabouts were studied in many countries. As a consequence, we now have a whole toolbox of different types of roundabouts. Figure 8.1 tries to illustrate diameters and range of traffic demand for roundabouts for different sizes:

- mini-roundabout with a traversable central island and a diameter between 13 and 23 m
- single-lane roundabouts with a diameter between 26 m (minimum required for European trucks to make a full turn) and 35 m (urban) or 40 m (rural) and only single-lane entries and exits
- semi-two-lane roundabouts with a diameter of 45–60 m, a lane widths of 8–10 m (no lane marking on the circle) and single lane exits but 1- or 2-lane entries
- larger two-lane roundabouts (which are banned, e.g. by German guidelines due to their bad accident experience)
- turbo-roundabouts with 1- or two-lane segments on the circle. The entries and exits may have one or two lanes, where the two-lane solution needs a specific design to avoid undesired lane changes.

This is the kind of classification used in Germany. But in most countries on the European continent, the view on roundabouts is quite similar. It should be emphasized that all rules in design guidelines of the continental European countries are governed by a maximization of traffic safety as the first target. Capacity is only of secondary importance. Less safe roundabout constructions are not treated as state of the art.

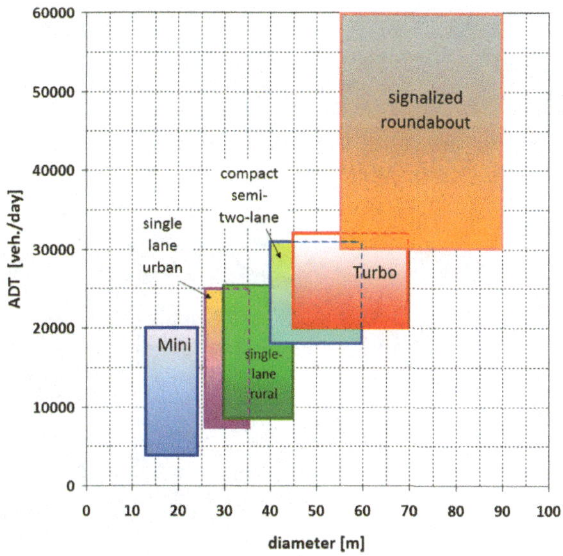

Fig. 8.1 Definition of types of roundabouts by their inscribed circle diameter and their potential range of applicability in terms of average daily traffic (ADT)

In the UK the situation seems to be different. There the design is not so much oriented in lanes. Instead, if capacity makes it necessary, the lanes are flared out near the roundabout to increase capacity. This is supported by the results of capacity investigations (Kimber 1979). This leads to a design which can differ considerably from European continental solutions.

The USA started rather late in the 1990s with experiments in modern roundabouts. They discovered the benefits of this type of intersection, and meanwhile there are many of them. However, as with everything, the roundabouts in the USA are larger than in Europe (Fig. 8.2).

8.3 Some Words Regarding Accident Statistics

An international comparison of traffic safety leads to some complications regarding accident statistics. Already the identification of an accident differs from country to country. Some countries count all accidents (including property damage only) which were reported to the police. Other countries take account of only accidents with personal injuries. Another difference concerns fatalities, e.g. in Germany a fatality is classified as such if the victim dies within 30 days after the accident. Other countries apply completely different definitions.

Also the researchers use different methods of evaluation, e.g. the distance on the approaching arms where accidents are treated as intersection-related varies and is not

Fig. 8.2 Typical examples for an urban single-lane roundabout in Germany

Fig. 8.3 Examples for a mini-roundabout

explained in most of the publications. The most serious way in intersection accident analysis is to define relevant parameters to describe accident risk. The following variables seem to be most characteristic:

$$\text{accident rate} = \frac{\text{number of accidents}}{N_T}$$

$$\text{accident } \cos t \text{ rate} = \frac{\text{damage by accidents}}{N_T}$$

where N_T = no. of vehicles travelled through the intersection, usually estimated by the average daily traffic (ADT) damage by accidents: evaluated in currency units.

For the calculation of accident cost rates, the damage caused by accidents must be evaluated in currency units where the figures used are standardized on a national basis (e.g. in Germany: BASt 2014).

These measures of accident occurrence are relative to the exposure to risk as it is represented by the number of vehicles travelling through the intersection. These parameters allow a more meaningful interpretation than absolute figures.

Unfortunately, all the publications on traffic safety apply different methods of analysis. Therefore, a definitive comparison of the results from different investigations is not easy.

8.4 Some Early Findings for Traffic Safety

Starting modern roundabouts in the 1980, studies about safety effects generated by this type of intersection have been performed in many countries. Each of these studies revealed a large potential of safety emerging from modern roundabouts. Among these early studies, we see investigations from France. Gambard (1989) found in a before/after study where conventional intersections had been converted into roundabouts a reduction of fatalities by 88%. Alphand et al. (1991) studied more than 500 roundabouts. 90% of these remained without any personal injury for the year of investigation. Roundabouts experienced only half the number of accidents as signalized intersections. Also two-wheelers had 77% less accidents at roundabouts than at signalized intersections. These analyses are based on a report by CETE (1986). Here we can also find that large roundabouts had a significantly higher accident rate. The results also supported higher safety for circles with a cross-fall (slope) to the outside.

Also Switzerland started experiments with modern roundabouts in the 1980s. Several studies by Buehlmann and Huber (1994), Buehlmann and Spacek (1997) testified the high level of traffic safety at roundabouts. In a before/after study at 113 intersections, most of them in urban environments, he found significant improvements in safety by converting intersections into roundabouts—especially at single-lane roundabouts. Multi-lane roundabouts, however, did also lead to worse safety conditions compared with the previous situation. The most important effect was the

Table 8.1 Reduction in accident rates in Germany

Roundabouts				Conventional intersections			
Size	AR	ACR	MC		AR	ACR	MC
Large	6.6	24.9	3.8	With traffic signal	3.6	21.7	6.5
1-lane	1.2	4.7	3.8	Two-way-stop	1.0	12.0	12.0

Source Brilon and Stuwe (1993)
AR Accident rate in acc./10^6 veh; *ACR* accident cost rate in DM/10^3 veh; *DM* German mark = valid currency until 2002; *MC* mean costs per accident in 10^3 DM/acc.

significant reduction in accident severity. The largest gains in safety were achieved for pedestrians.

Reports on roundabout safety in the Netherlands had been published by van Minnen (1992, 1995a, b). The study included 46 (1990) and 177 (1995) roundabouts—most of them in urban areas. Also here roundabouts turned out to be much safer than other types of intersections. After converting conventional intersections into roundabouts, the number of recorded accidents decreased by 47%, the number of victims by 71%. The biggest reduction was achieved for car passengers (−95%) and pedestrians (−89%), whereas cyclists benefited only by −30%. Accidents were considerably reduced by converting intersections into roundabouts. Similar results were reported by Schoon and van Minnen (1994) on the basis of 201 places which were converted into roundabouts. Here the number of accidents per intersection was cut into half, and the number of injury accidents was reduced by 70%.

Several studies—however with rather small samples—have also been made in Germany. Brilon and Stuwe (1993) report on their studies showing the advantage of single-lane roundabouts. Results from before/after studies are summarized in Table 8.1.

The before/after comparison showed a reduction of accident costs from 34.5 DM/10^3 veh to 5.1 DM/10^3 veh (i.e. −88%) as a consequence of intersection conversions into single-lane roundabouts. This effect is achieved by a strong reduction in the number of severe injuries. These results were a clear indication of an unexpectedly high level of traffic safety at small roundabouts.

Other reports provided very similar results come from Norway (Giaever 1992), Denmark (Jorgensen and Jorgensen 1994), USA (Flannery and Elefteriadou 1999), and Australia (Tudge 1990; Troutbeck 1993). A good international overview on roundabout safety in the earlier times is also given by Jaquemart (1998). Table 8.2 gives an overview of the published effects of roundabout implementation at street junctions from the earlier days obtained from various publications, e.g. also Rodegerdts et al. (2010, pp. 5–16).

Elvik (2003) has reviewed results from 28 roundabout safety studies in a meta-analysis. He concluded that, on average, roundabouts reduce injury accidents by 30–50% and the amount of fatal accidents by 50–70% whereas the results for property-damage-only accidents varied quite a lot. Overall, also this study underlines the significant improvement of traffic safety by roundabouts.

Table 8.2 Rough figures for the reduction of the number of accidents in various countries obtained from earlier publications

	All	Injury
	Crashes	Accidents
Netherlands	−47	−72
Australia	−41 up to −61	−45 up to −87
France		−57 up to −82
USA	−35	−76
Germany	+22	−75

All these former findings of positive safety effects obtained from roundabouts are the basis for the extreme dissemination of modern roundabouts in many countries of the world.

8.5 Current Research Results on Roundabout Safety

The number of roundabouts has exploded in many countries during the last three decades. Reports estimate the numbers in France to be above 30,000, in the UK above 30,000, in the USA 9000 (in 2021, cf. Kittelson 2021), in Denmark 1450 (in 2010; Underlien-Jensen 2013) and in Germany a count in the summer of 2014 revealed 12,500. As a consequence, against this broader background most countries have updated their experiences with modern roundabouts. Among the wide diversity of reports we concentrate only on some characteristic results.

8.5.1 USA

Bill and Khan (2014) report that accident research at 30 roundabouts in Wisconsin over 3–4 years each before and after a conversion of conventional intersections into roundabouts. Although the effects were not always as good as expected the general conclusion was positive, e.g. the total number of crashes increased by 12%. The crash severity, however, did significantly decrease. The number of accidents with personal injuries decreased by 38% and no fatality was reported. The best effects were achieved at unsignalized intersections.

An investigation about roundabout safety on a national platform had been performed by Rodegerdts et al. (2007) which is also documented in the second US roundabout guide (Rodegerdts et al. 2010). Fifty-five sites at different locations in the US had been investigated in a before and after study. The conversion into roundabouts achieved a reduction in the number of all accidents of 35% and for injury

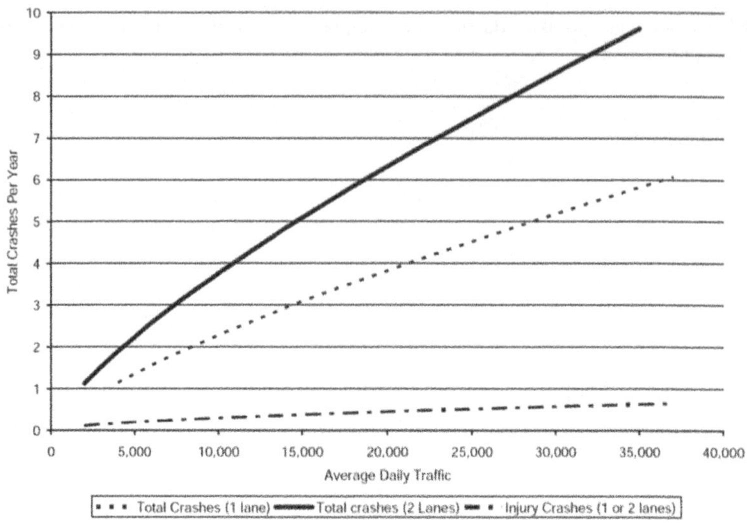

Fig. 8.4 Expected number of crashes per year for one- and two-lane roundabouts (Rodegerdts et al. 2010)

accidents of 77%—which again provides an indication that especially severe accidents are avoided by roundabouts. The study did also demonstrate that the single-lane accidents provide a better level in safety than the two-lane circles (Fig. 8.4).

8.5.2 Denmark

One rather careful investigation on roundabout safety has been performed in Denmark (Underlien-Jensen 2015). He has analysed 332 sites which were converted from conventional intersections into roundabouts during a period from 1995 until 2009. Empirical data were collected for general trends and for "regression to the mean". The number of injury crashes reduced by 47% with a decrease of injuries by 60%. The estimate is that the conversions prevented 20 fatalities. The improvements, regarding all accidents, were better at single-lane roundabouts than at multi-lane circles. Safety effects were the largest at 4-arm intersections, for a large percentage of left-turning traffic, and when the approach speeds were high. However, at 3-arm intersections under urban speed conditions, the accident number did increase. Special attention was directed on bicycles. The number of cycle accidents increased from 113 to 246% when the new roundabout had a cycle lane at the outer margin of the roundabout—a clear indication that this facility is extremely dangerous—this has also been found in other countries several years ago. On the other side, roundabouts with separate cycle paths and no priority to cyclists at the crossings lead to a reduction in the number of accidents by −81%. The study could also demonstrate that the long-term improvement effects were even better than the improvements in the first years. The problem with this study is that all investigated roundabouts were treated the same regardless of whether they were designed in line with the valid guidelines or they suffered from known errors in design (e.g. bicycle lanes).

8.5.3 Germany

The general situation regarding roundabouts in Germany has been described by Brilon (2014). Moreover, recently a rather well-sophisticated accident analysis at roundabouts was published in Germany (Bondzio et al. 2012). The investigation is concentrated on 100 single-lane urban roundabouts which comply with the current design guidelines. These sites were distributed over ten German states in towns and cities of all sizes. The traffic volumes were between 5000 and 25,000 veh/day, 0 to 800 pedestrians/2 h, and 0 to 7000 cycles/day. The research considered accident data from the police records from 2008 until 2010. Figure 8.5 as an example shows a plan from one of the investigated roundabouts with an inscribed diameter of 35 m, 24,000 veh/day, 720 cycles/day, and 120 ped/2 h. The circular roadway consists of a 4.5-m-wide asphalted roadway and an inner 3-m-wide truck apron. Bicycles run on the circular roadway.

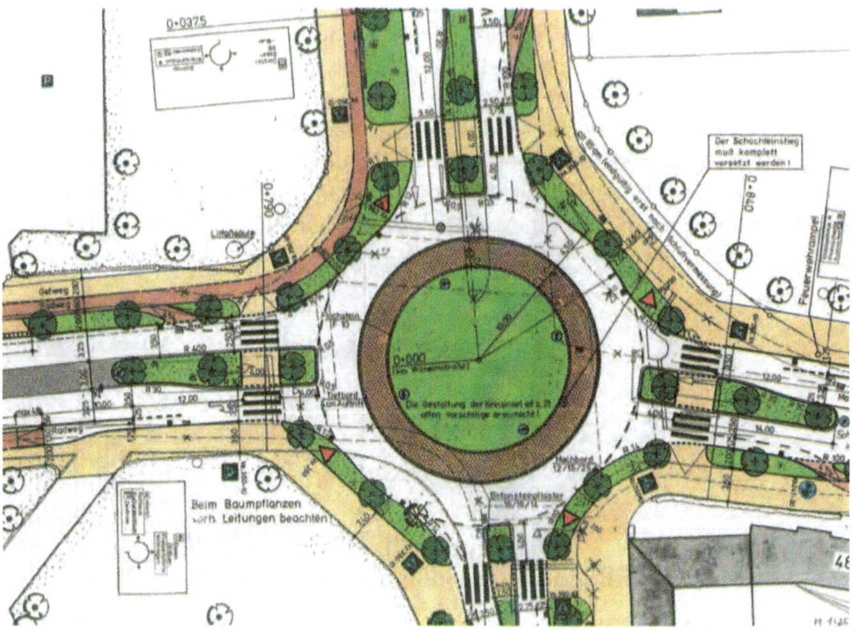

Fig. 8.5 Plan of one of the roundabouts

There was no fatality in the sample, 8 hospitalized injured (5 were pedestrians) and 92 slightly injured persons. Cars were mostly involved in the accidents (80%), cyclists with 10%, and pedestrians only with 1% (see Fig. 8.7). The picture of involvement becomes different if we analyse the traffic involvement of different road users only for accidents with personal injuries. Then we see the higher vulnerability of motor cyclists, bicycle users, and pedestrians. From the rather low involvement of pedestrians, we can obtain the high degree of safety of all roundabouts for pedestrians. Here it should also be noted that of a total of 15 accidents involving pedestrians 5 were severely injured persons and 11 had slight injuries, 9 from 11 accidents on crosswalks happened on zebra crossings (zebra crossing gives an absolute priority to pedestrians).

The average accident cost rate was 6.3 €/1000 veh which is much lower than the 10 €/1000 veh which are treated as a good result for an intersection. Therefore, on average these roundabouts proved to be a very safe solution for an intersection. The accident cost rate as it is distributed from extremely low to rather high values is illustrated in Fig. 8.6. We see that the average of 6.3 included points with an extremely low figure, but also roundabouts with intolerable severe accident occurrence. This picture of a concentration of severe accidents on few sites was even worse when focusing on bicycle accidents. The reasons for these differences have not been analysed.

Looking on the influence of design parameters: In a range from 25 m until 50 m the diameter did not influence the accident risk. However, a number of arms larger than 4 had a negative effect on safety (Fig. 8.7).

Table 8.3 provides a view on the distribution by types of accidents. We see that the most frequent type was a collision of an entering vehicle with a circulation road user (31%). Looking on accident severity, we see, however, the important role of bicycle accidents.

A special focus was set on bicycles at these roundabouts. The designs under test were: (A) cycles mixed with other traffic on the circle (e.g. Fig. 8.5), (B1) bicycle paths with priority to cyclists at the crossings (e.g. Fig. 8.8), (B2) bicycle paths together with pedestrian facilities, (B3) bicycle paths with no priority to cyclists

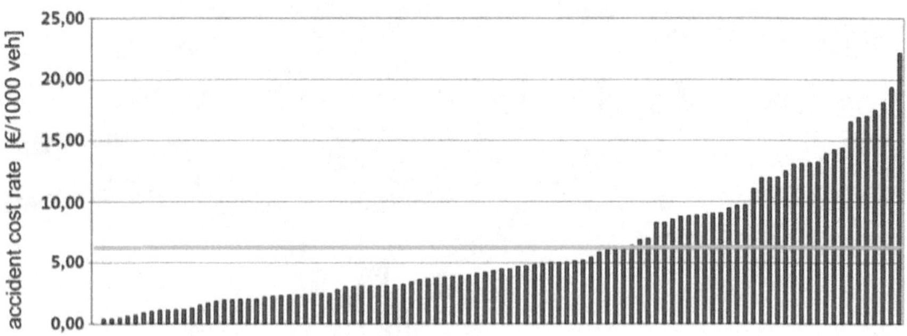

Fig. 8.6 Distribution of accident cost rate over the 100 roundabouts under investigation (Bondzio et al. 2012) (vertical lines: individual values of accident cost rate at the 100 sites in the study; horizontal green line: average)

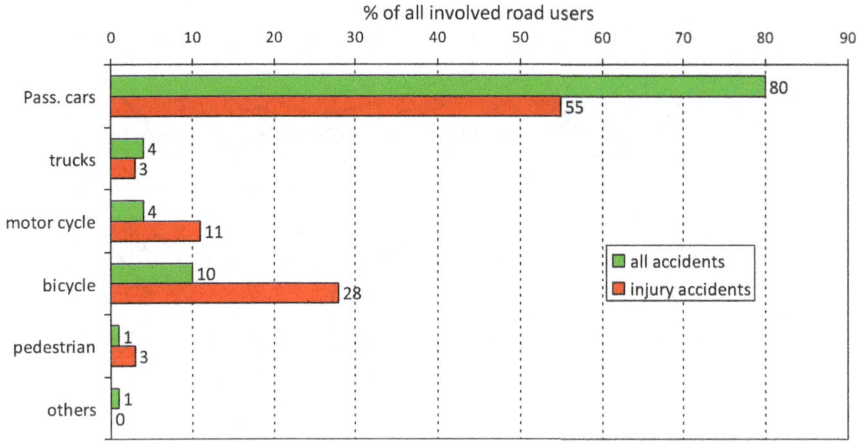

Fig. 8.7 Involvement of different road users in all accidents and in injury accidents (in %)

Table 8.3 Frequency of different types of accidents

	All Accidents	Injury Accidents
Collision of approaching with circulation vehicle	34	31
Single vehicle accident	13	10
Rear end collision on the approach	17	7
Rear end collision on the circle	19	8
Collision of an entering vehicle and a bicycle (on the cycle path crossing) at the entry	7	10
Collision of an existing vehicle and a bicycle (on the cycle path crossing) at the exit	5	15

at the crossings. In Fig. 8.9, we see that the safest solution is to provide separate paths guiding cyclists around the roundabout where the cyclist have to care for the priority of motor vehicles at the crossings. The two solutions with priority to cyclists were associated with a larger accident cost rate. The answer was the same when the exposure of the intersection to cyclists was taken into account. This result is well in line with results from other countries like Denmark (see above).

A comparison with other types of intersections or with prior situations was not a topic in this investigation. It was, however, complimented by an observation of road user behaviour at some of the analysed intersections. Here, among others it was found that at cycle paths (see e.g. Fig. 8.8), the bicyclists up to 50% use the crossings in the wrong direction which imposes a significant risk.

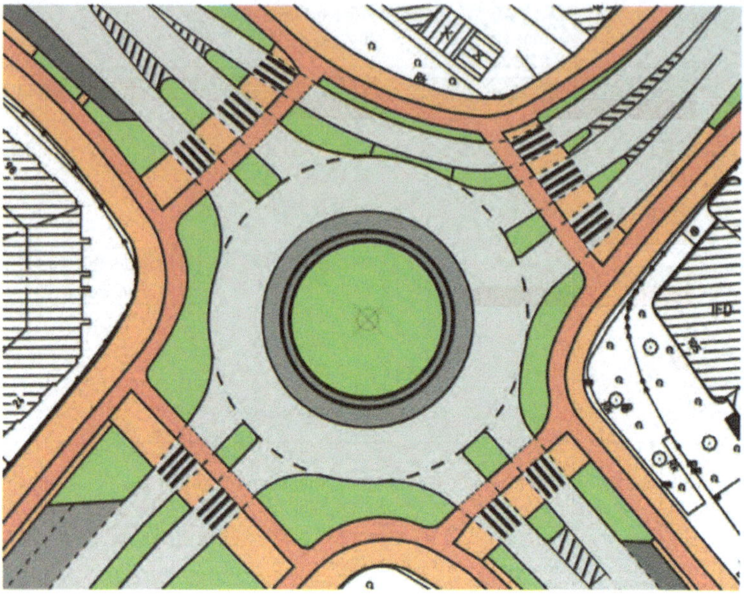

Fig. 8.8 Sketch of a roundabout with cycle paths which cross the entries and exits next to the pedestrian crossings (here with zebra marking). This site also has a bypass lane (on the top of the picture)

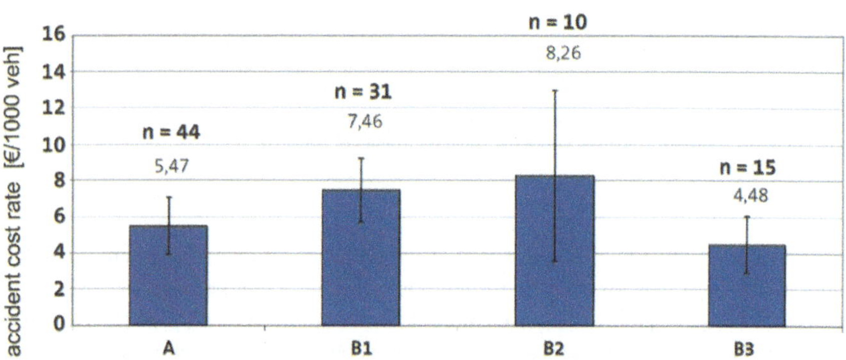

Fig. 8.9 Accident cost rates for the various designs for cyclists (Bondzio et al. 2012; types A, B1, … see text)

8.5.4 Mini Roundabouts

Mini-roundabouts are small circular junctions with a traversable central island (see Fig. 8.3). Small cars, have to drive around the central island, whereas trucks are forced to cross this island with their rear wheels. These types have first been introduced in the

Table 8.4 Effect of the conversion into mini-roundabouts

	Before (conventional intersection)	After (mini roundabout)
Average accident rate (acc./10^6 veh)	0.8	0.4
Average accident cost rate (DM/1000 veh)	29.3	3.5

Table 8.5 Accident cost rates (€/1000 veh) of mini-roundabouts compared to other types of junctions

	Mini roundabout	Unsignalized intersection	Signalized intersection
3 arms	2.02	4.68	6.60
4 arms	5.66	13.39	8.40

UK under the leadership of Frank Blackmore in 1968. Other countries have imitated this example rather late, e.g. Germany started to experiment with this form in 1997 (see Brilon and Bondzio 1999).

The safety of mini roundabouts in Germany has been studied first by Brilon and Bondzio (1999) and recently by Baier et al. (2014). Brilon and Bondzio report on the pioneering (for Germany) experiment of a conversion of 10 intersections into mini-roundabouts. As a result, the accident risk was significantly reduced (see Table 8.4). The accident cost rate was much lower than at a very safe conventional intersection (i.e. $\approx$10 DM/1000 veh in 1999).

Also the analysis by Baier et al. of 26 mini-roundabouts and 309 conventional intersections confirmed an extraordinary low accident cost rate (see Table 8.5).

Of course, mini-roundabouts are only allowed in urban areas with a general speed limit of 50 km/h. Details of a reasonable design are described in the guidelines (FGSV 2006) and in the report by Baier et al. (2014). The most important aspect is that the central island should consist of a paved circle which must be elevated by 4 or 5 cm above the asphalted circle.

8.5.5 *Turbo-Roundabouts*

A turbo-roundabout is a kind of circular intersection where the number of lanes on the circle varies between 1 and 2 and where the traffic through the intersection is strictly channelized by lanes (Fig. 8.10). This type does also allow a safe operation of two-lane exits. Usually, such a roundabout has a diameter of 50 or more meters. Application is useful if one or two of the movements have periods with very large traffic volumes. In Germany, traffic guidance is achieved just by lane markings, whereas in the Netherlands vertical lane dividers (similar to kerbs in the middle of

(a)

(b)

Fig. 8.10 Typical exit (**a**) and entry (**b**) of a turbo-roundabout in Germany

the roadway) are in use. Even if these circles are quite space-consuming their capacity is limited to around 35,000 veh/day.

Traffic safety of turbo-roundabouts has recently been investigated by Brilon and Geppert (2015), based on a limited sample size. The accident rate on average was 1.0 acc/10^6 veh, and the accident cost rate was 7.7 €/1000 veh which corresponds to the level of single-lane roundabouts and is better that the risk at conventional intersections.

Therefore, this type of roundabout provides an adequate level of traffic safety and, thus, is a useful instrument of traffic design in urban areas. However, it is not compatible with any kind of bicycle operation which means that for bicycles other kinds of traffic guidance (e.g. bridges) must be applied at the relevant sites.

8.6 Accident Prediction

Accident prediction at roundabouts has first been proposed by Maycock and Hall (1984). Based on the analysis of 84 4-arm at-grade roundabouts, they developed a linear model to predict the number of accidents (frequency of all crashes + pedestrian accidents) based on the entry path curvature, the roundabout diameter, entry width, angle of the approach relative to the circle, and traffic volumes. The equations can primarily be used to compare alternative designs according to the British style of roundabout design.

For the USA, an accident prediction model has been formulated by the NCHRP 572-report (Rodegerdts et al. 2007) which is also mentioned in the US roundabout guide (Rodegerdts et al. 2010; Exhibit 5-19, 5-20). Here the expected number of crashes per year is estimated by an exponential function of the ADT (annual average daily traffic), e.g.

$$\text{crashes/year} = 0.0038 \cdot ADT^{0.749} \quad \text{for a 2-lane 4-arm round about}$$

(see cited literature for other parameters)

Other parts of the model concern accident prediction for each approach. Here the ADTs of the approach, of the circle, and of the exit are of predominant importance. In addition, geometric parameters like entry radius, entry width, diameter, and others are used for accident prediction. These models estimate the number of accidents on the approach, entering-circulating, and exiting-circulating separately. These equations have the potential to compare several alternatives for the geometric design regarding safety. It must, however, be mentioned that the equations have a relatively small empirical background and that they are only based on the US background (e.g. definition of crashes, design style).

8.7 Cyclists at Roundabouts

All investigations underline the fact that cyclists at roundabouts constitute a specific problem. Usually, they also get some improvement in safety by a roundabout. However, these improvements are not as significant for them as they are for the other road users like car passengers or pedestrians. As a consequence, cyclists at roundabouts face the largest risks.

All the studies come to very similar conclusions. They distinguish between the following kinds of bicycle treatment. Moreover, the reports propose the following actions. These points apply only to single lane roundabouts.

- Bicycles in mixed traffic on the circular roadway together with cars: this is a very safe solution for lower traffic volumes. It should be favoured up to a total traffic volume of 15,000 veh/day (Germany, FGSV 2006; Haller et al. 2000) or 8000 veh/day (Netherlands, van Minnen 1995a, b).

- Bicycling lane on the outer margin of the circular roadway. This is the most dangerous solution. It must be absolutely banned (see, e.g., Brilon and Stuwe 1993; Schoon and van Minnen 1994; Daniels et al. 2009; Underlien-Jensen 2013).
- Bicycle paths separated from the roundabout. This is the recommended solution for larger traffic volumes. The crossings of the exits and the entries must be separated from the circle by 4 m, better by 1 car length, i.e. 5 m. The cycle paths should approach the crossings vertical to the direction of the roadway. It is evident that a priority for cyclists at these cross points induces a higher risk than a regulation where cyclist have to yield to motor vehicles (Bondzio et al. 2012; Underlien-Jensen 2013).

These recommendations apply for single-lane roundabouts. At multi-lane roundabouts, cyclists cannot be allowed on the same roadway as motor vehicles. Also bicycle crossings at multi-lane entries—and especially exits—are a significant risk. Thus, multi-lane roundabouts should only be implemented where the occurrence of cyclists can be completely excluded. Tunnels or bridges for cyclists are a must at these larger roundabouts.

For the mini-roundabouts, separate cycle facilities are not recommended. Here, if cyclists cannot be operated on the roundabout itself, then a mini is not a good solution for the relevant situation.

8.8 Conclusion

The paper tries to provide an overview about research results on safety at roundabouts with a focus on urban intersections. Even if it is written from a German perspective, it includes results from several other countries.

As a conclusion from all studies, there is no doubt that roundabouts are the safest type of intersection. Especially the single-lane roundabouts reveal the highest level of safety. Also mini-roundabouts have an extraordinary good safety record. Turbo-roundabouts—regarding safety—are on the same level as the single-laned. This high degree of traffic safety depends on the speed-reducing design of the whole intersection.

In comparison to conventional types of intersections like signalized or two-way-stop intersections, the car occupants and the pedestrians enjoy the highest gains from roundabout safety. On the other hand, bicyclists can become a problem for traffic safety at roundabouts. However, also cyclists can be accommodated with a sufficient degree of safety—but only if the requirements for design are strictly obeyed.

It should be emphasized that the high degree of safety is coherent to road user discipline and to the acceptance of the existing traffic rules. This acceptance should be strengthened by an adequate intersection design. Therefore, the favourable safety effects of roundabouts can only be achieved if the rules for modern roundabout design, as they are documented in many national design guidelines, are strictly applied.

It must also be ascertained that roundabouts are not the optimal solution in each situation. Following the continental European guidelines, there are limits in capacity (cf. Fig. 8.1) which in detail have to be figured out for each single case by adequate capacity models (see, e.g., Brilon 2014). Beyond these limits, signalized intersections remain to be useful solutions to manage traffic at urban intersections with a very large traffic demand. Also due to limited space, alternatives to roundabouts must be applied in many cases.

Overall, it can clearly be stated that the adequate use of roundabouts may be a real boon for traffic safety—especially in urban areas.

References

Alphand F, Noelle U, Guichet B (1991) Roundabouts and safety—state of the art in France. In: Intersections without traffic signals (II). Springer, Berlin, pp 107–125. ISBN 3-540-54180-2

Baier R, Leu P, Klemps-Kohnen A, Reinartz A, Maier R, Schmotz M (2014) Minikreisverkehre – Ableitung ihrer Einsatzbereiche und Einsatzgrenzen. Berichte der BASt, no. V240 (free download)

BASt (2014) Bast-info: Volkswirtschaftliche Kosten durch Strassenverkehrsunfaelle in Deutschland 2005–2012. Federal Highway Research Institute (BASt), Germany (free download) (see also: https://de.wikipedia.org/wiki/Unfallkosten)

Bill A, Khan G (2014) Safety evaluation of Wisconsin roundabouts. In: 4th International roundabout conference, Seattle, TRB

Bondzio L, Ortlepp J, Scheit M, Voss H, Weinert R (2012) Verkehrssicherheit innerörtlicher Kreisverkehre (Traffic safety of urban roundabouts). Report no. VI05 of the association of German insurance companies GDI (free download)

Brilon W (2014) Roundabouts: a state of the art in Germany. In: 3rd International roundabout conference, Seattle (free download)

Brilon W, Bondzio L (1999) Untersuchung von Mini-Kreisverkehrsplätzen in Nordrhein-Westfalen (Investigations on mini-roundabouts). Straßenverkehrstechnik 9:428–434

Brilon W, Geppert A (2015) Kapazitaet von großen Kreisverkehren: Turbo-Kreisverkehre (Capacity of larger roundabouts: turbo-roundabouts) Strassenverkehrstechnik 2(2015):81–90

Brilon W, Stuwe B (1993) Capacity and design of traffic circles in Germany. Transp Res Rec 1398:61–67

Buehlmann F, Huber CA (1994) Sicherheit von Kreiselanlagen – Erfahrungen und vorläufige Empfehlungen. Bfu-Bericht, Bern

Buehlmann F, Spacek P (1997) Unfallgeschehen und Geometrie der Kreiselanlagen. Forschungsauftrag Nr. 17/93 des Eidgenossischen Verkehrs- und Energiewirtschaftsdepartementes, Zurich

CETE (1986) Evolution de la securite sur les carrefours giratoires. CETE-Ouest, Nantes, France

Daniels S, Brijs T, Nuyts E, Wets G (2009) Injury crashes with bicyclists at roundabouts: influence of some location characteristics and the design of cycle facilities. J Safety Res 40(2009):141–148

Elvik R (2003) Effects on road safety of converting Intersections to Roundabouts: Review of Evidence from Non-U.S. Studies. Transportation Research Record no. 1847

FGSV (2006) Merkblatt fuer die Anlage von Kreisverkehren. Forschungsgesellschaft fuer Strassen- und Verkehrswesen, Cologne

Flannery A, Elefteriadou L (1999) A review of roundabout safety performance in the United States. In: ITE international conference, 28–31 Mar 1999 (free download)

Gambard JM (1989) Safety and design of unsignalized intersections in France. In: Intersections without traffic signals (I). Springer, Berlin. ISBN 3-540-18890-8

Giaever T (1992) Application, design, and safety of roundabouts in Norway. In: Proceedings of the roundabout conference in Nantes, France, Oct 1992. CETE, 92220, Bagneux, France, pp 83–92

GOV.UK: Highway Code, here: rules 184–190. See: https://www.gov.uk/using-the-road-159-to-203/roundabouts-184-to-190

Haller W, Lange J, Alrutz D, Stellmacher-Hein J (2000) Fusgaenger- und Radverkehrsfuehrung an Kreisverkehrsplaetzen. Schriftenreihe Strassenbau und Strassenverkehrstechnik, no. 793, Bonn

Jaquemart PE (1998) Modern roundabout practice in the United States. Synthesis of highway practice, 264. National Academy Press, Washington D.C. (free download)

Jorgensen E, Jorgensen NO (1994) Traffiksikkerhed I 82 danske rundkoersler - anlagt efter 1985. Institut for Veje, Trafik og Byplan

Kimber R (1979) The traffic capacity of roundabouts. TRRL-report LR942, Crowthorne

Kittelson (2021) https://roundabouts.kittelson.com/roundabouts

Maycock G, Hall RD (1984) Accidents at 4-arm roundabouts. TRRL report LR1120, Crowthorne, UK

Rodegerdts L, and many other authors (2007) Roundabouts in the United States. NCHRP report 572, TRB, Washington D.C. (free download)

Rodegerdts L, and many other authors (2010) Roundabouts: an informational guide, 2nd edn. NCHRP report 672, TRB, Washington D.C. (free download)

Schoon C, van Minnen J (1994) The safety of roundabouts in The Netherlands. Traffic Eng Control 142–148

Todd K (1988) A history of roundabouts in the United States and France. Transp Q 42(4):599–623

Todd K (1991) A history of roundabouts in Britain. Transp Q 45(1):143–155

Troutbeck R (1993) Capacity and design of roundabouts in Australia. Transp Res Rec

Tudge RT (1990) Accidents at roundabouts in New South Wales. In: 15th ARRB conference, proceedings, pp 341–349

Underlien-Jensen S (2013) Safety effects of converting intersections to roundabouts. Transp Res Rec 2389:22–29

Underlien-Jensen S (2015) Safety effects of height of central islands, sight distances, markings, and signage on single-lane roundabouts (unpublished paper)

van Minnen J (1992) Experiences with new roundabouts in the Netherlands. In: Proceedings of the roundabout conference in Nantes, France, Oct 1992. CETE, 92220, Bagneux, France, pp 153–162

van Minnen J (1995a) Rotondes en voorrangsregelingen. R-95-58. Stichting Wetenschappelijk Onderzoek Verkeersveiligheid SWOV, Leidschendam

van Minnen J (1995b) Rotondes en voorrangsregelingen II. R-98-12. Stichting Wetenschappelijk Onderzoek Verkeersveiligheid SWOV, Leidschendam

Chapter 9
Pedestrian Safety Versus Traffic Flow: Finding the Balance

Robert B. Noland

9.1 Introduction

Modern and developing cities are characterized by the pervasiveness of motorized traffic. Western nations began this process over a century ago when motorized vehicles originally became commonplace and ownership and use grew rapidly. Over the last 20–30 years, many rapidly developing nations are facing the same growth in motorized vehicle usage and are facing many of the challenges that the developed world tackled in the past. While there are well-known benefits to motorization, there are also costs. Many countries are making the same mistakes that western nations have been trying to fix over the last few decades. The promotion of mobility and traffic flow over other goals is now seen as having damaged the well-being of cities mainly by taking space away from people with consequences on the ability of people to walk and engage in activities without motorized transport.

This chapter provides an overview of the conflict between pedestrian safety and designing and building streets to maintain efficient traffic flow. I start with an historical perspective on the early conflicts over the use of streets in both Great Britain and the USA. This is followed by a discussion and critique of the guidance developed by traffic engineers to improve traffic flow and more recently traffic safety. The effectiveness of various approaches to improving both traffic and pedestrian safety is then discussed, based on recent research, while noting that in the USA, our safety data still has many problems. Conclusions focus on solutions with a spotlight on shifting more street space to pedestrians as a way of improving both safety and the livability of cities. The results of a cost/benefit analysis of one specific approach, a road diet, that reduces vehicle capacity is presented and shows overwhelming positive benefits.

R. B. Noland (✉)
Alan M. Voorhees Transportation Center, Edward J. Bloustein School of Planning and Public Policy, Rutgers University, New Brunswick, USA
e-mail: rnoland@rutgers.edu

G. Tiwari and D. Mohan (eds.), *Transport and Safety*, Springer Transactions in Civil and Environmental Engineering,
https://doi.org/10.1007/978-981-16-1115-5_9

I close with some discussion of how developing nations can tackle these issues before it is too late.

9.2 A Brief History Lesson

Starting in the mid-nineteenth century, before motorized vehicles, conflicts began to develop over the use of city streets in Britain. Horse-drawn carriages (also known as omnibuses) became much more common as incomes grew with the industrial revolution. Pedestrians dominated city streets and these new forms of transport were a hazard resulting in both fatal and injury crashes. As early as 1840, there were about 1000 deaths recorded and fatality rates averaged about 50 per million (Hair 1971; Ishaque and Noland 2006). For comparison, the rate in 2013 was about 26.72 per million (based on 1713 road deaths) (Department for Transport 2013). While data records in the nineteenth century may not be accurate, this implies that the fatality rate associated with road transport has only been cut in half in 170 years; however, mobility is vastly improved and we will return to this trade-off later.

Those responsible for traffic casualties were liable to have their vehicle forfeited under British law. This was known as the Law of the Deodands, and objects involved in the death of a person were forfeited to the Crown. This law, however, was repealed in 1846, just at the time that deaths from traffic were increasing. The reason for this was that jurors often did not convict the vehicle owner, partly because they owned vehicles themselves and they saw the forfeited items ending up with "lords and churches instead of with the family of the deceased." (Ishaque and Noland 2006; Farr 1870). This shows how those who were more mobile (and normally more privileged people sat on juries) sought to minimize the penalties associated with the benefits of mobility. This persists to this day, in that penalties for vehicle drivers who kill pedestrians are often trivial.

It also suggests that pedestrians were considered to be more blame-worthy for their fate than the drivers of carriages and omnibuses. The right to the mobility of these vehicles was seen as more important than the safety of those who were walking. One jurist wrote in 1869: "Accidents happen because the drivers do not believe, or at any rate will not admit, that foot passengers had as much right to cross a street or thoroughfare as persons driving has to pass along it." (Ishaque and Noland 2006; The Times 1869). This sounds very familiar to debates that we hear today over the use of road space. Recent research suggests that attitudes towards the deaths of vulnerable road users have not changed (Goddard et al. 2019; Ralph et al. 2019).

As traffic increased, policy and regulations were developed to both protect pedestrians and guarantee the ability for traffic to flow unimpeded. Three approaches in particular were followed in Britain. These were the requirement that footpaths (sidewalks) be built along streets, the development of traffic signals and ways for pedestrians to cross streets, and the building of guardrails along sidewalks to channel pedestrians to designated crossing points.

Sidewalks are undoubtedly useful for protecting pedestrians from fast-moving traffic, but also allow that traffic to move in the street by removing the right of pedestrians to walk in the street. A 1906 letter published in the *Proceedings of the Institute of Civil Engineers* sums up the attitude: "It is unfair to the motor car driver, as well as to the pedestrian, to allow any important road to be without a footpath" (Ishaque and Noland 2006; Lindsay 1905–06). Design standards for footpaths in London were quite liberal, requiring that they have a width that is 1/6th of that of the total carriageway. The latter were often 24 ft (7.3 m) in width; thus, the footpaths were at least 4 ft (1.2 m) in width on both sides of the street (Ishaque and Noland 2006). Footpaths were also seen as a sanitary measure to improve drainage and to bring more order to city streets, especially in the slums (Ishaque and Noland 2006).

The first traffic signal started operation in 1869 in front of the Houses of Parliament in London. This was a gas-fired semaphore, similar to railroad signals. It provided a 30 s window for pedestrians to cross every five minutes. This made compliance by pedestrians unlikely as the wait time was long and most carriage drivers also did not stop, given their unfamiliarity with the new technology. The signal itself suffered from several gas explosions. In any event, the experiment was discontinued in 1870 as many Members of Parliament protested (Ishaque and Noland 2006).

After the failure of this experiment, debate turned to other means for pedestrians to safely cross streets while allowing traffic to flow unimpeded. Police officers often guided pedestrians across streets, but this was seen as costly. Subway tunnels and footbridges were considered. The latter were considered to be unsightly and land acquisition costs would be high. The first pedestrian subway was opened in 1870 (near the failed semaphore). Many of these were built in London in subsequent years, but debates over the best means for pedestrians to safely cross streets continued. There was a recognition that many people would still cross at street level even if a subway tunnel or footbridge was present. Well into the twentieth century, there was little consideration of simply stopping traffic at key pedestrian crossings, suggesting that traffic flow was more important than providing safe pedestrian crossings (Ishaque and Noland 2006). Adding to the debate was the concern expressed in this statement before the House of Lords in 1938: "We do feel that if subways and bridges were put into general operation it would only confirm the view of the motorist that the public highway was a motor speed track and would lead to further accidents." (Ishaque and Noland 2006; House of Lords Sessional Papers 1938).

The other engineering measure taken was the installation of guardrails along sidewalks and also within median refuges. These served the purpose of keeping pedestrians from entering the street and also channeled them to designated crossing points. By assuring that pedestrians could no longer enter the street, this facilitated the free flow of traffic. These were widely built in the 1930s as motorized traffic increased, some as much as three miles long on some East London streets. These were very effective at keeping pedestrians out of the street. By the 1930s, the conflict between which mode dominates the street and what the purpose of the street was had clearly been decided in favor of the motorcar at the expense of the pedestrian.

A similar story was playing out in the USA, as documented by Peter Norton in his seminal book, *Fighting Traffic, The Dawn of the Motor Age in the American City*

Fig. 9.1 Dedication of children's memorial, Baltimore, MD, 1922 (*source* Norton 2008)

(Norton 2008). In the 1920s, the main victim of increased motorization was children. The vast majority of traffic fatalities in cities were pedestrians (e.g., in Philadelphia, pedestrians accounted for 75% of total traffic fatalities) and about half of these were children (Norton 2008). Children and their parents were used to the streets being places where children could play and wander freely. But with the danger introduced by motorized vehicles, there was outrage at the carnage that was occurring. In some cities, monuments to children slain by vehicles were erected as shown in Fig. 9.1 (Norton 2008).[1] As parents protested the conditions, the motor vehicle industry fought back and defended the "rights" of the population to be mobile. Some cities considered implementing speed restrictions. In 1923, Cincinnati, Ohio, debated an initiative that would have required speed governors in all vehicles. In response, the motor industry developed a strategy of shifting blame to pedestrians and children, implying that parents were irresponsible to let their children play in the streets. They created the term "jay-walker" to mock pedestrians who did not cross at designated crossing points; in fact, one year after the Cincinnati initiative failed, Los Angeles

[1]These monuments were forgotten over the decades, which in itself indicates that the motor car won this battle (Norton 2008).

passed the first law against jay-walking. As in Great Britain, by the 1930s, the battle over the use of street space had been decided with the motorcar firmly in control (Norton 2008).

9.3 Engineering Guidelines

In the 1920s, as motor vehicle traffic rapidly grew, cities were confronted with the problem of vehicles congesting central business districts. In response, and frequently at the behest of the business community, municipal engineers (the forerunner to today's traffic engineers) sought methods to improve the efficiency of traffic flow. As discussed above, one means was controlling pedestrian movements. Many engineers also saw the provision of more electrical transit systems as one way to improve efficiency. But other engineering approaches were also tried, including coordinated traffic signals (in Chicago), eliminating on-street parking, and ultimately reconfiguring the city itself to provide more road capacity for vehicles (Norton 2008). As early as 1925, the first formal engineering guidebook, *Street Traffic Control*, was produced (Norton 2008).

After the Second World War, growth in vehicle ownership accelerated. The Highway Research Board (forerunner to the Transportation Research Board) produced the first version of the *Highway Capacity Manual* in 1950. Updates have been produced every few years (the current being 2010). In 1965, the concept of "level of service" was introduced. This was a means of classifying the travel delay associated with highways and intersections. Underlying the concept is detailed calculations of traffic flow, queueing, and signal timings, but the output produces a simple A to F ranking of the level of service (A being the best, and F being the worst). In practice, C provides a reasonably stable flow of vehicles, and most traffic engineers become concerned only at levels of D and lower. For intersection level of service, the rankings are linked to estimated delays for both signalized and unsignalized intersections (see Table 9.1).

Level of service (LOS) requirements have been extremely influential. In most, if not all, major cities and towns in the USA, any new development requires an analysis of LOS. If a new development (or redevelopment) leads to a degradation of LOS,

Table 9.1 Level of service (LOS) at intersections (Transportation Research Board 2010)

LOS	Signalized intersection (s)	Unsignalized intersection (s)
A	≤ 10	≤ 10
B	10–20	10–15
C	20–35	15–25
D	35–55	25–35
E	55–80	35–50
F	≥ 80	≥ 50

usually below C, then the developer must either scale back the size of the development or fund mitigation measures such that there is no reduction in level of service. These measures may include increases in the width of the road, changes to the turning lanes at intersections, or installation of traffic signals. The incentive for the developer is frequently to simply build in an area with minimal existing traffic on the edge of the urbanized area; in other words, this encourages sprawling development patterns. But mitigation measures in developed areas make the pedestrian environment less friendly, such as increasing the time needed to cross streets, or by encouraging faster traffic speeds from widening the road, or installing turning lanes.

Thus, engineering guidelines as implemented via LOS requirements became the foundation for how cities grew after the mid-1960s. In 2015, some fifty years after their promulgation in the Highway Capacity Manual, there is debate over the value of using this metric as the only performance criteria for both changes in street design and land use. In California, environmental legislation required a LOS analysis for all new developments, leading in most cases to worse environmental outcomes as new development located in areas where there would be no impact on existing traffic. In 2008, legislation was passed in California that allows development in "transit-rich" or infill areas to not be subject to this requirement.[2] Further legislation in 2013 formally removed the LOS requirement and many cities have moved to metrics based on vehicle-kilometers of travel (Lee and Handy 2018).[3] While the impact of this change is still being assessed, it is seen as a way to reduce road design changes that benefit the motorist, while making the streets more walkable.

Another important guidance document is "*A Policy on Geometric Design of Highways and Streets,*" aka "the Green Book," produced by the American Association of State Highway and Transportation Officials (AASHTO 2011). This guidance document sets standards for how highways should be built, such as the width of traffic lanes, the curvature, and a large variety of other detailed design components. This has resulted in the design and construction of highways and streets that are "wider, straighter, and faster," primarily because these are seen as being both efficient and safe highways. As Dumbaugh and Gattis (2005) show, the concepts of rural road arterial design have been applied to urban streets, while forgetting that on urban streets, pedestrians are present, and the function of the road is very different than in a rural area. The Green Book has little consideration of the role of pedestrian movements, although the most recent guidance does recognize that pedestrians provide vitality to central business districts and thus should be catered to (AASHTO 2011). Much of the limited discussion on safety issues in the Green Book is dedicated to creating "clear zones," that is, a buffer whereby vehicles running off the road do not hit obstacles (e.g., trees). While the guidance recognizes this may be difficult to do in urban areas,

[2]The Sustainable Communities and Climate Protection Act of 2008 (Sustainable Communities Act, SB 375, Chapter 728, Statutes of 2008). Essentially, this act requires regions to develop plans to reduce greenhouse gas emissions via land use, housing, and transportation strategies.

[3]California Senate Bill 743 removed the requirement for LOS analysis. See: https://opr.ca.gov/ceqa/updates/sb-743/.

it still recommends doing as much as possible to remove roadside objects, precisely those that may protect pedestrians walking on sidewalks (AASHTO 2011).

The Green Book now defers much of the safety guidance to the newly developed *Highway Safety Manual*, published by AASHTO in 2010 (AASHTO 2010). The purpose of the manual is to provide practicing engineers with crash reduction factors (CRF) for different road types, so that these can be used in cost/benefit analysis of safety improvements. CRFs are basically coefficient estimates derived from statistical crash frequency models.

While it is commendable that there is now a concern for traffic safety, there are several problems with the approach being taken. The CRFs are assumed to be point estimates. That is, the uncertainty inherent in any statistical estimate is ignored. Another issue is that contextual issues are ignored; the CRF estimates may not include other factors associated with crash occurrence, possibly leading to biased estimates (Noland 2013; Mitra and Washington 2012). Research work we recently completed suggests that omitting variables that provide contextual information on an area increases the parameter estimates associated with various road design features. In other words, we are probably overestimating the benefits of design treatments relative to other risk reduction policies (Noland and Adediji 2018). The risk to cities is that once the HSM is put into widespread use, traffic engineers and decision makers will follow the guidance without considering its limitations and other contextual features of the urban environment.

The research underpinning the CRFs is also questionable. For example, Noland (2013) attempts to find the source of the CRF for rural two-lane roads, an area that has been researched to a large extent due to the prevalence of crashes on these roads. Tracing back the citations for the CRFs, the HSM refers to two studies: (Zegeer et al. 1981; Griffin and Mak 1987), but it is not transparent how these studies were used to derive the CRFs, and both suffer from major methodological flaws (Noland 2013). The CRFs for rural two-lane roads in the HSM are shown in Table 9.2 and as can be seen imply that wider roads are safer, especially for larger traffic flows. It is not reported whether there are any statistically significant differences between these estimates and it is also assumed that one can linearly interpolate between them, which (Persaud and Lyon 2007) suggests is not appropriate. Other work has questioned whether there is any safety benefit from wider lanes, even suggesting that crashes may increase (Noland 2003; Hauer 2005; Milton and Mannering 1998). Therefore,

Table 9.2 Crash modification factors from the highway safety manual (AASHTO 2010)

Lane width CMF	<400 vehicles/day	>2000 vehicles/day
9 ft or less	1.05	1.50
10 ft	1.02	1.30
11 ft	1.01	1.05
12 ft or more	1.00	1.00

developing strict criteria of this type may be counterproductive and actually undermine the goal of improving safety. Our research suggests that there are wide error bars for any mean estimate (Noland and Adediji 2018).

The development of the *Highway Safety Manual* poses a risk to planners and those concerned with reallocating space to pedestrians. By promulgating engineering guidelines that ignore contextual features, and which may not even be estimated correctly, decisions may be made that at best have no beneficial safety impact, and at worst, actually increase traffic risk, especially to pedestrians, and reduce the mobility of pedestrians.

9.4 Behavioral Adaptation

Efforts to improve the safety of vehicles and streets would seem like an uncontroversial topic. Everyone benefits when safety is improved, especially when the costs of those improvements are small. However, it is not clear that all safety improvements result in reductions in the number of injuries and fatalities. This is because people change their behavior in response to improved safety. This concept is known as "risk compensation" by economists (Peltzman 1975) and "risk homeostasis" by psychologists (Wilde 1982) and more generally defined as "behavioral adaptation" (Noland 2013). There is a long history of debate over the theory, with many safety researchers questioning the basic theory (Robertson 1981; Graham and Garber 1984). At the same time, there has been much research confirming the basic process and further theoretical extensions have incorporated "task homeostasis," that is, the difficulty of the task involved with maintaining safety (Fuller 2005).

Noland (2013) provides a full review of these issues and reframes the concept as a trade-off between safety and mobility. In this sense, the relative safety of travel or of any given mode is simply another attribute that people consider in their choice process. A key trade-off is the relative time devoted to a trip versus the safety of that trip; all else equal, faster speeds are riskier but reduce travel time. A pedestrian seeking to cross a busy road when traffic is approaching is merely seeking to minimize the travel time associated with his/her trip.

This framework is displayed in Fig. 9.2, reproduced from (Noland 2013). If a policy or technology is able to improve safety, the increased level of safety may lead to more mobility or more safety. That is, some of the safety improvement is converted to more driving which may offset some of the benefits of the improvement. Individual behavior may be seen as more risky, such as faster speeds, but if the vehicle is more crashworthy, the driver is simply making a rational decision to reduce his/her travel time. Other behavioral adaptations can also occur, such as driving less attentively, or engaging in other activities while driving (e.g., speaking or texting on a mobile phone). Referring back to Fig. 9.2, if one is initially at point A with a given level of mobility and safety, a change in policy or technology can improve safety for a given level of mobility (represented by the shifted curve). Point B would represent a case where all of the improvement is converted to increased safety, while

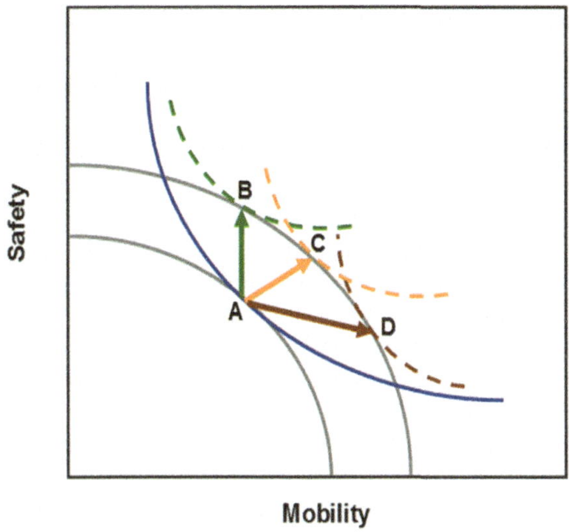

Fig. 9.2 Trade-offs between safety and mobility (*source* Noland 2013)

point C represents the case where some is taken as a mobility increase. Point D might represent the case where safety declines, perhaps because driver's have a high preference for increased mobility (or their perception of the safety improvement exceeds its actual level).

With these trade-offs in mind, it is easy to understand the decision making process of pedestrians. While the Green Book states that "pedestrian actions are less predictable than those of motorists" (AASHTO 2011), in reality pedestrians are making rational choices. Most do not take unnecessary risks but balance the costs and benefits of their actions. Seemingly risky choices are merely a decision to reduce travel time. When these risks are misperceived, whether due to fatigue or a cognitive deficit, then crashes are more likely to occur.

9.5 Data Issues

A major issue with safety data is its quality (O'Day 1993; Shinar et al. 1983). While it is well recognized that fatality data is generally of higher quality, less severe crashes typically are underreported or not fully investigated. Given the rhetoric that transportation agencies cite about the need to improve safety, it is somewhat astounding that there remain major data deficiencies in the USA. Recent work we have done indicates that there are likely major issues with pedestrian fatality data, ranging from how these are recorded and processed to how pedestrians are classified. There is also a lack of transparency with many State Departments of Transportation not making their crash data available to the public.

A recent analysis of New Jersey pedestrian fatality records using original police records suggests that out of 157 total reported fatalities, 13 are incorrectly reported based on the definition specified by the National Highway Traffic Safety Administration (NHTSA) (Noland et al. 2017). These include intentional homicides, incidents not on a public traffic way such as on private land, and one workplace accident. At least, nine of these incidents do not involve a *pedestrian*, if we assume that pedestrians are traveling on foot from one location to another. Another 22 incidents, all of which are within NHTSA's definition, do not involve pedestrians, but merely people not in a vehicle who are struck and killed. Most of these are cases where a driver was standing next to a vehicle or was walking away from a disabled vehicle along a high-speed road. As their primary mode of travel was a motor vehicle, they are more representative of the risk of driving and not the risk of being a pedestrian. Thus, over 20% of the total reported pedestrian fatalities in New Jersey for 2012 are not really pedestrians engaged in purposeful travel from one place to another (Noland et al. 2017).

Another issue is the inability to obtain raw data records from many states. While the federal government maintains the Fatality Analysis Reporting System (FARS), which is a publicly accessible dataset, injury data is not centrally collected and is generally unavailable. We explored obtaining data to assess the openness of state agencies with sharing this data for both analytical purposes and their willingness for us to make it easily accessible via online mapping tools. Some states are very open with their data repositories, and others have flatly refused to make their data available citing liability concerns; that is, they fear that should a crash occur then the data will be used to sue them. This shows a lack of agency accountability; one objective of open data is to make state agencies accountable to their constituencies.

With these data limitations in mind, some analysis of both nationwide data and New Jersey state data is presented in the next section.

9.6 What Policies Reduce Fatal and Injury Crashes?

Developed countries have seen enormous drops in fatalities associated with traffic crashes over the last 40 years (for the USA, see Fig. 9.3). This has occurred despite increases in vehicle usage and total populations. Recent research I have conducted has sought to identify some of the major policies associated with these trends. Of key interest is the role that the road network plays, especially given the large investment in trying to make roads safer.

Work that I conducted about 18 years ago (Noland 2003) sought to examine changes in the road network while controlling for other policies enacted by states. Using a cross-sectional time-series methodology of state-level data, it was found that most of the reduction (from 1984 to 1997) was associated with increased safety-belt use, reduced alcohol consumption, and better medical technology. Changes in demographics also played a role, as the fraction of the population below the age of 25 decreased, which is typically a group at higher risk of crashes. Various road network

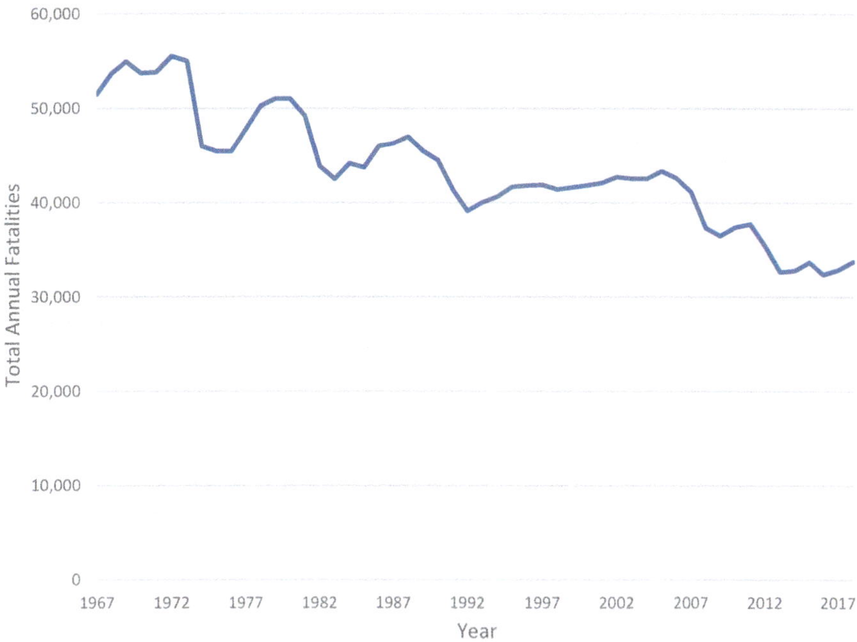

Fig. 9.3 Total traffic fatalities in the USA, 1967–2018 (*source* NHTSA)

features were found to have positive associations with fatalities and injuries, such as increases in arterial and collector roads, and overall lane mileage. While effects were weak, there was evidence that larger lane widths (of arterials and collectors) are associated with more fatalities and injuries.

An updated analysis (Noland and Zhou 2017) evaluates a broader set of policies, including those initiated in the prior 15 years. These include the implementation of graduated licensing policies, more motorcycle helmet laws, new laws that regulate mobile phone usage, reductions in alcohol consumption, and improved medical technology (as measured by a proxy variable). The economic climate has also had an impact, with the 2008 financial crisis and recession associated with drops in total fatalities, something seen in many other studies. An increase in the fraction of collector roads was associated with fewer fatalities, i.e., more arterials likely increase fatalities.

Those policies that tend to increase the cost of mobility, i.e., regulations on driver behavior such as motorcycle helmet requirements, mobile phone laws, and graduated licensing laws, all are effective. Those that reduce the cost of mobility, such as adding more arterial roads, tend to increase fatalities. Improvements in medical technology tend to reduce the likelihood of a fatality so while this reduces the cost of mobility, there is still a beneficial impact. This is consistent with what we would expect from behavioral adaptations.

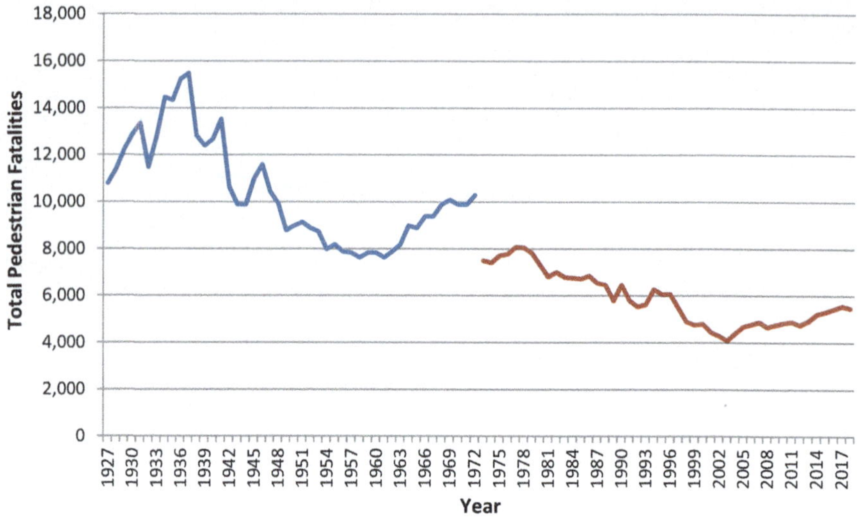

Fig. 9.4 Pedestrian fatalities in the USA, 1927–2018 (*sources* National Safety Council, 1927–1972; NHTSA, 1973–2018)

The record on pedestrian fatalities has been mixed, with a drop since about 1973, as seen in Fig. 9.4. Data prior to this time period was not collected by NHTSA, and there is a large discontinuity between 1972 and 1973, but there was likely a large reduction after the reported peak in the early 1930s. Since about 2003, however, there has been an increase in total pedestrian fatalities from a low of 4109 to 5489 in 2018.

New Jersey has had a fairly constant number of pedestrian fatalities since 1994 as shown in Fig. 9.5, and it is not possible to ascertain a trend. New Jersey has a relatively low incidence of traffic fatalities, but the proportion of those that are pedestrians is one of the highest in the nation (about 20%). This is largely because the state is densely populated, so exposure levels are higher. In Fig. 9.6, one can see the relative distribution of pedestrian casualties throughout the state; these roughly track the major population centers in the north and south of the state.

Analysis we conducted has examined both the probability of pedestrian casualties occurring throughout the state and a more detailed analysis of associations between road and pedestrian infrastructure and the level of severity of pedestrian crashes. The crash frequency analysis used spatial data from 2003 to 2008 to examine factors associated with pedestrian casualties (Noland et al. 2013). In general, results show that lower-income areas tend to have more casualties and areas with lower household vehicle ownership likewise have more casualties. These would be areas where people are more dependent on walking to engage in economic activities so exposure levels are likely higher. Alternatively, casualty rates are higher in areas with lower population density where streets usually have higher speed limits and fewer safe areas

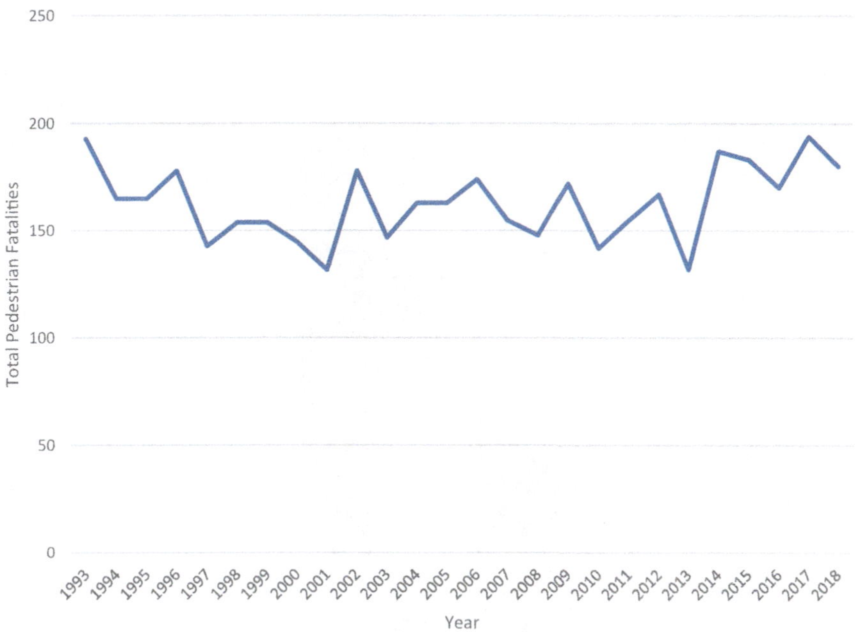

Fig. 9.5 New Jersey pedestrian deaths, 1993 to 2018 (*source* NHTSA, Traffic Safety Facts and FARS data)

for pedestrians to walk. A greater density of major highways, excluding controlled-access highways, is also associated with more pedestrian casualties (Noland et al. 2013).

To examine the relative severity of pedestrian casualties, we analyzed the road and pedestrian infrastructure for over 2500 crashes using imagery from Google Street View (Hanson et al. 2013). The features examined included the number of lanes, speed limits, presence of sidewalks and whether they had buffers along the street, crosswalk types and presence of medians. The results suggest that casualties are less severe when there are reduced speed limits, fewer lanes of traffic, and sidewalks with buffers (Hanson et al. 2013).

In general, these results suggest that a major source of pedestrian fatalities is large high-speed roads without adequate pedestrian infrastructure. These may also be more likely to cut through lower-income neighborhoods, partially explaining the frequency of crashes in those areas. In short, roads such as those in Fig. 9.7 are not safe for pedestrians, while the one pictured in Fig. 9.8 is relatively safe. All the images were the site of a pedestrian crash. The former were fatalities while the latter was a minor injury.

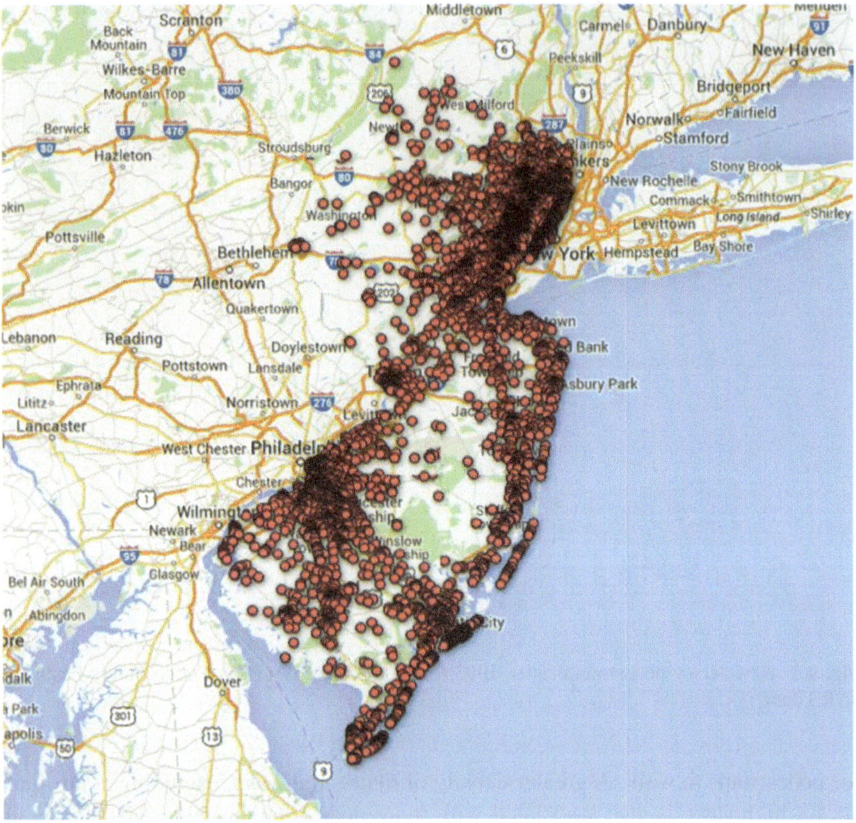

Fig. 9.6 Geo-coded pedestrian crashes in New Jersey, 2003–2013, representing about 35% of total pedestrian crashes (*source* Plan4Safety, Rutgers Center for Advanced Infrastructure and Transportation; Google Maps)

Fig. 9.7 High-speed arterials with no pedestrian infrastructure, (left image) US 130, Cinnaminson, Burlington County, NJ, and (right image) Bergen Boulevard (NJ 63), in Palisades Park Borough, Bergen County (*source* Google Street View)

Fig. 9.8 Low-speed two-lane street with buffered sidewalks, CR 655, Maplewood, Essex County, NJ (*source* Google Street View)

9.7 Solutions: Sharing the Street

Many cities in the USA and throughout the world are now recalibrating the balance of providing unimpeded traffic flow versus providing safety for pedestrians. These concepts range from traffic calming to slow traffic, mainly on residential streets, to shared space concepts, to reallocation of road space to reduce lanes for vehicles and provide more space for pedestrians.

Times Square in New York City, which is the heart of the theater district, was bisected by a major arterial road, Broadway. In 2009, Broadway was closed to traffic and the space reallocated to pedestrians (Fig. 9.9). The result has been a substantial increase in pedestrian activity and a much-enhanced environment for pedestrians, so much so that space is limited and some are complaining of too many pedestrians (Bagli 2015). If too many pedestrians are a problem, then one solution is to close additional streets that still traverse the square.

In London, a good example of a shared space concept was recently completed in South Kensington along Exhibition Road. This street runs from Hyde Park in the north down to the South Kensington Underground Station, passing by a major university and three large museums that attract substantial pedestrian traffic. During peak periods, the sidewalks were previously overcrowded and the vehicles on the street tended to speed through the area, making crossing dangerous. The concept of shared space is to send a message to motorists that they will share the road with pedestrians; this is done by eliminating curbs and putting a textured pavement on the street. In this case, the speed limit was lowered from 30 to 20 mph, which some might argue is still too high. The images in Figs. 9.10 and 9.11 show before and after

Fig. 9.9 Times Square, New York City, before and after pedestrianization (*source* NYC DOT)

Fig. 9.10 Exhibition Road, South Kensington, London, before shared space (*source* Royal Borough of Kensington and Chelsea)

Fig. 9.11 Shared space, Exhibition Road, South Kensington, London (*source* author)

images of Exhibition Road and clearly show how the capacity for pedestrians has been expanded.

Another concept being employed in the USA is a Complete Streets policy. This policy aims to reconfigure existing street design so that all road users are accommodated, pedestrians, cyclists, motorists, and transit. One implementation of this is the concept of a "road diet," which involves reducing the number of lanes for streets that were over-designed for vehicle capacity. This commonly involves taking a road with two lanes for traffic in each direction and converting it to one lane in each direction with a shared turning lane separating the traffic lanes. Normally, there is then additional road space to also add a bicycle lane on either side of the street, although not all implementations include this.

We conducted research to evaluate the costs and benefits of implementing a road diet along Livingston Avenue, a 1.5 mile arterial corridor in New Brunswick, New Jersey (Noland et al. 2015). This street runs from a major intercity arterial to the center of New Brunswick, traversing mainly a residential neighborhood but with some commercial activity and several schools along its length. Figure 9.12 shows the street as it is currently configured while Fig. 9.13 shows an overlay image of what a road diet would look like.

The costs associated with the road diet mainly consist of any additional travel time delay associated with reducing the capacity and subsequent vehicle throughput, as well as the construction costs (mostly restriping the pavement) associated with the conversion. The benefits are the expected reductions in vehicle crashes. There are

Fig. 9.12 Livingston Avenue as it is currently configured (*source* Noland et al. 2015)

Fig. 9.13 Photo simulation of Livingston Avenue with a road diet (*source* Noland et al. 2015)

Table 9.3 Relative disutility factors by injury severity level (AIS)

AIS level	Severity	Fraction of VSL
AIS 1	Minor	0.003
AIS 2	Moderate	0.047
AIS 3	Serious	0.105
AIS 4	Severe	0.266
AIS 5	Critical	0.593
AIS 6	Unsurvivable	1.000

Source Trottenberg and Rivkin (2013)
Note AIS is Abbreviated Injury Scale

also potential benefits of improving pedestrian mobility, although we did not attempt to assess these.

To estimate the travel delay, we ran several simulations using VISSIM microsimulation software (Noland et al. 2015). The value of travel time applied to the delay was based on formal guidance provided by the US DOT (Trottenberg and Belenky 2011), equivalent to 50% of the median income of $12.75 per hour; for business travel, it is 100% of the median income or $25.50 per hour. Values were indexed to 2014. To estimate the cost of crashes, estimates of the value of statistical life were used, again based on US DOT guidance (Trottenberg and Rivkin 2013). The 2014 value ranges from $5,311,875 to $13,177,537, with a mean of $9,295,782. As most crashes do not involve a fatality, factors are applied based on the level of severity of a crash as shown in Table 9.3. The number of crashes per year is about 38, on average with 17% involving pedestrians and 43% involving an injury. As we do not know precisely what the reduction in crashes will be, we assumed a 19% reduction based on a review of the crash reduction potential of similar road diet conversions (Thomas 2013).

Given the uncertainties about some of our assumptions, we conducted a detailed scenario analysis over a 20-year time frame. In all cases, the net present value of the road diet conversion was overwhelmingly positive. Benefits ranged from about $8 million to over $40 million.

Most of the political debate surrounding road diets involves a fear of creating congestion or reducing the level of service of the street. This was the case in New Brunswick as there was a desire to complete the study well before an upcoming election. However, shortly after the study was completed, three children were hit by a vehicle and one suffered serious injuries. After political protests over the safety of the street, the city began more detailed engineering work and committed to doing the conversion. Some initial restriping was done in front of schools along with a few blocks of the street, shortly after the crash. A photo is shown in Fig. 9.14. Work began on the project in early 2020, some six years after the decision to do so.[4]

[4]This long delay was due to regulatory requirements that required a traffic impact analysis, engineering design work, and the installation of a new signaling system. The road restriping could have

Fig. 9.14 Restriping of Livingston Avenue in May 2014 (*source* author)

9.8 Conclusions

In this chapter, I have reviewed the historical conflicts over road space and their resolution in favor of the motorcar in both Great Britain and the USA. While these conflicts were suppressed for many years, they have recently reemerged, recognizing that engineering guidelines on how to build roads ignored the pedestrian and damaged the fabric of many cities. The viewpoint that insists on reducing traffic congestion and facilitating traffic flow at any cost has been challenged.

From a research perspective, it is important to continue to evaluate the relative costs and benefits of traffic flow versus pedestrian safety. Benefits of pedestrianization can be more difficult to measure than travel time savings for vehicles. While safety improvements are one measure, other quality of life improvements are difficult to quantify. Safety data systems also need to be up to the task, they clearly are not in the USA, and developing countries often have even less reliable data systems. Making this data publicly available, both for researchers and to enforce government accountability, is crucial.

What are the lessons for developing countries, especially those that are rapidly motorizing? While motorization can improve mobility and economic linkages, it must be managed in such a way that the city is not destroyed as a place where pedestrian activity can thrive. Large high-speed arterial roads are chasms that pedestrians

---

been quickly and cheaply done without these requirements and thus avoid additional crashes over the time period.

cannot cross and damage the linkages between neighborhoods. Traffic engineers and transport planners need to be thoughtful about following engineering guidance documents on how to design roads and improve safety. The locational context of where road infrastructure is placed must not be ignored. The main lesson is to find the right balance, and many cities have gone too far in catering only to motorized traffic at the expense of pedestrians.

There are growing examples throughout the world of successful reallocation of space from vehicles to pedestrians. These increase the vibrancy of neighborhoods and reduce many of the negative costs associated with an overreliance on motorized transport. A good example in India is Raaghiri Day.[5] This is a weekly event where selected streets are shut down to traffic and opened to all for entertainment and recreation. This is part of the Open Streets movement that seeks to reclaim streets for all users. Originating in Bogota, Columbia, with Cyclovia, these have proliferated throughout the world.[6] These serve to educate the public about how streets can be used for other purposes and hopefully will lead to broader institutional changes.

Acknowledgements Thanks go to Geetam Tiwari for inviting me to present this research at the Annual TRIPP lecture without which I would not have written this. The lecture series was funded by the Volvo Research and Educational Foundation. I would also like to thank the many students and researchers who collaborated with me on this work over the years, including: Dr. Nicholas Klein, Dr. Christopher Hanson, Prof. Eric Gonzales, Nicholas Tulach, Charles Brown, Dong Gao, Feiyang Sun, Yuhan Zhou, Bahareh Sehatzedeh, Dr. Moazzam Ishaque, James Sinclair, Elizabeth Harvey, James van Schoick, and Patrick Brennan. Funding sources for the projects cited have included the Rees Jeffreys Road Fund, Transport for London, the Region 2 University Transportation Research Center, the Rutgers Community-University Research Partnership Grant for New Brunswick and the New Jersey Department of Transportation. All omissions and errors are my own.

References

AASHTO (2010) Highway safety manual. American Association of State Highway and Transportation Officials, Washington, DC

AASHTO (2011) A policy on geometric design of highways and streets, 6th edn. American Association of Highway and Transportation Officials, Washington, DC

Bagli CV (2015) Times square's crushing success raises questions about its future. New York Times, Jan 26

Department for Transport (2013) Reported road casualties Great Britain: 2013 Annual Report

Dumbaugh E, Gattis J (2005) Safe streets, livable streets. J Am Plann Assoc 71(3):283–300

Farr W (1870) Persons killed in the streets of London, thirty-first Annual Report of the Registrar General of Births, Deaths and Marriages in England (England and Wales). London

Fuller R (2005) Towards a general theory of driver behaviour. Accid Anal Prev 37(3):461–472

Goddard T, Ralph K, Thigpen CG, Iacobucci E (2019) Does news coverage of traffic crashes affect perceived blame and preferred solutions? Evidence from an experiment. Transp Res Interdiscip Perspect 3:100073

[5] https://raahgiriday.com/.

[6] https://en.wikipedia.org/wiki/Ciclovia.

Graham JD, Garber S (1984) Evaluating the effects of automobile safety regulation. J Policy Anal Manage 3(2):206–224
Griffin LI, Mak KK (1987) Benefits to be achieved from widening rural, two-lane, farm-to-market roads in Texas, paper #87-0233, 1987, National Research Council, Transportation Research Board
Hair PEH (1971) Deaths from violence in Britain: a tentative secular survey. Popul Stud 25(1):5–24
Hanson CS, Noland RB, Brown C (2013) The severity of pedestrian crashes: an analysis using Google Street View imagery. J Transp Geogr 33:42–53
Hauer E (2005) Fishing for safety information in murky waters. J Transp Eng 131(5):340–344
House of Lords Sessional Papers (1938) Minutes of Evidence taken before the Select Committee on Prevention of Accidents, London
Ishaque MM, Noland RB (2006) Making roads safe for pedestrians or keeping them out of the way? An historical perspective on pedestrian policies in Britain. J Transp Hist 27(1):115–137
Lee AE, Handy SL (2018) Leaving level-of-service behind: The implications of a shift to VMT impact metrics. Res Transp Bus Manag 29:14–25
Lindsay C (1905–06) Correspondence on country roads. Minutes of Proceedings of the Institution of Civil Engineers, vol 165, p 67
Milton J, Mannering F (1998) The relationship among highway geometrics, traffic-related elements and motor-vehicle accident frequencies. Transportation 25(4):395–413
Mitra S, Washington S (2012) On the significance of omitted variables in intersection crash modeling. Accid Anal Prev 49:439–448
Noland RB (2003) Traffic fatalities and injuries: the effect of changes in infrastructure and other trends. Accid Anal Prev 35(4):599–611
Noland RB (2013) From theory to practice in road safety policy: understanding risk versus mobility. Res Transp Econ 43:71–84
Noland RB, Adediji Y (2018) Are estimates of crash modification factors mis-specified? Accid Anal Prev 118:29–37
Noland RB, Gao D, Gonzales EJ, Brown C (2015) Costs and benefits of a road diet conversion. Case Stud Transp Policy 3(4):449–458
Noland RB, Klein NJ, Tulach NK (2013) Do lower income areas have more pedestrian casualties? Accid Anal Prev 59:337–345
Noland RB, Sinclair JA, Klein NJ, Brown C (2017) How good is pedestrian fatality data? J Transp Health 7:3–9
Noland RB, Zhou Y (2017) Has the great recession and its aftermath reduced traffic fatalities? Accid Anal Prev 98:130–138
Norton PD (2008) Fighting traffic, the dawn of the motor age in the American City. The MIT Press, Cambridge, MA
O'Day J (1993) Accident data quality, NCHRP Synthesis 192. Transportation Research Board
Peltzman S (1975) The effects of automobile safety regulation. J Polit Econ, 677–725
Persaud B, Lyon C (2007) Empirical Bayes before–after safety studies: lessons learned from two decades of experience and future directions. Accid Anal Prev 39(3):546–555
Ralph K, Iacobucci E, Thigpen CG, Goddard, T. (2019). Editorial patterns in bicyclist and pedestrian crash reporting. Transp Res Rec 2673(2):663–671
Robertson LS (1981) Automobile safety regulations and death reductions in the United States. Am J Public Health 71(8):818–822
Shinar D, Treat JR, Mcdonald ST (1983) The validity of police reported accident data. Accid Anal Prev 15(3):175–191
The Times (1869) Editorial, 11 June
Thomas L (2013) Road diet conversions: a synthesis of safety research, for Federal Highway Administration, DTFH61-11-H-00024. Pedestrian and Bicycle Information Center, University of North Carolina, Highway Safety Research Center, Chapel Hill, NC
Transportation Research Board (2010) Highway capacity manual. Washington, DC
Trottenberg P, Belenky P (2011) Revised Departmental Guidance on Valuation of Travel Time in Economic Analysis

Trottenberg, P. and Rivkin, R.S., 2013. Guidance on Treatment of the Economic Value of a Statistical Life in U.S. Department of Transportation Analyses .

Wilde GJ (1982) The theory of risk homeostasis: implications for safety and health. Risk Anal 2(4):209–225

Zegeer CV, Deen RC, Mayes JG (1981) The effect of lane and shoulder widths on accident reductions on rural, two-lane roads. Transp Res Rec 806:33–43

Chapter 10
Current Vehicle Safety Technologies and Future Directions

Murray Mackay

10.1 Introduction

In 2010, the European Union sets a target of reducing road traffic deaths by 50% by the year 2020. This target follows a previous target set in 2001 to halve such deaths by 2010. That earlier target was in fact missed by two years (E.T.S.C. 2011). However, the incentives to meet specific targets for member states and to compete against others have driven governments towards better, science-based policies. Reductions in deaths have come in roughly equal measures from behavioural changes, environmental changes and vehicle design improvements (E.T.S.C. 2013). The main behavioural improvements have been reduced drinking and driving, reduced speeds and increased seat belt and helmet use. The environmental changes have been traffic calming measures such as speed humps, chicanes and pedestrianisation in urban areas, speed management and improvements in highway design. Vehicle safety improvements have come from improved crash performance, largely driven by the EuroNCAP rating process, seat belt reminders and electronic stability control systems.

At the level of individual countries, targets for reductions in serious injuries have been used successfully. At the European Union level, however, no target has been set for a reduction in serious injuries because of large differences of definition as to what constitutes a serious casualty across the 28 countries (Mackay 2005).

Vehicle safety in high-income countries over the last two decades has undergone a profound change. Until the late 1990s, most of the advances in vehicle safety came from improved crash performance, with front and side airbags and side curtains, pre-tensioning and load-limiting seat belts, improved seat and head restraint design, enhanced structural performance in collisions which reduces intrusions into the passenger compartment and diminishes the maximum decelerations

M. Mackay (✉)
University of Birmingham, Birmingham, UK
e-mail: mmackay@manx.net

G. Tiwari and D. Mohan (eds.), *Transport and Safety*, Springer Transactions in Civil and Environmental Engineering,
https://doi.org/10.1007/978-981-16-1115-5_10

applied to the occupants. In low- and middle-income countries however, many of these crashworthiness techniques have yet to be widely adopted.

More recently in high-income countries, because of electronic developments in sensing traffic conditions and driver performance, emphasis has moved towards crash avoidance. Some enthusiasts in this area of vehicle safety claim that all that can be done in better crash performance has been done. Research money and effort have shifted to focus on crash avoidance almost exclusively. This has led to the introduction of electronic stability control, autonomous emergency braking, adaptive cruise control, road/lane departure warnings, pedestrian detection/warning and blind spot detection. Radio detection and ranging (Radar), laser imaging detection and ranging (lidar) and vision technologies are being applied to the traffic environment. This is leading towards real-time vehicle-to-vehicle and vehicle-to-environment communication. This is leading ultimately towards the driverless car.

The effectiveness of these various technologies is discussed in more detail below. However, the issues of how these technologies can or should be applied in the context of traffic in low- and middle-income countries are, I believe, a subject to which more research needs to be directed.

10.2 Before the Crash Systems and Their Effectiveness

There is a timeline for before and after a crash which is as follows:

Normal driving, abnormal driving, developing situation, crash unavoidable, crash, post-crash.

There are technologies which can be applied to each of these conditions which can break or modify this chain of events.

1. *Normal Driving*—There are now a number of systems which aid the driver under normal traffic conditions.

   Blind spot detection—A radar system which lights a warning indicator when there is a vehicle in the blind zone on the appropriate side.

   The effectiveness of BSD in actual traffic, in terms of reductions in crashes, has not yet been adequately assessed. Experimental studies of different traffic situations indicate that radar systems are effective (Forkenbrock et al. 2014), but some comparative work suggests that external mirrors with wide-angle sections for the outer part of the mirror, as are fitted to many cars today, are almost as good as the more complex radar systems.

   Traffic sign recognition—A visual system which displays, for example, the current speed limit or other relevant conditions such as road works either in the instrument cluster or in head-up display on the inside of the windscreen.

   Doubtless, this system has an influence of the driver's choice of speed, and it improves the response to unusual changes in the road conditions ahead. However, its role in reducing crash involvement has yet to be assessed.

High/low beam assist—This automatically dips the headlights at the appropriate time when there is a vehicle approaching.
The actual safety benefits, if there are any, are unknown.
Adaptive cruise control—A radar system which automatically adjusts the vehicle's speed so that a safe distance is maintained behind the vehicle in front. ACC is undoubtedly a significant safety aid, especially for motorway driving, as it reduces the vigilance required for what is a relatively boring task. Its real-life acceptance is tempered by the fact that, in moderate and heavy traffic flows, the driver with ACC leaves such a gap between himself and the car in front so that other traffic quickly moves into that gap. As a result, the ACC user finds himself continuously moving rearward relative to other traffic. Effectiveness studies show that ACC systems reduce the stress on the driver and diminish the frequency of error and heavy braking (Lee et al. 2008).
There are some situations where ACC may not detect small, narrow vehicles or other vehicles emerging from an adjacent lane or from a junction. It may also react incorrectly to vehicles in an adjacent lane on a bend or in the lane which carries traffic travelling in the opposite direction (see Fig. 10.1).
The technology of ACC blends into an autonomous braking system, the effectiveness of which is discussed below.
Intelligent Speed Assistance—This system increases the force required on the accelerator if the speed limit is exceeded. Alternatively, it provides either audible or visual warnings that the current speed limit is being exceeded. More complicated systems modify the speed limit shown in a head-up display in the windscreen by considering weather, darkness and traffic conditions. Yet more sophisticated systems change the speed limit according to the presence of schools, pedestrian crossings, road works or other hazards. The system can either be switched off voluntarily or incorporated into the car's instrumentation so that it cannot be switched off (European Transport Safety Council 2005). A number

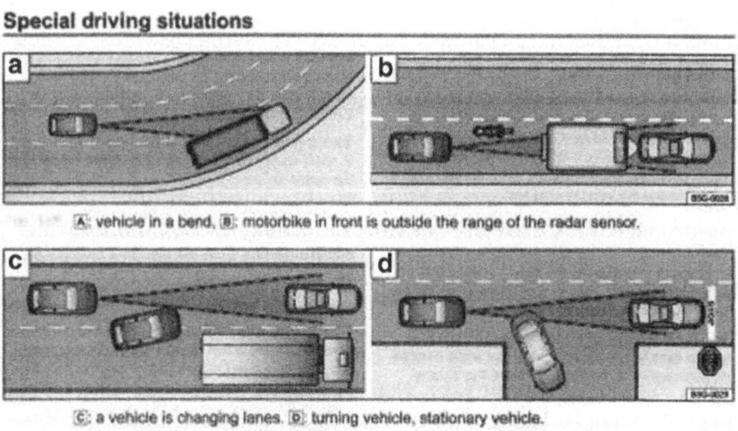

Fig. 10.1 Some limitations of adaptive cruise control

of studies (S.W.O.V. 2015) show that speeds and speed differentials are reduced when the system is widely used. In terms of casualty reductions, estimates have been made based on the relationships between known reductions in speeds and the frequency of crashes (Carsten and Tate 2005). These estimate, for example, that in crashes in which someone is injured, those crashes are reduced by 10–20% depending on whether the ISA system is voluntary or mandatory. For serious crashes and fatalities, the reductions are estimated to be at least double those of slight injury crashes.

Night Driving Assistance—An image generated by a heat-sensing device detects the outline of pedestrians, cyclists and animals at distances up to twice that of typical headlights. It gives a warning to the driver if the object comes within the range of being hit. This is done by projecting an enhanced visual picture in a head-up display on the windscreen. A subset of this is the dynamic spot light which selectively illuminates pedestrians and animals with a separate marking headlight. Man–machine interface studies show substantial benefits in terms of early and better responses under experimental conditions. However, the systems are not yet adopted widely enough for the effect on actual nighttime accidents to be assessed. If this night vision assistance system is combined with autonomous emergency braking, then the crash reduction benefits could be substantial.

2. *Abnormal Driving*—This covers the situations where a driver does something unusual or does not do something which should have been done during normal driving.

   Electronic Stability Control—This covers a range of systems. If a vehicle is rounding a curve at an excessive speed, traction control systems govern the braking effect on each individual wheel of a vehicle which discourages brake away and sliding. Further systems also control the throttle setting. An additional function is to assess the roll rate of a vehicle and adjust the suspension characteristics to reduce rollover risk. All these functions are integrated in a central processing unit which combines to greatly reduce the risk of loss of control by the driver and a possible rollover.

   Numerous studies have examined the effectiveness of ESC. Most of such studies have demonstrated very significant benefits. In terms of all police-reported crashes, ESC reduces the number of crashes by 20–40%. The benefits are greater for single-vehicle crashes and for SUVs, but less on wet/icy roads (Thomas and Frampton 2007).

   Fatigue Detection—Fatigue is a well-known cause of crashes, particularly for long-distance truck drivers. A number of techniques are on offer which senses that a driver is fatigued and in danger of falling asleep. These include the assessment of reduced minor movements of the steering wheel, drooping eyelids, reduced eye movements and changed posture. Eye mark cameras and steering wheel sensors can trigger acoustic and/or light signals to stimulate the sleepy driver. Monitoring the driver's electrocardiogram is a possible technique to anticipate when he is beginning to fall asleep. Quantifying the benefits of such systems has not been made, but preliminary studies of truck drivers suggest that they could be substantial. Naturalistic driving studies show how these benefits can be quantified in the future (Bonnard et al. 2014).

3. *Developing Situation*—This covers the period of some 10 s or less before a crash.

   Lane Departure Warning (LDW) and Lane Change Assistance (LCA)—These systems sense when a vehicle is about to or is crossing the edge line or lane markings using either cameras or infrared systems. A visual, audible or tactile warning is given to the driver. The effectiveness estimates vary widely (Visvikus et al. 2008) from 10% for cars and 30% for heavy trucks in reducing the crossing of lane and edge lines. The USA estimates for reductions in off-road crashes for heavy trucks range from 22 to 53% depending on the types of off-road crashes (Houser et al. 2009).Visual signals are the least effective and the audible ones most effective. For LDW, the actual effect on casualties has been estimated as a reduction of 5% for slight casualties, 9% for serious casualties and 12% for fatalities (Visvikus et al. 2008). More recent research in the US environment suggests that some 29% of all road departure crashes are caused by the driver drifting out of his or her lane (Kusano et al. 2014).

   Tactile Edge Lines or Rumble Strips—These alert the driver to the fact that he has deviated laterally from the road or from a lane. In a review of the available before and after studies, Corkle et al. (2001) found reductions of between 20 and 72% for run-off-road crashes.

   Warning Bars—Usually used on high-speed roads, these are transverse bars of paint, raised above the road surface slightly. They are positioned before some change in the road ahead such as a roundabout, a traffic signal or a junction. They start around some 200–300 m before the hazard and become progressively closer as the hazard is approached following a logarithmically decreasing arrangement (Denton 1971). They alert the driver and encourage early braking. Before and after studies at 42 roundabouts in the UK showed a 57% reduction in crashes and an 11–18% reduction when used at the approach to slip roads (Martindale and Urlich 2010).

   Autonomous Emergency Braking—AEB is in a sense a development of ACC in that it automatically applies emergency braking. For speeds up to 30 mph, if the driver is approaching a stationary car, it can bring a car to rest before contact without any braking by the driver. Beyond that approach speed, it can markedly reduce the severity of the impact. It depends on inputs from short-range radar and visual signals. In recent developments, it can be operated when there are pedestrians, cyclists or large animals in the danger zone.

   Effectiveness studies of Volvo CX40 with and without AEB show a reduction in front to rear crashes of 23% (Isaksson-Hellman and Lindman 2012). Other studies give reductions of 9–33% depending on the road environment and traffic conditions. AEB is also effective in reducing crash involvement and injury severity in other collision types such as serious and fatal injuries in front to side impacts by around 15% (Grover et al. 2008).

   Automated braking can also be enhanced by the application of the Torricelli vacuum braking system. This consists of a shallow medal plate with the edges curved downwards, fitted to the underside of the vehicle near the rear axle line. When the AEB is triggered, the plate is lowered to the road surface and a vacuum is applied to the inside of the curved plate, generating a downforce of around 15kN. This can reduce the braking distance by as much as 40%.

4. *Unavoidable Crash*—This covers the period of about one second or less before impact.
   Pre-Pre-tensioner—This is an electric motor within the seat belt system which is triggered when the car is subjected to high decelerations such as emergency braking or severe lateral cornering forces. It tightens the seat belt, removing any slack in the system caused by bulky clothing, poor sitting position or, to an extent, excessive fatty tissue.
   Improved Seat Position—Combined with pre-pre-tensioning, the seat base, seat back and head restraint are moved to optimal positions for a collision. Quantifying the benefits of these various systems has yet to be done.
   Autonomous Emergency Braking—Even with only a second available, emergency braking can significantly reduce the severity of a collision; for example, at 40 mph, when braking at 0.7 g for one second reduces the impact speed to some 25 mph.

10.3 In the Crash

In high-income countries, the crashworthiness of cars and trucks has reached an advanced stage of development. Coherent crashworthiness was first introduced in the late 1960s with the promulgation of the FMVSS200 series of standards in the USA. It has now matured some 60 years later. In essence, there are three areas which encompass the crashworthiness concept; structures, restraints and sensing.

Structures—Crashworthiness design aims at limiting intrusions into the passenger compartment and optimising the load paths of crash forces so that peak decelerations of the passenger compartment are minimised and those forces are absorbed without major structural failures. Meeting those requirements for the various collision types is a challenge. The main crash types are the distributed frontal, offset frontal, side impact for both distributed and narrow objects, together with rear impact and rollover conditions all present major design issues. Material strength and weight require a range of different metal alloys and other composites, with costs always holding back the choices.

Restraints—The heart of good occupant protection is the seat belt. By connecting the occupant to the car's structure, the occupant can "ride down" a collision, undergoing the change in velocity over as long a distance as possible when a collision occurs. In addition, the seat belt itself stretches and thus provides extra stopping distance. Seat belt performance is enhanced by pre-tensioning, which involves sensing that a collision is beginning and either quickly winding some webbing back into the retractor or pulling the buckle at the junction of the shoulder and lap sections downward. That takes place within around 15 ms of the collision occurring. Any slack in the seat belt is removed and the overall effect is to reduce the crash forces applied to the occupant.

Airbags should always be considered as being additional to the seat belt. Inflatable systems cover frontal and side airbags, side and roof curtains, knee bags, anti-submarining seat base bags and a variety of such inflatable systems for rear-seat occupants. Recently, central airbags, which inflate between the two front seats, are triggered in lateral impacts when a far-side occupant is present. This limits the sideways motion of that occupant who otherwise would come out from under the shoulder section of the seatbelt and strike the nearside occupant or his seat back.

Sensing—Seat belt technology and airbags both depend on detecting that a collision is occurring very early, that is, necessary so that inflatable systems are deployed before occupants move significantly around the passenger compartment under the influence of the collision forces. This is especially the case for lateral impacts because there is little structure between the occupant and the striking object so a side airbag must be inflated before that structure closes the gap needed for a side bag to inflate. That means that the gas generator must be ignited within 5 ms of the beginning of an impact.

Different problems present themselves in the sensing of a rollover in time for inflatable curtains and roof bags to be deployed before the occupants move upwards and outwards (relative to the vehicle interior). Such sensing systems measure the onset of angular accelerations and also the angle of the vehicle as a potential roll is initiated. These two variables are combined together in a sensor and if a critical value is exceeded the gas generators are fired.

Rear Impacts—Good crashworthiness in rear impacts depends predominantly on the structure and geometry of the seat. The aim is to limit the relative motion of the head relative to the torso of the occupant. This is achieved by the head restraint being relatively close to the occupant's head and by coordinating compliance of the seat back and the head restraint to minimise the head-to-torso relative motion. In addition, the deflection of the seat back is controlled to limit the forces applied to the occupant but also to keep the occupant in place so that ramping up the seat back does not occur. Active head restraints, which move the head restraint closer to the occupant's head, are triggered when the occupant loads the seat back above a threshold level.

Varying collision severities and a range of occupant weights and sitting heights mean that these population issues have also to be considered.

Glass—High penetration-resistant laminated glass has been well established as the best type of glass since the 1970s (Mackay 1973). More recently laminated glass is finding favour in side windows, especially in SUVs and people carriers where the risks of partial or complete ejection are greater than for standard cars.

Pedestrians—In terms of crash protection for pedestrians, as well as cyclists and motorcyclists, test procedures have been developed which greatly influence the front end shape of today's cars and the strength and compliance of particularly the bumper region, the front edge of the bonnet, the top of the bonnet and the cowl at the base

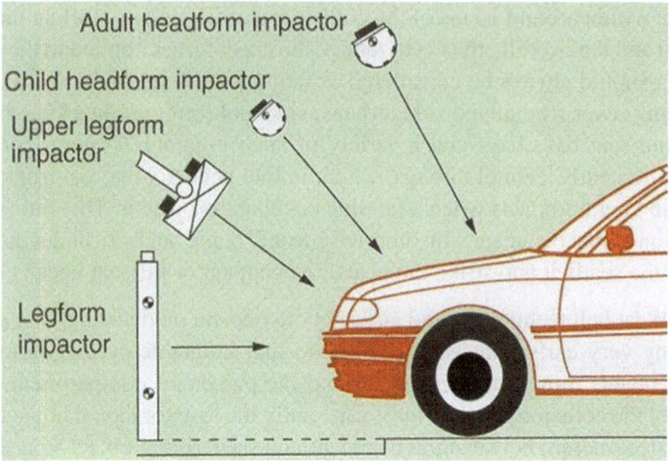

Fig. 10.2 Sub-system testing for pedestrian protection in EuroNCAP

of the windscreen. The EuroNCAP process tests these parameters with sub-system impactors (www.euroncap.com) (see Fig. 10.2).

More advanced technologies involve sensing that a pedestrian is being struck. That can be done either with a contact sensor in the bumper or by an active radar/visual sensor. The latter has the advantage that emergency braking could be operated before the impact, reducing the severity of the collision. Such signals can be used to trigger a pyrotechnic hood-lifter which increases the clearance of the hood from rigid components such as the top of the engine. This therefore presents a softer, more compliant structure when the pedestrian's head makes contact. In addition, airbag systems can be activated to cover the base of the windscreen frame and the A pillars, particularly hostile components in the pedestrian to car collision.

Compatibility Issues—All over the world one of the most important crash configurations which involve car occupant fatalities is the collision between a car and a heavy truck. Technical standards have been developed for front under-run, rear under-run and side impacts with trucks but meeting those standards in many high-income countries is not a necessary requirement.

One encouraging development within the European Union is the relaxation of the rules governing the overall length of single unit or tractor–trailer combinations of heavy trucks. These changes will allow an extension of the vehicle at the centreline of the truck whilst maintaining the existing rules of length measured from the front corner of the vehicle. This allows a more curved, aerodynamic shape for the front of the cabin with significant savings of up to 3% in fuel consumption. This will also allow the forward structures to be designed to provide better energy management for car-to-truck frontal collisions. It will also allow better protection for vulnerable road users in terms of the compliance of the exterior of the front structures and the some

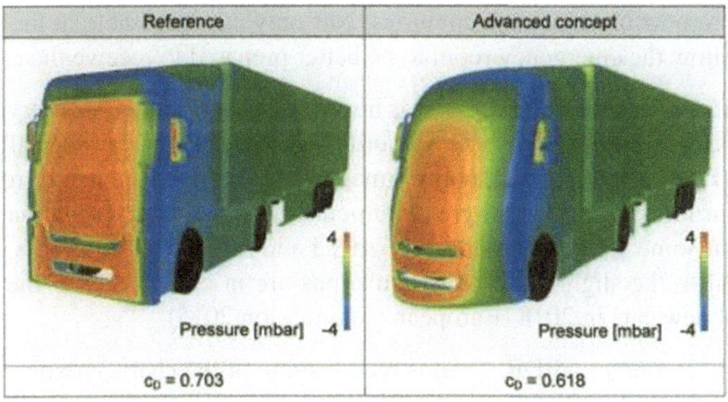

Fig. 10.3 Aerodynamic load on the front end of a standard slab fronted truck and an aerodynamically efficient truck

deflection of the VRUs so that they do not go so readily under the truck or its wheels, as happens with the current slab front of today's heavy trucks (see Fig. 10.3).

New Car Assessment Programmes (NCAP)—The initiation of a system which provides safety information and crash test rating of specific car models began in 1978 in the USA. Its logic is to inform consumers of the safety of the cars which come onto the market. It goes beyond the statutory requirements of government regulations and allows car models to be compared. A similar programme began within Europe almost 20 years later in 1997, driven in part because of the time required to establish Europe-wide regulations. Such regulations take so long to be agreed upon that when they are finalised, the technical requirements may well be a decade out of date. The success of the NCAP systems has been such that manufacturers compete for the best rating because that has a significant effect on future sales.

Since then, many national and regional NCAP systems have been put in place such as the ASEAN NCAP. A Global NCAP has been established with the aims of encouraging the flow of better vehicle safety information and better vehicle safety design. It has been shown to be a powerful way of encouraging car manufacturers to improve their products. Testing goes well beyond the regulatory requirements and more recently includes evaluating the fitting of active safety technologies such as electronic stability control and autonomous emergency braking.

10.4 After the Crash

Triage Systems—With the increasing power of telecommunication systems, structured triage at the scene has evolved beyond physiological and anatomical parameters such as pulse > 100/min, systolic blood pressure <90 mmHg, Glasgow Coma Score, > 65 years. Crash characteristics such as front, side and rollover, extent of intrusion

(>12 in.) are now used in some countries. Not only can these help at the scene but they also allow the emergency room to be better prepared to receive the casualty.

eCall—In Europe, an automatic signal is transmitted to the emergency services when a crash occurs. It provides the GPS coordinates and whether airbag inflation has occurred. Trials suggest the response times of arrival at the scene are reduced by some 5 min in urban areas and greater amounts in rural ones. Preliminary studies suggest that some 2–5% of fatalities are saved and some 10% of serious casualties are reduced to the slight category. Regulations are in place to make the fitting of eCall to all new cars in 2018 (European Commission 2015).

Event Data Recorders—EDR systems record some of the characteristics of a collision. These can include the direction of the crash forces on a vehicle, the peak acceleration experienced, the change in velocity, the presence of braking or not, the speed of the vehicle, the settings of the gear, the lights and the time to fire for the inflatable devices. The decoding of EDRs in some jurisdictions is common and is used in subsequent insurance claims or legal disputes. Privacy issues arise but some insurance companies will reduce the premiums if the insured chooses to have these devices used. For research, they provide a most useful amount of data which would otherwise be unobtainable after the fact. As with flight data recorders in commercial aircraft, EDRs are becoming a mainstream technology in traffic accident reconstruction.

Virtually all of the technologies described above, both pre-crash and crash related, apply to cars and trucks. Very little work has been conducted on motorcycles, motor scooters and electric-powered bicycles. With adaptive cruise control, for example, because of the narrow radar beam scanning, the road ahead will not detect narrow vehicles such as motorcycles which are ahead of the car but outside the lateral limits of the radar. Figure 10.1 illustrates such a situation.

Motorcycle use in a number of high-income countries, notably Spain, Italy, Germany, Sweden and Greece has increased by around 30–50% since 2005. This probably occurs because of increasing urban congestion and increased running costs for cars. Hence, motorcycle safety needs to be improved across the world, not just in low- and middle-income countries.

10.5 The Application of Advanced Safety Technologies in Low- and Middle-Income Countries

The application and the appropriateness of the above technologies to low- and middle-income countries raises many issues relating to the traffic mix, the physical environment and the social conditions.

In India, for example, Mohan (2015) has presented data on the growth of motor vehicle sales from 1972 to 2013, and the likely growth in cars and motorised two-wheelers (MTWs) to 2030 assuming growth rates of 10% per annum (see Fig. 10.4).

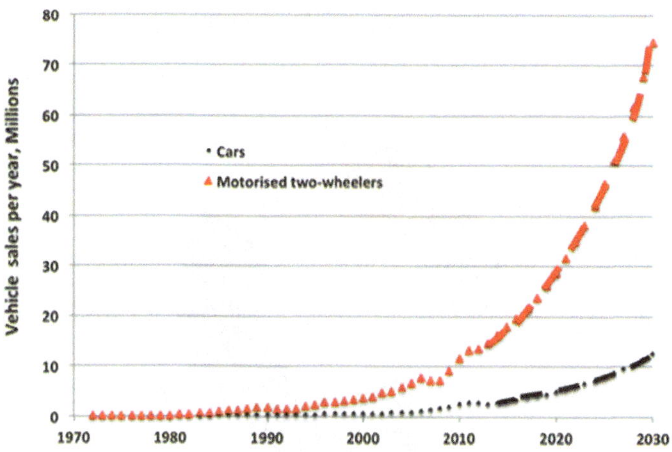

Fig. 10.4 Personal motor vehicle sales in India 1972–2013 (*Source* Society for Indian Automobile Manufacturers) and projections to 2030 at growth rate of 10% per year

If these predictions are even approximately correct, they indicate that the vehicle mix in India will continue to be dominated by motorcycles for the next 15 years. The interaction of cars with each other and with MTWs in this traffic mix is very different from that in high-income countries. For example, the considerable gaps that separate cars from the ones in front in cities in Europe would be, in Delhi, immediately filled by MTWs. Hence, sensing systems such as adaptive cruise control or autonomous braking will be unable to operate in cars in their present form in such an environment.

Other systems, such as road and lane departure warnings which depend on predictable roadway edge marking or lane marking, will not function if there are no such markings in place; a commonplace condition in low- and middle-income countries.

Beyond such obvious differences in the physical infrastructure, the non-verbal communication which goes on between different road users in various countries is very marked; compare pedestrians waiting at a signal-controlled crossing in say Stockholm or Hamburg when there is not a single motor vehicle in sight, with pedestrians crossing the road in Istanbul, Delhi or Bangkok. Such differences need to be considered when the technologies described briefly above were to be applied in low- and middle-income countries.

Countermeasures for Motorised Two-Wheelers—Fig. 10.5 shows the proportions of traffic deaths by road user category for India (WHO 2013).

The category of "other" is of interest. It is described as rickshaws, cycle rickshaws, hand carts and animal-drawn vehicles. Therefore, the overall category of vulnerable road users, which conventionally include MTWs, cyclists and pedestrians, should include most of the deaths in the "other" category, that results in the VRUs in India being some 63% of all traffic deaths. Of that group, the largest are

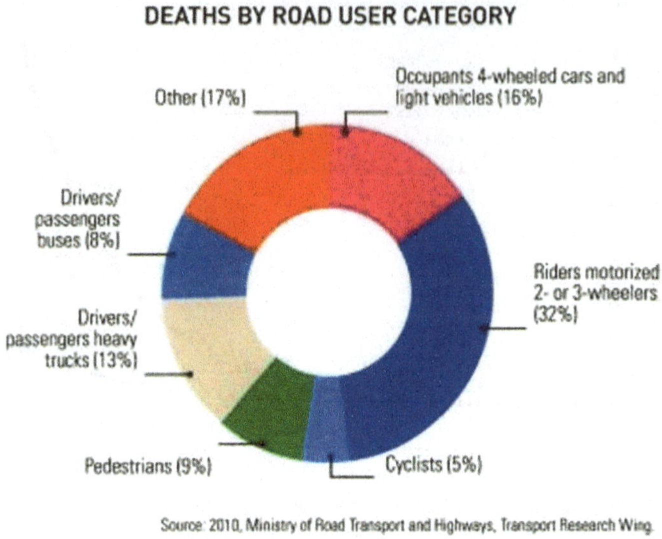

Fig. 10.5 Proportion of road traffic deaths by road user category

riders of MTWs. There are some well-researched countermeasures which could be applied immediately.

Daylight Running Lights—Although somewhat controversial, daylight running lights on motorcycles, at least in high-income countries, have been demonstrated to be effective. The effectiveness varies between studies, but a reasonable figure is a reduction in motorcycle/other vehicle injury crashes during daylight of around 16% (Bijleveld 1997). The effect on fuel consumption is trivial. To maximise the effectiveness, the lights should be so wired that they come on automatically when the machine is started.

Anti-lock Brakes on Motorcycles—Studies of insurance claims in the USA show that there were 21% less claims for motorcycles equipped with anti-lock brakes compared with motorcycles without ALB (H.L.D.I. 2014). Anti-lock brakes are particularly effective in wet weather and single-vehicle loss of control accidents.

Leg Protection—It has long been established that limb injuries, and particularly lower limb trauma, are the most frequent injuries to riders who go to casualty departments (Pedder et al. 1981). Many riders choose to fit crash bars, presumably on the assumption that they will protect the legs when they are struck by an opposing vehicle. In reality, such bars generate serious injuries themselves see below (see Fig. 10.6). Some manufacturers now fit contoured energy-absorbing structures, which under crash test conditions protect the legs very well.

Lower limb injuries are often serious because they affect load-bearing bones and joints. A specific injury to pillion riders is the heel flap injury. It occurs if the heel of the pillion passenger enters the plane of the rotating spokes of the rear wheel

Fig. 10.6 Crash bar causing lower limb injury and an energy-absorbing fairing to protect the legs

which is often very close to the rear footrest. The rotating spokes can fracture the calcaneus, a very disabling injury. Simple shields fitted to the frame can prevent such occurrences.

Airbags—Over a decade ago, an airbag and its sensing system was developed for a motorcycle. Currently, such a system is only fitted on one model, the large and expensive Honda Goldwing. In frontal crash testing, a motorcycle airbag provides good protection for the rider, but its effectiveness in the real world has not been evaluated. In the context of motorcycles for low- and middle-income countries, such a technology is economically unviable.

Airbag Jackets—Airbag technology applied to clothing is a more practical proposition. Sensing is a simple affair in that the rider connects himself to his machine and then primes the jacket. If he is thrown from his normal sitting position, the connection is cut and the jacket inflates. Such jackets are especially good in reducing the sliding abrasive injuries when riders fall off at speed and slide and roll substantial distances on the road surface. In tropical countries, such clothing clearly has acceptability problems but with fabric developments, there may be acceptable solutions.

Autonomous Emergency Braking for Motorcycles—This technology has been applied successfully to motorcycles but it is so far not yet available commercially. Theoretical assessments have been made by examining a sample of motorcycle accidents and applying the system to each individual case (Savino et al. 2014). The general result was that an AEB system would have operated in between 37 and 53% of the crashes. Impact speeds would have been reduced by up to 10%, depending on the initial speed and the crash configuration.

In the immediate future, vehicle technology for MTWs in low- and middle-income countries offers substantial benefits from relatively cheap and well-tested safety measures, especially daylight running lights and anti-lock brakes.

Car Safety in Low- and Middle-Income Countries—The introduction of the Global NCAP rating system promises to improve car crash performance substantially. Series of Global NCAP tests have shown that car models produced for these countries are inferior to the same models sold in high-income countries. Given the success of the NCAP process in Europe and North America, where manufacturers compete for a five-star rating because it has a direct impact on sales, the same incentive will probably work in south-east Asia. There are however major problems of compatibility between cars and heavy trucks particularly on the networks of motorways now being constructed in many countries (JP Research 2013). Front, rear and side under-run crashes are all important; the solutions are well established with specifications for guard structures for trucks. There are strong economic disincentives in terms of cost and weight. Hence, legislation and enforcement is a necessary policy.

10.6 Conclusions

In high-income countries, from 1950 to 1970, traffic safety policies were largely based around an irrational belief that behaviour change would occur largely through publicity campaigns and road user education. The application of rational, science-based countermeasures over the last 30 years has produced major reductions in traffic deaths and injuries. In low- and middle-income countries with very rapid growth in powered vehicles of all kinds, the need to develop appropriate, science-based policies is great. Improved vehicle safety will be a major contribution towards that end.

References

Bijleveld FD (1997) Effectiveness of daylight motorcycle headlights in the European Union. Report R-97–9. S.W.O.V., Leidschendam, Netherlands

Bonnard A, Hugot M, Brusque C (2014) Development of innovative safety performance indicators from naturalistic data. In: Transport research arena, Paris, France

Carsten OMJ, Tate FN (2005) Intelligent speed assistance: accident savings and cost-benefit analysis. Acc Anal Prev 37(3):407–416

Corkle J, Marti M, Montebello D (2001) Synthesis on the effectiveness of edge line rumble strips. Report MN/RC-2002–07, Minnesota Dept. Transportation, St Paul, Minnesota, October

Denton G (1971) The influence of visual pattern on perceived speed. LR 409, Road Research Laboratory, Crowthorne, UK

E.T.S.C. (2005) Intelligent speed assistance—myths and reality. European Transport Safety Council, Brussels, Belgium, September

E.T.S.C. (2011) Road safety target outcome: 100,000 fewer deaths since 2001. 5th Road safety PIN report, European Transport Safety Council, Brussels, Belgium, June

E.T.S.C. (2013) Back on track to reach the EU 2020 road safety target? 7th Road safety PIN report, European Transport Safety Council, Brussels, Belgium, June

European Commission (2015) eCall to be installed in new cars by 2018. eu.europa.eu., 28 April

Forkenbrock G, Hoover RL, Gerdus E, Van Bushkirk TR, Heinz M (2014) Blind spot monitoring in light vehicles—system performance. DOT HS 812 045. N.H.T.S.A., Washington DC, July

Grover C, Knight I, Okoro F, Simmons I, Couper G, Massie P, Smith B (2008) Automated emergency braking systems: technical requirements, costs and benefits. PPR 227, TRL Ltd., Crowthorne, UK
Hershman LL (2001) N.C.A.P.—past, present and future. SAE paper 2001-06-0245. Int. Congress, Detroit, Michigan. March
H.L.D.I. (2014) Evaluation of motorcycle anti-lock braking systems. Bulletin 31(11) (Insurance Institute for Highway Safety, Arlington, Virginia)
Houser A, Murray D, Shackelford S, Kreeb R (2009) Analysis of benefits and costs of lane departure warning systems for the trucking industry. F.M.C.S.A.-RRT-09–022, American Transportation Research Institute, Arlington, Virginia, February
Insurance Institute for Highway Safety (2013) New research adds to evidence that motorcycle ABS prevents crashes. Status report, vol 48, no 4, Arlington, Virginia, May
Isaksson-Hellman I, Lindman M (2012) The effect of a low-speed automatic braking system: estimated from real life data. In: Proceedings of 56th conference, annals of advances in automotive medicine, pp 231–240, October
JP Research India Ltd. (2013) Mumbai-Pune expressway road accident study. JP Research India PVT Ltd., Pune, Maharashtra, India
Kusano K, Gorman T, Sherony R, Gabler H (2014) Potential injury reduction in the US vehicle fleet for lane departure warning-equipped vehicles in single vehicle crashes. Traffic Injury Prev 15:157–164
Lee JD, McGehee DV, Brown TL, Marshall DC (2008) Rear-end crash avoidance systems: effects of adaptive cruise control. DOT HS 810 981, N.H.T.S.A., Washington DC, August
Mackay M (1973) The effectiveness of vehicle safety design Changesin accident and injury reduction. In: Proceedings of conference on vehicle safety legislation, Institution of Mechanical Engineers, London, pp 65–78, July
Mackay M (2005) Quirks of mass accident data bases. J Traffic Injury Prev 6(4):308–311
Martindale A, Urlich C (2010) Effectiveness of transverse road markings. New Zealand transport agency research report 423, Wellington, NZ, October
Mohan D (2015) Autonomous vehicles and their future in low and middle-income countries. In: Conference on driverless technology and its urban impact. NYU Marron Institute of Urban Management, New York, May. https://tripp.iitd.ernet.in/publications/paper/planning/2015%20Autonomous%20Vehicles%20DMohan%20woe.pdf
Savino G, Rizzi M, Brown J, Piantini S, Meredith L, Albanese B, Pierini M, Fitzharris M (2014) Further development of motorcycle autonomous emergency braking. What can in-depth studies tell us? Traffic Injury Prev 15:165–172
S.W.O.V. (2015) Intelligent speed assistance. SWOV fact sheet. Institute for Road Safety Research, Leidschendam, The Netherlands, January
Thomas P, Frampton R (2007) Real world assessment of relative crash involvement rates of cars equipped with electronic stability control. In: Paper 07-0184, Proceedings of 20th experimental safety vehicle conference, N.H.T.S.A., Washington DC
Visvikus C, Smith TL, Pitcher M, Smith R (2008) Study of lane departure warning and lane change assistance systems. Project report 374, Transport Research Laboratory, Crowthorne, UK, November
W.H.O. (2013) Global status report on road safety 2013. World Health Organisation, Geneva, CH, p 137

Chapter 11
Possible Futures of Vehicle Safety

Yves Page

Human Kind is condemned to progress. Till eternity
Alfred Sauvy (French economist)

11.1 Introduction

In more and more countries in the world, we can observe a proliferation of road safety planning strategies which present very clear ways and measures to struggle against road crashes in the countries where these plans are initiated. They are basically a response to a very detailed diagnosis of traffic safety issues, generally broken down into three categories: risk factors (e.g., speed, alcohol, vigilance, distraction, etc.); vulnerable users or group at (over)-risk (e.g., young drivers, motorized two-wheelers, etc.) and accident types (e.g., loss of control, intersection crashes, night crashes, etc.). Recommended deterrence actions are obviously responses to the safety issues but, above this, they belong to a paradigm, e.g., the 'Safe System', whose basic principles are the following:

- A human being has limited biomechanical capabilities to withstand impacts. Reduction or avoidance of impacts is therefore inevitable, noticeably by a management of impact energies and the limitation of people exposed to forces likely to provoke injuries.
- A human being is fallible, and therefore makes errors. The whole system of traffic and land transport must be designed and maintained taking these errors into consideration.
- A comprehensive diagnosis of safety issues must be conducted in order to determine safety measures that have high safety benefit potentials and a high efficiency/cost ratio.

Y. Page (✉)
Renault Group, Guyancourt, France
e-mail: yves.page@renault.com

G. Tiwari and D. Mohan (eds.), *Transport and Safety*, Springer Transactions in Civil and Environmental Engineering,
https://doi.org/10.1007/978-981-16-1115-5_11

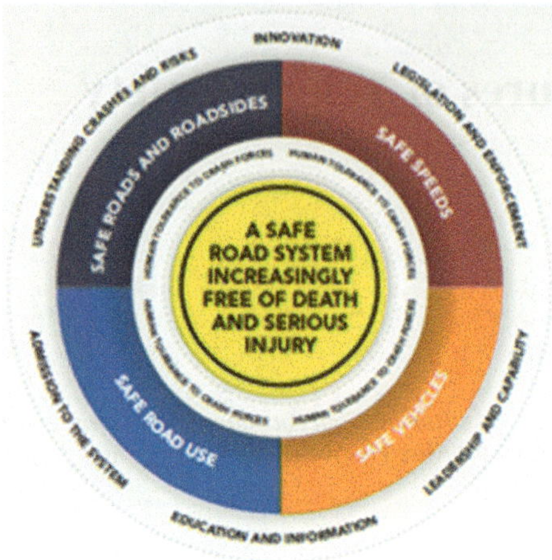

Fig. 11.1 General principles of the 'Safe System'

- Measures must be consistent with choices for society (economic, human and environmental), with positive economic impact on providers of products and services useful for safety on land transport.
- The responsibility of safety actions must be shared among all public and private players and not by only one player (e.g., the public authorities).
- Safety is first generated by the reduction of crashes that are easily avoidable, and then by sustaining a transport system (not only roads) compatible with the safety of future generations.

These general principles are often presented in these programs, according to 4 (prosaic) chapters which put forward systemic, holistic and integrated aspects or road safety policies (Fig. 11.1). These four chapters are:

- Safe speeds
- Safe vehicles
- Safe roads, streets and roadsides
- Safer road usages

And, of course rescue services and medical treatment.

11.2 Safe Vehicles

The focus of this paper is safe motorization particularly passenger cars which count for approximately 83% of all motorized vehicles in France. They are also involved in 78% of injury crashes and in 75% of fatal crashes. These figures may vary a lot between regions, especially in emerging countries where two or three motorized two-wheelers are predominant as well as trucks, buses and coaches. Technologies that apply to passenger cars can in most cases also apply to other motorized vehicles except some that are very specific (e.g., car structure, restraint systems, dynamic control, etc.).

The question is: What is a safe vehicle? The answer to that question claims will help us understand the extent to which the automobile design contributes to road safety.

Safety consists of a series of measures which ensure that a task or a whole set of tasks are conducted without any property damage or injury or any kind of other harm (moral harm, economic loss, social harm or esthetic harm), to the one who conducts the tasks and to others as well (Page and Coz 2003).

Safety can be envisaged from to several angles:

- Primary safety aims to prevent a harm. The difference between prevention and avoidance is a bit vague. Say, for the sake of simplicity, that any measure that targets attitudes and behaviors of road users as well as other players of road safety (in charge of design, maintenance and control of vehicles, road, road equipment and traffic management) are preventive actions, whereas measures that correct a driving situation which is critical or is about to be critical are avoidance measures. We do not elaborate here on the term 'precaution', which consists in taking measures against a new risk, unknown and not well-documented, which generally leads to very restrictive measures.
- Secondary safety aims to reduce the consequences of a harm (severity).
- Tertiary safety aims to bring the best and fastest care to victims.
- Quaternary safety aims to reduce the physical and psychological sequelae after a harm.

Therefore, vehicle safety consists of a series of measures which ensure, via a vehicle (or, nowadays, the so-called extended vehicle taking also into consideration the connectivity between vehicles or between vehicles and environment), automobile trips with minimum harm and external effects.

Road risk prevention is currently being coordinated by public authorities which encourage associations and the private sector to develop, each in its own field of expertise, actions that can help in preventing or avoiding crashes or, at least, mitigating injury severity in the case of a crash. However, since decades, for the sake of innovation in safety or under regulatory or consumerism pressure, vehicle manufacturers and automotive suppliers have developed systems to protect vehicle occupants (e.g., safety belt was patented in 1903 and the concept of bag filled in with air is dated 1941) or external users (e.g., pedestrian-friendly bonnets) to improve safety inside or outside the vehicle in reducing the consequences of an impact. Active safety (accident avoidance) has been improved via steering, dynamic control and braking performances.

A vehicle manufacturer contributes to efforts in road crash and injury prevention by undertaking and applying many safety measures via technology and by conducting safety actions and social responsibility (driver training, education, communication, research and corporate sponsoring). OEM's contribute to national and international road safety actions by acting specifically in vehicle safety, i.e., in designing and fitting vehicles with primary safety systems (assisting drivers in his/her navigation/guidance/control tasks), secondary safety systems (i.e., able to optimize occupants protection as well as external users protection in case of a crash), tertiary safety systems (e.g., automatic crash notification; rescue code, which is a QR code pasted on the front and back windshields allowing the rescue services to quickly get a vehicle identification card of the vehicle, which helps in cutting it at the right places in case of necessary extrication).

These systems have various origins: either the availability of technologies, or a brilliant engineering idea or inspiration, or the existence of such a system by a competitor, or a specific strategy of a vehicle manufacturer which makes safety a brand identity, or an economic interest, or a regulation, or a norm, or a specific consumer test (such as New Car Assessment Programs), or, a particular accident analysis that reveals that such and such a safety issue could be a priori tackled by these systems.

11.3 Secondary Safety of 'Safety of Protection'

Sequentially, an automobile impact induces an impact between an occupant (or an external user) and a part of the vehicle (or another element of the environment if the occupant is ejected) and then impacts between different internal organs of the victim's body (Page 2012).

First impact is, for an occupant, the consequence of one of these three injury mechanisms:

- Intrusion in the passenger compartment.
- Projection (or deceleration) of an occupant against a rigid part of the compartment, or interaction between an occupant and the restraint system (belt or airbag).
- Ejection of the body (or part(s) of the body outside the compartment).

There are other ways to be injured or die on the roads, such as carbonization or intoxication by smoke.

Injury severity depends on, for each of these mechanisms, on the violence of impact, on impact configuration (frontal, side, rear, rollover, etc.), on stiffness of the obstacle and on biomechanical tolerances of the human body. These three mechanisms should be targeted by secondary safety and it is usually done by using two options:

- Vehicle structure.
- Restraint systems.

Vehicle structure must dissipate energies released during the impact while maintaining the integrity of the compartment, i.e., avoiding/reducing intrusion in also limiting the pulses and efforts sustained by the occupant. Prosaically, a soft structure would properly dissipate the energy when the vehicle is deformed but would induce intrusion. A rigid structure would reduce intrusion but would put much more pressure on the occupant's body.

Vehicles being very different from one to another, one must also ensure a good compatibility between them (mass compatibility, stiffness compatibility and geometric compatibility) for some vehicles must not be too aggressive to the others. This is the reason why, for example, trucks are in most countries fitted with anti-underrun devices or heavier passenger cars must not be too stiff otherwise too aggressive against smaller and less stiff cars (even though they still have to be stiff enough to appropriately protect their occupants in case of a crash against an stiff obstacle).

Restraint systems (e.g., seat belt with pretensioners and load limiters tuned for persons biomechanically fragile; frontal airbags; side airbags for thorax and head; seat bossage; etc.) are essential complements to a stiff vehicle structure, which is the current choice. Occupants must absorb their own kinetic energy residuals: the pretensioner (seatbelt tension) couples the occupant to his/her seat. In a frontal impact, for instance, load limiter unrolls a few centimeters of the belt for it does not provoke lesions to the thorax organs, and then the airbag 'welcomes' the occupant head and thorax and diffuses more efforts while, additionally, preventing head and thorax to smash into rigid parts of the wheel or the dashboard. Obviously, these features work at reasonable impact violence. Above a certain threshold (like, for example, a frontal impact against a deformable obstacle above 70/75 km/h), the laws of physics make full protection hardly achievable at a reasonable automotive cost. In that case, the vehicle sustains a heavy deformation and intrusion can no longer be prevented and the restraint system can no longer be functional.

Lots of improvements in protection safety (usually known as passive safety) was done at the end of the 1990s, with voluntarist policies of a few OEM's which can be considered as pioneers in the field, and with the development of regulations concerning frontal impact (Directive 96/79/CEE et ECE.R94) and side impact (Directive 96/27/CEE et ECE.R95), as well as development in consumer testing such as EuroNCAP (and other kinds of NCAP's around the world) which assign points and stars (1 up to 5) to new cars. The first car which was ever awarded with Euro NCAP 5 stars is a French vehicle, in 2001. Ever since, a lot of passenger cars have been awarded 5 stars, even though with the continuous improvements, hardening and broadening of tests since 2009 and planned up to 2020 (see Road Map EuroNCAP released in 2014, which specifically hardens tests in passive safety and adds tests for the presence of preventive/active safety devices). We must here underline that EuroNCAP does not only target car occupant protection. Tests also target pedestrian safety and soon pedal cyclist safety in crashes against passenger cars. Other kinds of NCAP's, under the supervision of global NCAP, also have roadmaps for enlarging their testing.

11.4 Primary Safety

The analysis of a road traffic system reveals that it is composed of motorized and non-motorized road users, who drive or walk on roads/streets. They move in a general environment that they do not generally control or monitor (e.g., traffic conditions, temporary signals, road works, etc.). These trips are governed by traffic laws. Each and every road user is therefore supposed to monitor and adapt his trip/driving according to rules/laws that he/she is also supposed to know since he/she has got a driving license (not all pedestrians I admit), according to the situation he/she faces (road, trip motivation, type of driven vehicle, etc.) and the presence of other users at the same time and place.

The diversity of components of a traffic system obviously shows that the user is not responsible for everything. He/she does not conceive/maintain roads, he/she does not select the weather, he/she does not select traffic density, road works occurrence, missing signals, etc. On the other hand, he/she has to take the right decisions against what he/she encounters on the road. In other words, he/she is the last regulator of his/her trip. Vehicles that are sold, roads that he/she drives on, road and traffic maintenance, must increase his/her safety and must optimize his/her decisions/actions (even though, in such an automated task like driving, decisions are often implicit).

Therefore, if we set apart design, the conception and effectiveness of conditions in which transport of persons and goods are performed (e.g., failures in urban management, inefficient transport and land planning, missing alternative transport, etc.), if we set apart problems in designing and maintaining roads and infrastructure (i.e., visibility, clarity, adequacy to characteristics of vehicle dynamics, potential for forgiveness, etc.) as well as vehicles, traffic safety issues develop around drivers who violate basic safety rules (excessive drinking and driving, excessive speed, drug use, aggressive driving, risky driving, etc.), or make mistakes and errors (perception, cognition, vehicle control, bad or insufficient skills) often due to altered states (alcohol, inattention, distraction, stress, fatigue, lack of sleep), or due to inexperience of driving, or due to specific trip conditions.

Consequently, OEM's find another way in traffic safety via driving assistance systems. These systems have two interests: they make it easier to perform some driving tasks and help drivers not to enter into dangerous or critical situations… or to get out of these situations. There are a lot of these systems on the market (mainly in highly industrialized countries), with a lot of variants, but they often demand environment sensors that are often costly and not yet sufficiently robust.

There are currently a few taxonomies of driving assistance systems. We propose 4 of them, according to the assistance type, according to the level of influence of the assistance and according to the active or passive participation of the driver.

According to the assistance type:

- The assistance can be of a strategic type. It then targets the itinerary planning and the navigation. Navigation systems are typical examples of this type.

- The assistance can be of a tactical type. It targets the selection and the performance of the manoeuver adapted to the encountered situation. Blind spot helps detection is a typical example of a tactical aid.
- The assistance can be of an operational type. It consists of controlling the vehicle trajectory. Emergency braking or electronic stability control is typical examples.

According to the assistance influence:

- In a first step, the assistance brings an information to the driver (e.g., an information about density of traffic or tire pressure).
- In a second step, the vehicle activates an alarm (e.g., a tone if the seat belt is not buckled up).
- The vehicle can also activate an enhanced information (e.g., a long and high tone if the belt is still not buckled up after a few seconds).
- The vehicle can perform a corrective action on the manoeuver (e.g., electronic stability control).
- The vehicle can, at last, take control, the driver being fully out the loop (automatic braking for instance).

According to the driver participation:

- Without driver intervention (e.g., automatic emergency braking)
- An assistance which takes part of the driving task (e.g., autonomous cruise control)
- A driving assistance under control, which accompanies an action by the driver (e.g., electronic stability control).

Apart from automotive parts (e.g., steering column, hydraulic brakes and tire rubber) that are reliable and safe (compliance to rules and laws as well as general safety of products), and apart from the considerable progress over time on steering, braking and dynamic control, and specific primary safety systems are not that much fitted in most vehicles nowadays and are often available in premium passenger cars or as an option in mid-class cars. Speed limiters or cruise controls are largely disseminated (in any case by French OEM's) but not the adaptive intelligent cruise control (maintaining a speed compatible with the vehicle pace ahead). Systems such as ABS, ESC, emergency braking systems, navigation systems, automatic head lights, automatic commuting low beam–high beams, tire pressure monitoring systems, automatic wipers, blind spot help detection, lane departure warnings, lane keeping assist, driver alert systems, night vision systems and variants of intelligent speed adaptation are on their way, in the pre-market introduction, at a reasonable pace. On the other hand, systems such as anti-collision at junction radars or anti-head-on collisions radars, alert of incidents/accidents ahead of the trip, or systems informing the drivers about risky sites or black spots are still at the research or advanced engineering phases.

The field for driving aids is henceforth extremely huge. Sensors detecting the environment (ultrasonics, cameras, radars, lidars, navigation maps, GPS, etc.) are more and more mature and algorithms more and more powerful: they detect obstacles around the vehicle, lanes on the roads, junctions, insertion access, line markings, road and traffic signals and therefore assist the driver in his/her driving tasks (navigation/guidance/longitudinal and lateral control).

The question is: to what extent OEM's and suppliers are able to propose protection systems and driving assistance systems that are economically accessible to mass production and mass commercialization. Recent history shows that it is definitely possible when we, for example, look at the large dissemination of ABS and ESC. Of course, often, sophisticated systems are released first on high-end vehicles before being tried on entry level vehicles. The fast renewal of the automobile fleet is one of the most promising safety measures though. To that end, the affordability of such systems is crucial.

11.5 Tertiary Safety

Development of portable devices (smartphones, tablets, etc.) including the e-call (most well known as automatic crash notification in the USA) will inevitably reduce delays in the intervention by rescue services. Although the European Commission considers that generalization of e-call could save up to 5–15% of fatalities in Europe, the latest studies show lower estimates, around 2–3% in France for instance.

In addition to the alert, cooperation between OEM's and rescue services has recently come up with a Standard of Extrication Card which allows fire brigades to cut vehicles more efficiently (if necessary, of course), in order to extricate occupants trapped after an impact. It is very likely that, in future, intelligent e-calls (able to provide information about the crash like impact speed, occupant's presence and belt age status, etc.), and e-health may also contribute to help rescuers and hospitals in their injury and injury severity diagnosis and therefore help in triaging the victims to the appropriate hospitals and trauma centers.

11.6 Technology and Safety Benefits

Past and current studies show high potential safety benefits of existing systems (protection systems as well as driving assistance systems already largely disseminated onto the market) (Sferco et al. 2001; Forêt-Bruno et al. 2001; Page et al. 2005, 2006a, 2009b; Page and Cuny 2006; Kassaagi et al. 2006; Couturier et al. 2007; Zangmeister et al. 2007; Page and Labrousse 2007; Cuny et al. 2008; Zangmeister et al. 2009; Page 2011, b; Fildes et al. 2015). Passive safety systems coupled with collision prevention system and/or injury mitigation systems such as ESC or EBA already show an unprecedented effectiveness value ever since seat belt effectiveness, first speed limit settings or automatic speed camera settings (in France at least) came into being: a front seat occupant of a vehicle which got 5 stars at the EuroNCAP (old rating) and which also has the ESC and the EBA has reduced the risk of severe fatal injury by 70% compared to an occupant in a vehicle without ESC nor EBA.

A recent study showed that, if we consider the declining trend of fatalities between 2000 and 2010 in France (−48%), 6 points are due to the vehicle safety improvements over this period (Page 2010). This seems very small but if we also consider that the rate of deployment of these technologies in the vehicle fleet is very low and slow (vehicle fleet is renewed every 15 years or more), and these systems have started being deployed in early 2002–2003 for passive safety and even later for active/preventive safety, and mainly in high-end vehicles first, then we are forced to conclude that the promise in safety benefits is actually high.

Perspectives for systems that are not largely disseminated, or still under development, are also positive (Page et al. 2006b, 2009a; Page and Hermitte 2007, 2009; Driscoll et al. 2007; Chauvel et al. 2013; Hynd et al. 2015). Each system has an expected effectiveness (in terms of savable lives or avoidable severe injured people) relatively low (often between 2 and 5%, a bit more for some of them, if the whole fleet is fitted) according to available studies. Therefore, a combination of systems that address various safety issues (loss of control, loss of guidance, blind spot, late braking, night vision, etc.) is to be preferred. Expected safety gains are potentially high and effectiveness studies research (today embedded as 'stand-alone' systems and tomorrow connected with one another) must be encouraged to identify this promise in greater detail (Page et al. 2015).

It is indeed difficult to establish a 'Top 10' rating of the most promising systems which for a few reasons (Page and Hermitte 2009): Effectiveness studies are of three kinds: the ones which simulate the expected effectiveness of systems not yet or poorly on the market, the ones which observe the actual effectiveness of systems in the market according to their penetration rate in the fleet, and the ones which extrapolate the safety gains that would be observed if existing systems would be disseminated 100% in the fleet. We thus have a problem of consistency among different estimates.

Whatever their types, available studies vary in the effectiveness indicators they use. It can be reduction in injury crashes, reduction in all kinds of crashes, reduction in fatalities, in severe injuries, in crash risk, in injury risk, taking into consideration (or not) the penetration rate, etc. (Possibly depending on accident or impact types such as loss of control, frontal impact and pedestrian collisions). As a consequence, they are not exactly comparable.

Similarly, methods and techniques of evaluation as well as simulation assumptions (noticeably concerning the use of driving aids by drivers) vary a lot too. Furthermore, some sensitivity studies establish effectiveness estimates depending on different values of a set of parameters entering into consideration to make the function work. Subsequently, effectiveness estimates might be quite different between variants of the same system.

Actually, numerous systems present numerous variants. Variants may concern the function itself: for example, a lane departure warning can have different triggering thresholds, possibly selected by the driver, when the car is about to leave the lane, when it crosses the lane line, or long before crossing the line; or an AEB can detect only moving obstacles in the same direction, or can detect any kind of moving or stopped obstacles.

OEM's and suppliers continuously improve the systems, from time to time and new systems are continuously released. Therefore, the long list of functions and variants is always evolving. A Top 10 would shed the light on a few fashioned systems at a given time and possibly hide promising functions not properly analyzed yet. Moreover, some functions improve better than others overtime by, for example, extending their coverage (AEB against moving vehicle, than against stopped or fixed obstacles, than against pedestrians, at low speed and speeds) or by strengthening the technology.

Primary, secondary or tertiary safety systems must not be considered in competition with one another but rather like different opportunities to solve similar problems. For example, an intelligent speed adaptation system can reduce the driving speed, an automatic braking system can reduce the impact speed, a reinforced car structure combined with restraint system can be even better at lowering impact speed and an automatic crash notification can reduce intervention delays by rescue services. Rating them all in a Top 10 would mean ignoring their additive impacts.

If systems are sometimes complementary or additive, they are seldom fitted individually in a vehicle, which would demand the establishment of a Top 10 of the 'packages of systems' rather than a Top 10 of isolated functions. Given their high number, classifying hundreds of combinations or packages is impracticable.

To our knowledge, there is no unique Top 10, accepted by the scientific community as absolutely irrevocable.

Some systems might have a restrictive target population (e.g., blind spot detection address between 4 and 6% of the injury crashes) but a high effectiveness potential (e.g., 50% out of the 4–6%) and a low cost. It would be unfair to disqualify these systems by underlying the low effectiveness if it can be reached at a lesser cost. The 'Top 10' should therefore be established on the basis of the effectiveness as well as on the basis of the effectiveness/cost ratio, which would, in the end, make it fully undecipherable.

Some systems, highly effective ex-ante in theory, can be fully rejected by drivers/users because of whatever reasons. They might feel like they (the systems) are inefficient, useless, intrusive, and non-adapted to driving. This is, for instance, the case of lane departure warning in the USA, which is often disconnected because it is considered too intrusive in daily driving. This drives us back to real usage of the systems and their parametrization, of which knowledge is still poor even though it is highly important in the estimation of their effectiveness.

As a consequence, we do not mean to establish a kind of 'Top 10' so far, but recommend to multiply effectiveness studies in a private/public partnership to consolidate effectiveness estimates available by now, and disqualify ambiguities.

11.7 Current Trends and Possible Future

A lot is said, written, discussed, wrongly or badly, about connected vehicles and automated driving (Pajon et al. 2012). In both cases, connectivity and automation make people dream about better days for road safety (an automated or autonomous vehicle is supposed to eradicate human driving errors and connected driving assistance systems are supposed to be better than current stand-alone driving aids) but they also frighten people (is technology relevant, reliable, robust? What about human beings in a highly or fully automated world, what about transmission of personal data? etc.)

It is therefore useful to recall the basics of connectivity and automation to avoid any confusion or misunderstanding. In the current real-world, what is all this about precisely?

11.7.1 Connected Vehicle

Professionals and the public now seem to be aware of what we call 'the connected vehicle,' meaning vehicles connected to other vehicles or to infrastructure or to … whatever. This is actually a technical definition that hides two different functional definitions of connectivity (Road Safety & Connected Mobility (Collectif) 2014):

- A driver or a passenger can be connected with the external world via a nomadic device (e.g., a smartphone or a tablet) which has nothing to do with the vehicle. He (or she) just uses the device while driving (or as a passenger) as he or she would use these devices outside the car. This is just the general continuation in the car of the 'connected user.'
- A driver or a passenger can be connected via an integrated device which is embedded in the vehicle and can offer different types of services. In this case, the vehicle offers some services, which could by the way be redundant to the services available with a nomadic device. Let us call it the 'connected vehicle.' Of course, the connected vehicle mediates between the driver (and the passengers) and the external world. The connected vehicle can also give information to the rest of the world in case it is itself a sensor (e.g., if it detects slippery road and sends the information to the surrounding traffic).

In both cases, the services provided by connectivity (whatever the technologies behind and whatever the medium, nomadic or integrated) can be classified according to the following taxonomy:

- Safety systems: the service has a primary objective to prevent crashes and injuries. For example, car-to-car communications can help in preventing crashes at intersections where visibility is reduced by buildings, trees, bus stops, whatever kind of fixed or mobile masks to visibility.

- Driving assistance: the service has a primary objective to help the drivers in performing a driving task (navigation, guidance or control). For example, a navigation system helps the driver in choosing his (her) route and to follow directions that are proposed by the system.

These two categories can easily be grouped together since driving assistance systems often have a safety aspect too.

- Traffic information: the service has a primary objective to help the driver knowing more about the traffic ahead, e.g., whenever a route is congested, road works are present ahead of the trip or whether a route is closed for whatever reasons—services related to transport, usually called intelligent transport systems.
- Services not related to transport, often called infotainment (Internet in the car, watching or downloading videos and many other applications currently available on smartphones and tablets…).

The connection is ensured by whatever kinds of technologies (3G, 4G, 5G, DSRC, etc.), which are beyond the scope of this paper but which present high performances as well as limits. Therefore, especially for connected safety systems and driving assistance systems, the functions work under particular circumstances called 'use cases' and not in any circumstances. For example, as connected technologies usually use GPS to localize a vehicle or a person somewhere on earth, this information is known to be not very accurate (a few meters accuracy) which prevents one from using it for impact avoidance for example (at least for the moment).

Moreover, international standards of principles allow some consensual rules for human–machine interaction (HMI) in order to properly design interfaces that are not distracting drivers. These apply for any kind of manipulation the driver is in charge of (radio tuning, navigation system use, etc.).

These systems are in full expansion, though in pre-deployment phase by now. They deserve a lot of attention, especially to hinder possible distractive effects, to select, amongst all systems, those which have the largest expected safety benefits, and of course to prevent cyber-crime.

Preliminary effectiveness studies about connected driving assistance systems or connected safety systems (and for functions such as alert of incidents, blind sport detection and information about status of traffic lights) show that expected benefits are positive but minor. These effects are even lower if they are considered to be 'in addition' to stand-alone driving assistance systems.

11.7.2 Automated Driving

Autonomous vehicle, automated driving, self-driving cars, driverless vehicles, unmanned driving, automated car, etc., are expressions often used to name a vehicle (and not necessarily a car) which takes over all or part of the driving task which is today under the driver control during a trip or a fraction of the trip. The delegation from the driver to the system consists of longitudinal control and/or lateral control and/or environment monitoring, and especially obstacle detection.

Therefore, an automated car is not systematically automated continuously and automation is not systematically complete. Automation can be conditional, partial, high or full, depending on automation level and driving situations (so-called use cases). A use case depicts a function of delegation, in certain conditions (traffic, road and environment and delegation mode). For example, the 'Traffic Jam Assist' function often addresses the following use case: dense traffic, up to 40–50 km/h, the vehicle drives on its lane, without any possibility of automated lane change, with road markings (left and right), all weather conditions but fog (or lack of visibility), when the driver can have his/her hands off the driving wheel.

Various bodies (NHTSA, SAE, OICA, VDA5, etc..) defined automated driving levels (in general, from 0 to 5), starting from manual driving or driving assisted with some low level of assistance, i.e., information aids (level 0) to full automation in all circumstances (level 5). These levels are established according to the distribution of tasks and driving authority between the driver and the vehicle, especially in situations when an impact is predictable (Table 11.1).

Motivations for such vehicles are ecological (optimization of traffic flows and reductions of pollution), demographical (assistance to anxious drivers, increase of comfort, assistance to elderly drivers in some difficult/uncomfortable manoeuvers), safety (reduction in crashes and mainly injury crashes), economical (optimization of vehicle lifetime and use, optimization of land use) or related to quality of life (additional time to do something else in the vehicle) (Page 2014).

The big challenges are technological (performance, reliability and robustness of sensors and artificial intelligence), ethical and legal (are we ready to drive on

Table 11.1 Automation levels

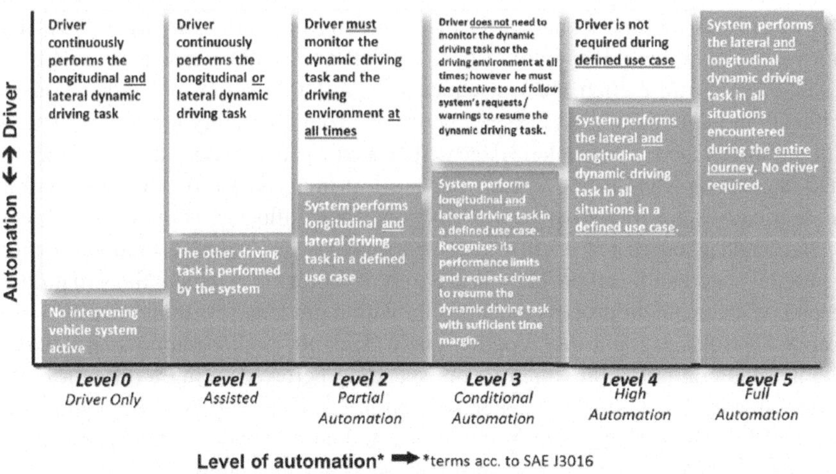

Source OICA, vehicle standards map 2014, Ministry of Transport, New Zealand

automated road and what is the appropriate legal framework that goes with it?) and ergonomics (how to design driver automaton relationships that manage driving situations, critical situations and crash risk situations better than today?).

Some driving tasks are already partially or fully automated, for example, longitudinal control in certain traffic conditions (e.g., adaptive cruise control), but automation will come progressively, starting with simple use cases and then with more complex ones (i.e., congested traffic on dual carriage ways with longitudinal control exclusively, and then on highways at higher speeds, and then in both cases with lateral control and possible lane change, and then in more complex environments such as urban areas, etc.). Of course, fully automated vehicles are also experimentally possible now in complex traffic, at very low speed, at a high cost, on very well-known routes that the system learns: shuttle on reserved lanes without any traffic will be possible sooner than usually expected.

Conditions for deploying such vehicles also concern technical certification (homologation requires reference to technical regulations or assumption of absence of danger related to the technology if no technical regulation exists yet) and their compatibility with usage regulations: indeed, if a vehicle is certified but cannot be used because of traffic laws, for example, deployment is impossible.

There are a lot of automated systems which work in certain traffic conditions, in certain modes (eyes-on, hands-on, eyes-off, etc.), for passenger cars, public transports, other types of vehicles and on dedicated roads or public roads. There are, by now, in the research phase or in the experiment phase, current vehicles on the market being fitted with driving assistance systems underneath level 3 (out of 5) of automation. In other words, these are assistance systems. The driver can always act as a supervisor (except for autonomous braking systems, but these kinds of 'last resort' systems are not classified in the SAE taxonomy).

Assessment of the expected safety benefits of automated driving system is starting and the few available studies are not really convincing since they usually assert that 90% of crashes are due to human errors (meaning implicitly the driver) and that removal of drivers would remove crashes... which is to a large extent questionable.... Driver error is most often the consequence of a combination of factors, human, technical and situational that prevent the driver from correcting a critical situation with which he/she is confronted (Page 2013). Error is a symptom of a malfunction (that sometimes leads to the crash), but not a primary cause of the malfunction: accident causes are found before the error and their influence is direct or indirect, depending on when we go back in the accident analysis. For example, the combination of fatigue, speed and grip problem is a group of factors that impact the situation at the end of the accident process that ends up with a guidance error and finally a loss of control. But, upstream this process, one can identify other intermediary causes such as road configuration which favors speeding before entering a difficult bend, lighting that does not allow a good bend visibility, etc.; as well as causes even more 'indirect' but also predominant in terms of accident prevention such as education, information, social culture toward speed, driving, risk and safety.

An additional way of presenting things consists of considering driver errors as limits of adaptation of drivers with critical situations to which the current traffic system confronts him/her. We have also to consider that 'human factors' are what actually make the driving system work, more or less efficiently, despite its shortcomings, thanks to the adaptation capabilities that are inherent in humans.

If the driver is removed from driving, driver adaptation is also removed...a loss that has to be compensated by automated systems.

11.7.3 Frugal Engineering and Frugal Safety

In high-end countries, short-term future and mid-term future are definitely related to automate driving and connected vehicles that allow different types of services, especially safety services. Highly or fully automated driving, even in some specific simple use cases such as traffic jam pilot or traffic jam chauffeur on motorways, will definitely not come very soon since there is still room for technical improvements to get safe and secure vehicles that will drive with non-automated vehicles on the same roads. The more eyes-off, the more difficult. If occupants are supposed to do something other than driving, they will of course be willing to sit in a vehicle comfortably and not behind at the wheel seat, and not necessarily belted. Therefore, we collectively have to think about new restraint systems compatible with an 'eye-off' traffic, i.e., 'out-of-regular-position'-occupants!

Beside this, the most predominant barriers to the deployment of safety systems are definitely their costs, and if most of them are considered to be effective in preventing injuries and extending years of life, they deserve to be encouraged.

One way of doing this is frugal engineering. Frugal engineering is the process of reducing the complexity and cost of a good and its production. Usually, this refers to removing nonessential features from a durable good, in order to sell it in developing countries. It also refers to make cost reductions during the process of innovation, engineering, production and commercialization. There are many ways of doing this, the biggest challenge is to avoid producing bad products/services at low cost in bad conditions, which could have long-term negative effects on the brand, the whole economy and the safety/security of products.

Current examples of low-cost safety systems can be found in smartphones, with a lot of free or cheap applications such as lane departure warning, drowsiness warning or forward collision warning, based on simple algorithm and the smartphone camera as sensor, for example (or blood pressure sensor as another example). They do not prove efficient and cannot be considered as frugal engineering as they are often nomadic devices produced outside the OEM's world. A step forward would be to integrate frugal engineering as a basic paradigm for safety, so that it is for low-end cars which present nevertheless quite good levels of quality so far.

Automated vehicles, connected vehicles and frugal safety engineering are the three likely pillars for the future of vehicle safety based on technology.

References

Chauvel C, Fildes B, Lahausse J, Page Y (2013) Automatic emergency braking for pedestrians. Effective target population and expected safety benefits. In: 23rd ESV conference, Seoul, June 2013

Couturier S, Faure J, Page Y, Labrousse M (2007) The benefits of double pretension in decreasing knees & lower legs injuries in frontal impacts. In: ESV conference, Lyon, 2007

Cuny S, Page Y, Zangmeister T (2008) Evaluation of the safety benefits of existing safety functions. Rapport D.4.2.2. du projet TRACE (Traffic Accident Causation in Europe), Juin 2008

Driscoll R, Page Y, Lassarre S, Ehrlich J (2007) An evaluation of the potential safety benefits of the French intelligent speed adaptation project. In: 51ème AAAM conference, Melbourne, Oct 2007

Fildes B, Keall M, Bos N, Lie A, Page Y, Pastor C, Pennisi L, Rizzi M, Thomas P, Tingvall C (2015) Effectiveness of low speed autonomous emergency braking in real-world rear-end crashes. Accid Anal Prev 81:24–29

Forêt-Bruno J-Y, Page Y et al (2001) Comparison of thoracic injury risk in frontal car crashes between occupant restrained without belt load limiter and those with 6 kN and 4 kN belt load lilmiters. Stapp Car Crash J 45

Hynd D, McCarthy M, Carroll J, Seidl M, Edwards M, Visvikis C, Tress M, Reed N, Stevens A (2015) Benefit and feasibility of a range of new technologies and unregulated measures in the fields of vehicle occupant safety and protection of vulnerable road users. Final report. TRL for the European Commission, Mar 2015

Kassaagi M, Bouslimi W, Val X, Page Y, Bersac J-M (2006) Effectiveness of emergency brake assist in rear-end accident scenarios. In: FISITA conference, paper F2006D062, Yokohama, Japan, 22–27 Oct 2006

Page Y (2010) How safe is vehicle safety? The contribution of preventive and passive safety to saving lives on European road and streets. Jubilee seminar on road traffic safety, Delhi, Dec 2010

Page Y (2011a) An overview of the safety benefits of (some) passenger car safety systems. Revue Méditerrannéenne de Medecine d'Urgence, Juin 2011, Numéro 7

Page Y (2011b) Thierry Hermitte et Sophie Cuny. How safe is vehicle safety? The contribution of vehicle technologies to the reduction in road casualties in France from 2000 to 2010. In: AAAM conference, Paris, Octobre 2011

Page Y (2012) A comprehensive overview of the frequency and the severity of injuries sustained by car occupants and subsequent implications in terms of injury prevention. In: AAAM conference, Seattle, USA, Octobre 2012

Page Y (2013) Do vehicle technologies address specific safety issues inside urban areas. Contribution à un ouvrage 'Safety, sustainability and future urban transport', New Delhi

Page Y (2014) Un véhicule autonome… Pourquoi? Congrès Médecine et Traffic. Automobile Club Médical de France, Paris, Novembre 2014

Page Y, Coz J-YL (2003) Démarche accidentologique et sécurité des véhicules. Revue de la Gendarmerie Nationale N° 207, Juin 2003

Page Y, Cuny S (2006) Is electronic stability program effective on French roads? Accid Anal Prev 38(2):357–364

Page Y, Hermitte T (2007) The TRACE Project, an initiative to update accident causation and evaluate the effectiveness of technologies. In: ESV conference, Lyon, 2007

Page Y, Hermitte T (2009) Ranking the safety benefits of road safety applications. Outcomes of the EU-funded TRACE project. In: SIA vehicle dynamics symposium, Lyon, France, Sept 2009

Page Y, Labrousse M (2007) An overview of the safety benefits of (some) passenger car safety systems. In: Ipass Conference, Bordeaux, Nov 2007

Page Y, Foret-Bruno J-Y, Cuny S (2005) Are expected and observed effectiveness of emergency brake assist in preventing road injury accidents consistent? In: 19th ESV conference, Washington, June 2005

Page Y, Thomas P, Hervé V, Kirk A (2006a) An estimation of side airbag effectiveness in preventing thoracic injuries using French, British and German crash data. In: 50ème AAAM annual conference, Chicago, IL, 17–20 Oct 2006

Page Y, Driscoll R, Lassarre S (2006b) LAVIA: Les effets potentiels du Limiteur s'Adaptant à la Vitesse Autorisée sur la Sécurité Routière Colloque LAVIA, Versailles, 9 Nov 2006

Page Y et al (2009a) Reconsidering accident causation analysis and evaluating the safety benefits of technology: final results of the TRACE project ESV conference, Stuttgart, June 2009

Page Y, Zangmeister T, Cuny S, Kreiss JP (2009b) The evaluation of the safety benefits of combined passive and on-board active safety applications. In: 53th AAAM conference, Baltimore, Oct 2009

Pajon M, Page Y et al (2012) Enlighting the future: from autonomous ADAS to autonomous driving. In: VISION conference, Versailles, Octobre 2012

Page Y et al (2015) A comprehensive and harmonized method for assessing the effectiveness of advanced driver assistance systems by virtual simulation: the P.E.A.R.S. initiative. In: ESV conference, Gothenburg, June 2015

Road Safety & Connected Mobility (Collectif) (2014) Rapport. Challenge Bibendum, Chengdu, Novembre 2014

Sferco R, Fay P, Page Y (2001) Potential effectiveness of enhanced stability programs (ESP). What European field studies tell us. In: ESV conference

Zangmeister T, Kreiss J-P, Page Y, Cuny S (2007) Simultaneous evaluation of multiple safety functions in passenger vehicles. In: ESV conference, Lyon, 2007

Zangmeister T, Kreiss J-P, Page Y, Cuny S (2009) Evaluation of the safety benefits of passive and/or on-board active safety applications with mass accident databases. In: ESV conference, Stuttgart, June 2009

Chapter 12
Head Trauma Biomechanics

Remy Willinger, Caroline Deck, and Nicolas Bourdet

12.1 Introduction

Traumatic brain injury is the leading cause of death and permanent impairment in accidents. In both the severe and mild TBI, diffuse axonal injury (DAI) is the most common pathology. Computation of axon elongation by using finite element head model in numerical simulation can enlighten the DAI mechanism and helps to establish advanced tissue level head injury criteria. The main objective of this research is to propose a brain injury criterion based on multiscale computation of axonal elongation under real-world head trauma.

A literature review of head trauma biomechanics is presented to introduce existing and recently proposed head injury criteria based on global head kinematics parameters, in terms of translational motion, rotational motion, and combined kinematics. The state of the art in the domain of head FE modeling is presented and also the model validation aspects. Finally, focus is on the development of the Strasbourg University FE head model (SUFEHM).

For this model, the modeling technic of the three-layered skull and the establishment of a skull fracture criteria based on real-world skull fracture data will be shown. Coming to the modeling of the brain, a new generation of models will be introduced. The implementation of new medical imaging data such as fractional anisotropy and axonal fiber orientation from DTI into the FE brain model was performed to improve the brain constitutive material law with more efficient heterogeneous anisotropic visco-hyper-elastic material law which enables it to compute axon elongation at the time of impact.

An extensive well-documented real-world head trauma simulation exercise was performed with this advanced head FE model including the computation of axonal

R. Willinger (✉) · C. Deck · N. Bourdet
Laboratoire Icube Equipe MMB, Universite de Strasbourg, Strasbourg, France
e-mail: remy.willinger@unistra.fr

G. Tiwari and D. Mohan (eds.), *Transport and Safety*, Springer Transactions in Civil and Environmental Engineering,
https://doi.org/10.1007/978-981-16-1115-5_12

elongation. Based on the statistical analysis, axonal strain was shown to be the most relevant metric to predict moderate DAI. It was found that the threshold value in terms of axonal strain for a 50% risk of moderate DAI (AIS2+) is 15% of axon strain.

The transfer of this novel head injury prediction tool toward industry and standard organization will be addressed including virtual testing in the automotive environment and new experimental versus numerical helmet test methods.

12.2 Existing Head Injury Criteria Based on Linear Acceleration: Presentation and Critics

12.2.1 Introduction

Over the past forty years, a slant has been put by the biomechanical research on the understanding of the head injury mechanisms. One of the main difficulties of this research field is that a functional deficiency is not necessarily directly linked to a damaged tissue. Nevertheless, an injury is always a consequence of an exceeded tissue tolerance to a specific loading. Even if local tissue tolerance has very early been investigated, the global acceleration of the impacted head and the impact duration is usually being used as impact severity metric. In some previous studies, numerous injury predictors based on translational head motions were proposed:

12.2.2 Maximum Resultant Head Acceleration

A head injury criterion which is often used because of its simplicity is the *maximum resultant head acceleration* (a_{max}). The threshold for a_{max} depends on its application, because of the time-dependent nature of the resultant acceleration with respect to head injury. Maximum linear acceleration is used for many years and continues to be used in many helmet standards $A_{max} < N$ with N a value which depends on the standard used. This criterion does not take into account the time duration of the impact, impact orientation, the type of injuries and the rotation…

A variation of this criterion is A_{3ms} value which refers to the maximum deceleration that lasts for 3 ms. Even if a "kind" of time duration is taking into account, same limitations can be done for this criterion. The 3 ms criterion is based on the WSTC. A_{3ms} should not exceed 80 g (Got et al. 1978).

12.2.3 Wayne State Tolerance Curve

The Wayne state tolerance curve is considered to be the foundation of research on human head injury criteria. This curve evolved from the work of Lissner et al. (1960), Gurdjian and Webster (1945), Gurdjian et al. (1961) and Patrick et al. (1963), and give the tolerable average acceleration in A–P direction (Anterior–Posterior) as function of the pulse duration. The curve is given in Fig. 12.1. Slight cerebral concussion without any permanent effects was considered to be within human tolerance. Only translational accelerations were used in the development of the curve which was obtained from different experiments with cadavers, animals, and volunteers. That substantial acceleration causes injury over short durations, while smaller accelerations require longer duration to cause injury is an assumption fundamental to the curve formulation. The short duration part of the curve ($2 < t < 6$ ms) was derived from cadaver tests in which skull fracture was chosen as injury criterion. Cadaver and animal tests were used for the intermediate pulse durations ($6 < t < 10$ ms). For this part of the curve, intracranial pressure was used as the injury criterion in the cadaver tests and concussion was chosen as the injury criterion in the animal tests. The long duration part of the curve ($t > 10$ ms) was obtained from volunteer tests. There was no head impact in these tests, and no injuries were observed. By assembling all these tests in one curve, it was assumed that skull fracture and concussion correlate. Lissner et al. maintained that for a given duration, accelerations above the curve lead to injury (survival hazards), while accelerations below the curve are tolerable and cause, at most, cerebral concussion without permanent effects. Except for the long duration accelerations, the WST-curve has never been validated for living human beings.

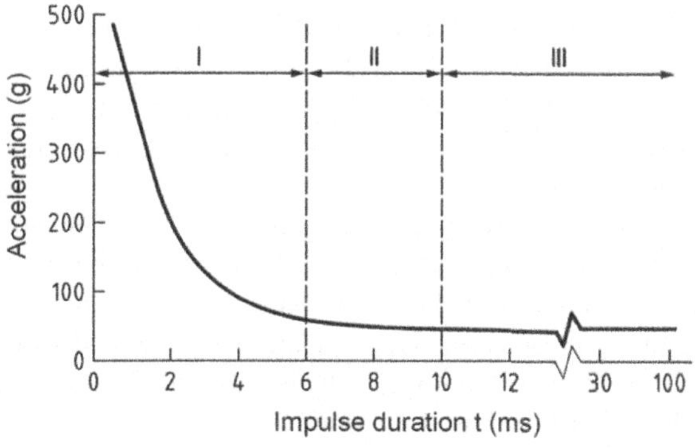

Fig. 12.1 Wayne state tolerance curve the figure is divided into three parts: (1) short duration area, obtained from cadaver experiments; (2) intermediate duration area, obtained from cadaver and animal experiments; (3) long duration area, obtained from volunteer tests. At a given duration, accelerations above the curve give injury, while accelerations below the curve do not lead to injury (Beusenberg 1991)

12.2.4 Head Injury Criterion, HIC

The Wayne state curve as described above led to the development of the Gadd Severity Index (GSI), proposed by Gadd in 1966, which was expressed in the form:

$$\mathrm{GSI} = \int_T a(t)^{2.5} \mathrm{d}t$$

where T = the total pulse duration, and $\alpha(t)$ = acceleration at the center of mass of the head, as a function of time.

This was described as the weighted impulse criteria for which a value of 1000 was considered unsafe. However, it can be shown that for irregular pulse shapes, there may exist within the pulse envelope which has a value greater than that for the whole pulse.

The GSI has received significant scientific criticism, because it deviates considerably from WSUTC (Slattenschek and Tauffkirchen 1970). Thus, it was decided that the maximum value within the pulse should be assumed to be the criterion for head injury. This became the Head Injury Criteria, HIC, which is given below:

$$\mathrm{HIC}\left[\left(\frac{1}{t_2 - t_1}\int_{t_1}^{t_2} a_{\mathrm{res}} \mathrm{d}t\right)^{2.5} (t_2 - t_1)\right]_{\max}$$

with: t_1 and t_2 (ms) any two points in time during any interval in the impact; a = resultant linear acceleration of the center of mass of the head.

After much discussion over many years, t_1 and t_2 were defined to be any two times during the entire impact duration for which HIC is a maximum value. Hodgson and Thomas (1975) suggested that the critical HIC interval should be less than 15 ms, even if the HIC value exceeded the threshold of 1000 over a longer interval. His finding was based on examination of events where the concussive outcomes were known or could be determined. The threshold of 1000 is still under discussion; because head injuries were found at HIC values of 500, while HIC values of 3000 were sustained without major injury. The benefit of HIC over peak linear acceleration is that HIC is related to time, and it is known that pulses with the same peak value but different duration can give a different injury outcome. Unfortunately, HIC and AIS values have never been satisfactorily correlated. Moreover, HIC is based on the linear accelerations of a one mass headform, HIC is based on skull fracture and not brain injury, HIC is not specific to direction of impact, no distinctions about injuries (SDH, SAH, skull failure, DAI), it is based on WSUTC and an important limitation of the HIC is that head rotational acceleration is not taken into account although rotation is debated to be the primary cause for various types of traumatic brain injury, in particular acute subdural haematoma and diffuse brain injury (Adams et al. 1983; Gennarelli et al. 1987; Holbourn 1943).

Moreover, Marjoux et al. (2006) demonstrated that this criterion was poor correlated with observed injuries. During APROSYS SP5 project, HIC prediction was evaluated based on Strasbourg accident database (i.e., 68 real-world accidents). In a very first step global (input), parameters as well as HIC value have been considered in order to evaluate the correlation of these parameters with the occurrence of head injury. When the binary logistical regression method is used (using SPSS software package), it appeared than HIC presents an acceptable correlation with severe neurological injury which means in most of the cases when victims are dead or in coma for a long time. Threshold parameter for a 50% injury risk obtained with the present set of accident is, respectively, 150 G for maximum acceleration and 1500 for HIC. However, correlation of HIC with moderate neurological injury as well as with SDH is poor (Marjoux et al. 2006).

12.2.5 Conclusion

Over the past years, several head injury assessment functions have evolved. Most of them are based on the Wayne state tolerance curve. The most commonly acknowledged and widely applied head injury criterion is the HIC which is based on the assumption that the translational resultant acceleration of the head is a valid indicator of head injury thresholds. This criterion has enabled vehicle safety to be improved. Nevertheless, it has shortcomings and does not take into account rotational acceleration and direction of impact. Furthermore, it is not clear how this injury criterion relates to the (unknown) injury mechanisms.

A final remark concerns the tolerance levels for the injury criteria. The choice of tolerance level depends on the headform, on the application, and on the level of injury risk allowed. For example, the tolerance level for HIC in the helmet standard ECE-R.22 is 2400 using a rigid headform, whereas the tolerance level for HIC in the car crash standard FMVSS 208 (NHTSA 1972) using a Hybrid III headform is 1000. Same maximum value (HIC = 1000) is also required in Circular AC25.562.1b "Evaluation of Seat Restraint Systems and Occupant Protection on Transport Airplanes" calculated with a Hybrid II dummy.

Brain injury is reported to correlate with stress, strain and strain rate (Lee and Haut 1989; Viano and Lövsund 1999). However, strains and strain rates inside the brain (during impact) are difficult to measure. Advancements in computational techniques have led to more accurate and more detailed numerical models of the human head. These models bring a detailed injury assessment closer to reality and at tissue level, since they enable stresses and strains to be examined.

In the next section, a state-of-the-art finite element head model is presented and will be used in order to assess head safety in aircraft at tissue level.

12.3 Head Modeling and Injury Criteria

12.3.1 Human Head Model

Strasbourg University Finite Element Head Model (SUFEHM), which is a 50th percentile FE model of the adult human head, is a state-of-the art finite element head model with enhanced brain and skull material laws (Sahoo et al. 2013a, b).

The main anatomical features included the brain, brainstem, skin, and cerebrospinal fluid (CSF), represented by brick elements, and the skull, face and two membranes (the falx and the tentorium) modeled with shell elements. The SUFEHM presents a continuous mesh that is made up of 13,208 elements, including 1797 shell elements to compose the skull and 5320 brick elements for the brain. The total mass of the head model is 4.7 kg. The geometry of the inner and outer surfaces of the skull was digitized from a human adult male skull to ensure anatomical accuracy. Isotropic, homogeneous, and elastic mechanical constitutive material models were applied to each of the SUFEHM parts except for the brain and skull.

The skull model was improved by using a composite material model which incorporates fracture (Sahoo et al. 2013b). The skull was modeled as a three-layered composite shell representing the inner table, diploe and outer table of the human cranial bone (thickness of 3 mm for the diploe layer and 2 mm each for the two cortical layers). More information about the constitutive law and failure modes is available in Sahoo et al. (2013b). To demonstrate the robustness of the enhanced skull model, various parametric studies were conducted and reported in Sahoo et al. (2015a, b).

The skull and brain mechanical parameters implemented under LS-DYNA is shown in Table 12.1. A detailed presentation of different parts of the SUFEHM is shown in Fig. 12.2.

The brain model was improved by implementing anisotropy based on fractional anisotropy and fiber orientation extracted from medical imaging (DTI) into the brain FE model in order to mimic the main axon bundles as detailled in Chatelin et al. (2013). The mechanical behavior of the brain model has been extensively validated for brain pressure against intracranial pressure data from at different impact locations and brain strain in accordance with experiments (Nahum et al. 1977; Trosseille et al. 1992; Hardy et al. 2001, 2007). The maximum difference of peak pressure was under 5% between simulation and experimental data. Further, this model was used for real-world head trauma simulation in order to derive brain injury criteria.

12.3.2 Accident Database

An important step in the current study is the presentation of the head trauma database which will be used for the simulation of the head impacts and from which brain injury criteria will be derived. The accident database used consists of 109 cases

Table 12.1 Skull and brain mechanical parameters of the SUFEHM implemented under LS-DYNA (Sahoo et al. 2013b, 2016a, b)

Skull mechanical parameters		
Parameters	Cortical bone	Diploe bone
Mass density (kg/m^3)	1900	1500
Young's modulus (MPa)	15,000	4665
Poisson's ratio	0.21	0.05
Longitudinal and transverse compressive strength (MPa)	132	24.8
Longitudinal and transverse tensile strength (MPa)	90	34.8

Brain mechanical parameters		
Matrix	$C_{10} = -1.034$ kPa	$C_{01} = 7.809$ kPa
Fibers	$C_3 = 13.646$ kPa	$C_4 = 4.64 * \text{FA}$
Viscoelasticity	$S_1 = 4.5$ kPa; $S_2 = 9.11$ kPa	$T_1 = 1 \times 10^9\ \text{s}^{-1}$; $T_2 = 6.8966\ \text{s}^{-1}$

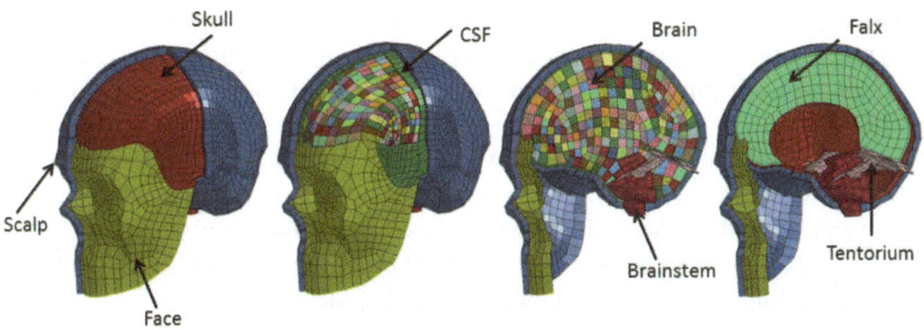

Fig. 12.2 Illustration of the different parts of Strasbourg University finite element head model, the different colors shown for brain represent the 5320 brick elements of brain and have different FA and anisotropy vector

collected from different existing accident databases and involving pedestrian, motorsport, American football player, and motorcycle accidents. In each database, the accident report consists of the final position of the pedestrian and vehicle after the accident, skid marks on the road and vehicle, type of vehicle, vehicle speed, impact position of the pedestrian on the vehicle, and the condition of the road at the scene. The medical report includes the victim's age, gender, height, weight, and details of injuries sustained by the victims. For each of these accidents, the head impact conditions were investigated in partnership with a number of institutions as briefly exposed here after and exposed in more details in Deck and Willinger (2008):

- Six motor sport accidents were collected from Formula 12.1 accidents.

- A total of 11 well-documented motorcycle accident cases were collected from European cooperation in science and technology (COST) 327 accident databases.
- Twenty-two American football impact events were collected.
- Seventy pedestrian accident cases including 15 well-documented pedestrian accidents collected from in-depth Investigation of the Vehicle Accidents in Changsha, China (IVAC) database, 28 pedestrian accidents from the "German In-depth Accident Study" (GIDAS) database, 7 pedestrian accidents from the "Centre for Automotive Safety Research" in Adelaide, South Australia database and finally 12 pedestrian accidents collected from the Tsinghua accident database plus 8 cases collected from the Crash Injury Research (CIREN) accident database.

The 70 pedestrian accident cases were reconstructed in previous studies by using MADYMO software. From this multi-body replication of the pedestrian kinematics, the information about the velocity of the head just before the impact, impact location, and orientation of the head were obtained.

Concerning motorcyclist accidents, American football player impacts and FIA cases, the 3D head acceleration fields were recorded during experimentations which have been done in the COST project for motorcyclist's cases, for American football player accident cases and FIA cases, respectively. For all these experimental accident reconstructions (39 cases), the 6 accelerations versus time curves (3 linear and 3 rotational accelerations) recorded at the center of gravity of the dummy were transferred to Strasbourg University and considered as input for the numerical simulation of the head trauma.

Finally, the 109 head trauma cases were divided into two groups, i.e., with DAI (27%) and without DAI (73%). The distribution of pedestrian head impact locations on the vehicle windscreen for all 70 pedestrian cases is illustrated in Fig. 12.3.

In Fig. 12.3, the different markers represent the injury severity sustained by the victims according to the Abbreviated Injury Scale (AIS).

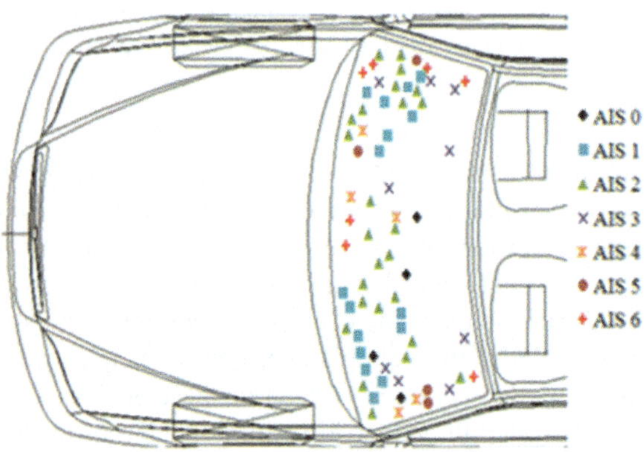

Fig. 12.3 Distribution of pedestrian impact locations on vehicle windscreen for the 70 well-documented pedestrian accident cases

12.3.3 *Head Trauma Simulations and Statistical Analysis*

The methodology for accident reconstruction is composed of several steps, as described in Fig. 12.4. The foremost step is to develop a well-documented database as exposed previously. The next step is to get the victim kinematics. The accident cases were divided into two categories: pedestrian accident cases in which there was a direct head impact against the windscreen and the other accidents cases (motor sports, American football, and Formula 12.1 accident) for which the experimental 6D head acceleration fields were implemented at the center of gravity of the head. For these latter cases, the skull was assumed to be rigid which is acceptable as only helmeted victims are in this group and none of them presented any skull fracture.

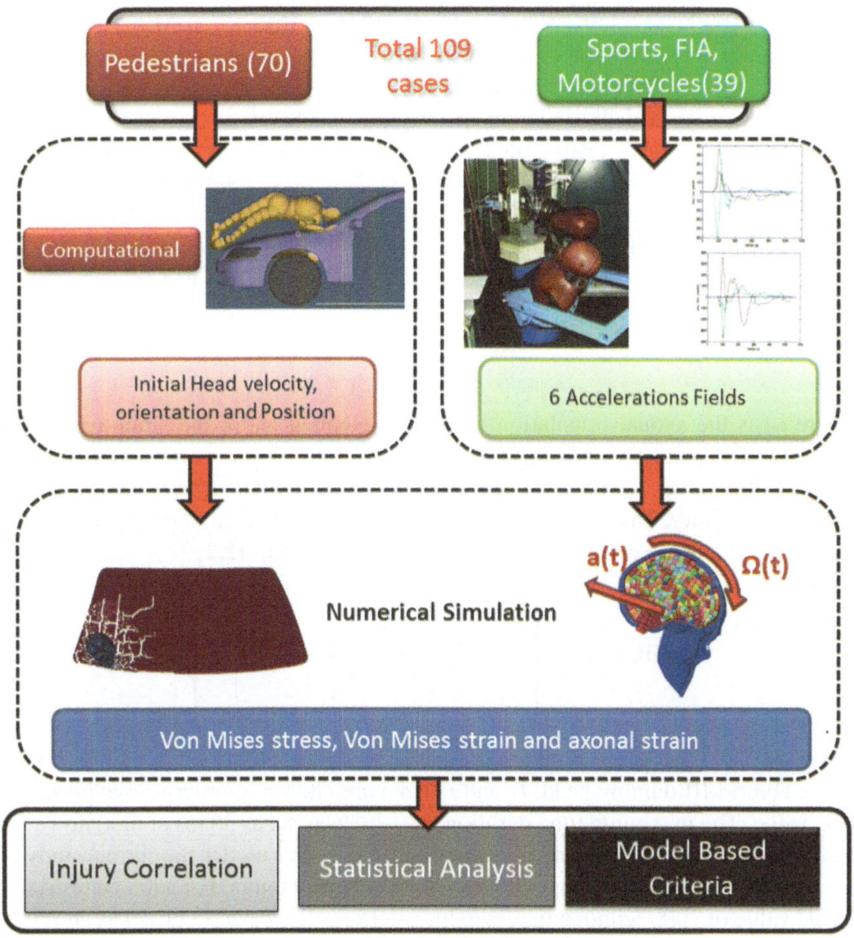

Fig. 12.4 Methodology for numerical head trauma simulation. Pedestrian cases are driven by head orientation and initial velocity, and the other cases are driven by the 6D acceleration curves

For the simulation of the 70 pedestrian head impacts, a realistic FE windscreen model was used as all of the victims impacted the head against the windscreen Deck and Willinger (2008). Proposed was a three-layer composite model (double-layered glass and PBV-tied model). Validation of the windscreen model was performed by comparing the numerical and experimental accelerations at the center of gravity of the headform under controlled experimental impacts as well as glass crack patterns. The head finite element model was propelled against the windscreen model at the same location observed on the accident's scene. The loading condition was the relative head position and the initial velocity between the head and the windscreen at the time just prior to the impact as computed from the victim kinematic simulation. For the cases which did not require a windscreen, the loading condition was obtained by implementation of the 6 head accelerations curves (3 linear and 3 angular accelerations). A total of 39 cases were reconstructed by this method.

For the 109 head trauma simulations, Von Mises stress, Von Mises strain, and axonal strain were computed. The Von Mises stress and Von Mises strain were extracted from the simulation results by LS-DYNA post processing tool. For axon strain calculation, a dedicated Python program was used to extract the Green—St-Venant strain tensor (also called the Green-Lagrangian strain tensor) for each brain element (5320 brick elements) from the LS-DYNA output data. Then these strain tensors are oriented along the main anisotropy direction $\vec{l}_{el}$ of each element to calculate the axonal strains during the total time duration of the simulation for each element. The axonal strain, noted ϵ_{axon}, is calculated based on Eq. 12.1.

$$\epsilon_{\text{axon}} = \left(\overline{\epsilon} \times \vec{l}_{\text{el}}\right) \cdot \vec{l}_{\text{el}} \tag{12.1}$$

where $\vec{l}_{el}$ is the axonal orientation vector per element and $\overline{\epsilon}$ the strain tensor per element. In the current study, head injury criteria (HIC) is also evaluated for all the 109 head trauma cases. HIC is a head kinematics-based injury criteria widely used to access head injury. The head injury criterion (HIC) was related to the work of Gadd (1966) as mentioned earlier and defined in equation in Eq. 12.2.

$$\text{HIC} = \max_{(t_1,t_2)}\left\{(t_2 - t_1)\left[\frac{1}{t_2 - t_1}\int_{t_1}^{t_2} a(t)\mathrm{d}t\right]^{2.5}\right\} \tag{12.2}$$

where a (m s^{-2}) is the resultant linear acceleration measured at the center of gravity of the Hybrid III dummy head. t_1 and t_2 (ms) are chosen in order to maximize the HIC value. The maximum time duration $(t_2 - t_1)$ was set as 36 ms at first; however, current standards use 15 ms and corresponding HIC_{15} was used in the current study. A HIC_{15} of 700 was estimated as a 5% risk of AIS4+ head injury. For adult pedestrians, a HIC value of 1000 within a time window of 15 ms has been proposed as an injury tolerance level for severe head injuries.

In order to define the best suitable parameter to predict DAI, statistical analysis was carried out for the mechanical parameters calculated numerically. The aim of the statistical analysis is to provide a means of assessing the capability of a number of variables to predict brain injury.

Binary logistical regression was used for this assessment and carried out using the version 18.0 release of the statistical software package (SPSS). This method involved fitting of a regression model between possible brain injury metrics (x = Von Mises stress, Von Mises strain, brain axon elongation, and HIC). With this method, the probability of brain injury (DAI) is defined as in Eq. 12.3.

$$P(x) = \frac{e^{a+bx}}{1 + e^{a+bx}} \tag{12.3}$$

where a and b are two parameters calculated by regression.

The candidate parameters (Von Mises stress, Von Mises strain, axon elongation, and HIC value) were then compared using the Nagelkerke R^2 statistic (where the limits for this measure are 0 for a poor fit and 1 for a good fit) to determine which head injury metric provides the best correlation with the occurrence of brain injury.

12.3.4 Derivation of Injury Criteria

Head trauma simulations of 109 well-documented accident cases were performed by using an advanced FEHM under LS-DYNA platform. The results in terms of maximum of Von Mises stress, Von Mises strain, and axonal strain with occurrence or not of brain injury are shown in Figs. 12.5, 12.6 and 12.7. The white columns represent the cases without DAI, and the black columns represent the cases with DAI. The min–max range for Von Mises stress, Von Mises strain, and axonal strain are 21–205 kPa, 0.006–2.47, and 0.02–0.43, respectively. It appears from Fig. 12.7 that

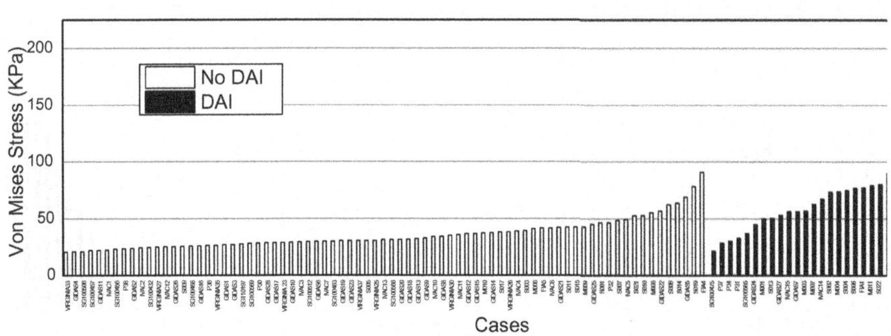

Fig. 12.5 Von Mises stress calculated for all the accident cases reconstructed using an advanced FEHM

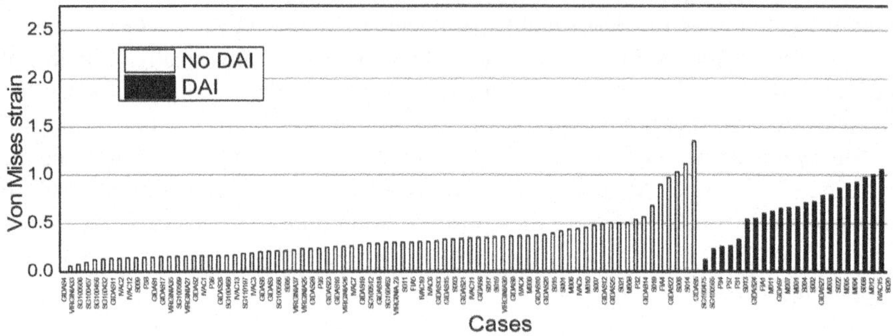

Fig. 12.6 Von Mises strain calculated for all the accident cases reconstructed using an advanced FEHM

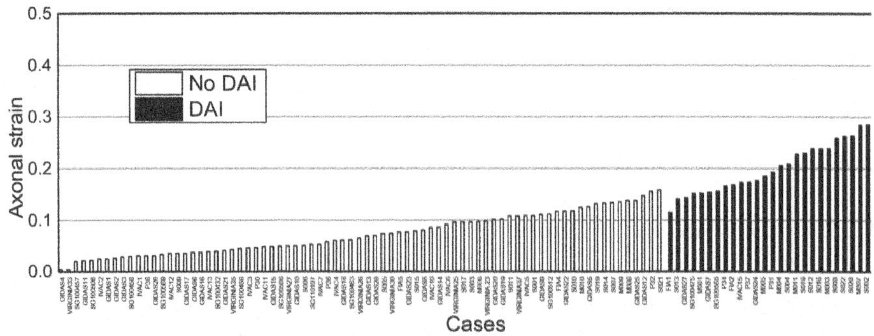

Fig. 12.7 Axonal strain calculated for all the accident cases reconstructed using an advanced FEHM

axon elongation presents low overlap of injured/non-injured cases. This observation will now be objectively confirmed by the statistical analysis.

The HIC was also calculated for 109 head trauma cases as shown in Fig. 12.8. The maximum time duration $(t_2 - t_1)$ was set as 15 ms, and corresponding HIC_{15} was computed according to Eq. 12.2. It is observed from Fig. 12.8 that there is a

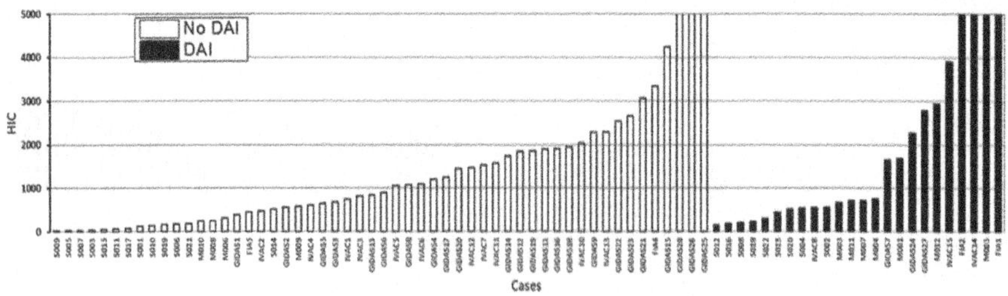

Fig. 12.8 HIC value calculated for all the head trauma accident cases

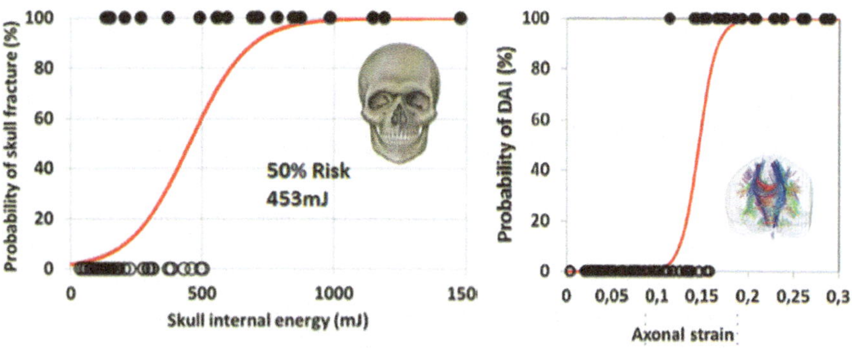

Fig. 12.9 Injury risk curves to predict probability of skull fracture (**a**, left) by addressing skull strain energy and moderate DAI (**b**, right) (moderate diffuse axonal injuries which means loss of consciousness AIS2+, possible reversible brain injury) by addressing axon strain

considerable overlap of injured/non-injured cases by taking in to account HIC for predicting DAI.

The statistical analysis for the 109 simulated head traumas is carried out by using binary logistical regression as exposed in the previous section. The candidate parameters for predicting injuries (DAI) along with their Nagelkerke R^2 value were calculated. It appears that axonal strain presents the highest R^2 value with a value of 0.876, whereas for Von Mises stress, Von Mises strain, and HIC, the Nagelkerke R^2 values are 0.502, 0.446, and 0.055, respectively. Hence, axonal strain is the most suitable parameter to predict DAI.

Based on this statistical analysis, the injury risk curve for predicting DAI is plotted as shown in Fig. 12.9b. The solid black circles represent the cases for which injury occurred, and the white circles are representative of non-injured cases. From this injury risk curve, the critical axon elongation for a 50% risk of DAI is obtained and corresponds to an axon strain of 0.1465. Injury risk curves to predict probability of skull fracture by addressing skull strain energy are obtained with a similar methodology and reported in Fig. 12.9a.

12.4 Conclusion and Applications

A state-of-the-art finite element head model with enhanced brain and skull material laws which enable it to compute axon elongation was used for numerical simulation of real-world head trauma to develop a robust head injury criterion. A total of 109 well-documented TBI cases were simulated, and axon elongations were computed to derive a brain injury tolerance curve. Based on statistical analysis, it was shown that axonal strain was the appropriate candidate parameter to predict DAI. The proposed brain injury tolerance limit for a 50% risk of DAI has been established at 14.65% of axonal strain. The proposed threshold in terms of axonal strain is in accordance with studies reported in the literature. Head kinematics-based metric (HIC) has also been computed for all head trauma cases to compare its capability to predict brain injury with the new axon strain-based metric. Results demonstrated very poor correlation of HIC with injury compared to axon strain. It is therefore concluded that the present study provides realistic methods and tools for advanced model-based head injury risk assessment and mitigations.

The developed head injury prediction tool can be applied in a "full FE approach" when a FE model of the protective system exists. In this case, the head–structure interaction can be simulated and it is possible to assess the head injury risk directly for this simulation. For protected head impacts, when the skull fracture is unlikely to appear it is possible also to implement a "coupled experimental versus numerical approach." In this case, the head 6D kinematic is recorded from the experience and considered as the initial condition of the head impact simulation as shown in Fig. 12.10. This coupled experimental versus numerical method permitted also to propose novel helmet test methods which consider oblique impacts with the recording of the 6D headform kinematic and followed by the numerical assessment of the brain injury risk as shown in Figs. 12.11 and 12.12.

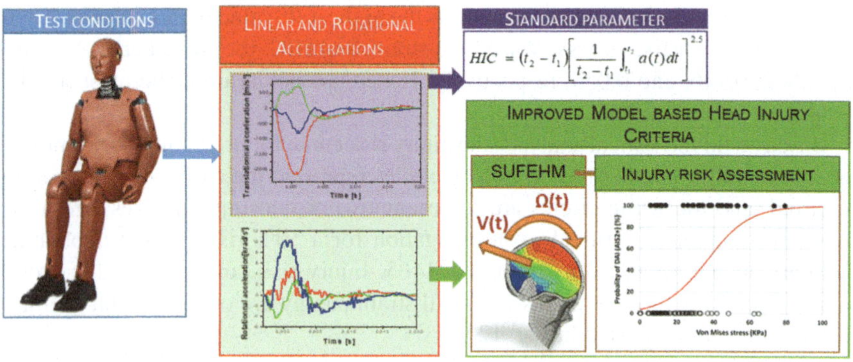

Fig. 12.10 Illustration of the coupled experimental versus numerical test method

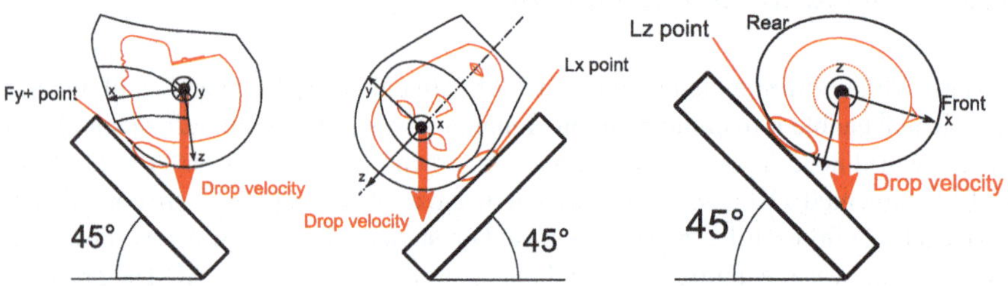

Fig. 12.11 Illustration of the three tangential impact conditions introducing angular acceleration around Y axes (called F_{y+} and F_{y-}), X axes (called L_x) and Z axes (called L_z)

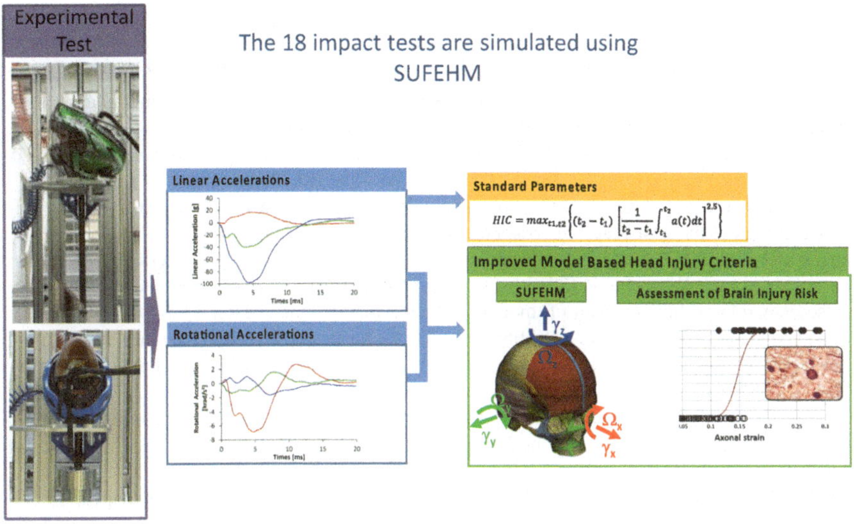

Fig. 12.12 Illustration of the new helmet test method: a total of 18 linear and oblique impacts are conducted and simulated with the model-based head injury prediction tool

References

Adams JH, Graham DI, Gennarelli TA (1983) Head injury in man and experimental animals: neuropathology. Acta Neurochir 32(15):30

Chatelin S, Deck C, Willinger R (2013) An anisotropic viscous hyperelastic constitutive law for brain material finite element modeling. J Biorheol 27(1–2):26–37

Deck C, Willinger R (2008) Improved head injury criteria based on head FE model. Int J Crashworthiness 13(6):667–678

Gadd CW (1966) Use of a weighted – impulse criterion for estimating injury hazard. In: Proceedings of the 10th STAPP car crash conference, pp 164–174

Gennarelli TA, Thibault LE, Tomei G, Wiser R, Graham D, Adams J (1987) Directional dependence of axonal brain injury due to centroidal and non-centroidal acceleration. In: Proceedings of the 31st Stapp car crash conference. SAE paper No. 872197, pp 49–53

Got C, Patel A, Tarrière C, Walfisch G (1978) Results of experimental head impacts on cadavers: the various data obtained and their relations to some measured physical parameters. In: Proceedings 22nd Stapp car crash conference, 24–26 Oct 1978, Ann Arbor, MI, Society of Automotive Engineers, Warrendale, PA, pp 57–99

Gurdjian ES, Webster JE (1945) Linear acceleration causing shear in the brain stem in trauma of the central nervous system. In: Mental advances in disease, pp 24–28

Gurdjian ES, Lissner HR, Evans FG (1961) Intracranial pressure and acceleration accompanying head impacts in human cadavers. Surg Gynecol Obstet 112:185–190

Hardy WN, Foster CD, Mason MJ, Yang KH, King AI, Tashman S (2001) Investigation of head injury mechanisms using neutral density technology and high-speed biplanar X-ray. Stapp Car Crash J 45:337–368

Hardy WN, Mason MJ, Foster CD, Shah CS, Kopacz JM, Yang KH, King AI, Bishop J, Bey M, Anderst W, Tashman S (2007) A study of the response of the human cadaver head to impact. Stapp Car Crash J 51:17–80

Hodgson VR, Thomas LM (1975) Head impact response. VRI report 7.2 SAE

Holbourn AHS (1943) Mechanics of head injuries. Lancet 2:438–441
Lee MC, Haut RC (1989) Insensitivity of tensile failure properties of human bridging veins to strain rate: implications in biomechanics of subdural hematoma. J Biomech 22(6/7):537–542
Lissner HR et al (1960) Experimental studies on the relation between acceleration and intracranial pressure changes in man. Surg Gynaecol Obstet 111:329–338
Marjoux D, Baumgartner D, Deck C, Willinger R (2006) Head injury prediction capability of the HIC, HIP, SIMon and ULP criteria. In: Proceedings of the IRCOBI conference, pp 143–158
Nahum A, Smith R, Ward C (1977) Intracranial pressure dynamics during head impact. In: Proceedings of the 21st Stapp car crash conference. SAE paper no. 770922
Patrick LM et al (1963) Survival by design - head protection. In: 7th Stapp car crash conference
Sahoo D, Deck C, Willinger R (2013a) Development and validation of an advanced anisotropic viscohyperelastic human brain FE model. J Mech Behav Biomed Mater 33:24–42
Sahoo D, Deck C, Yoganandan N, Willinger R (2013b) Anisotropic composite human skull model and skull fracture validation against temporo-parietal skull fracture. J Mech Behav Biomed Mater 28:340–353
Sahoo D, Deck C, Yoganandan N, Willinger R (2015a) Influence of head mass on temporo-parietal skull impact using finite element modelling. Med Biol Eng Comput 53(9):869–878. https://doi.org/10.1007/s11517-015-1295-6
Sahoo D, Deck C, Yoganandan N, Willinger R (2015b) Influence of stiffness and shape of contact surface on skull fractures and biomechanical metrics of the human head of different population under lateral impacts. Accid Anal Prev 80:97–105
Sahoo D, Deck C, Willinger R (2016a) Brain injury tolerance limit based on computation of axonal strain. J Accid Anal Prev 92:53–70
Sahoo D, Deck C, Yoganandan N, Willinger R (2016b) Development of skull fracture criterion based on real world head trauma simulations using finite element head model. J Mech Behav Biomed Mater 57:24–41. https://doi.org/10.1016/j.jmbbm.2015.11.014
Slattenschek A, Tauffkirchen W (1970) Critical evaluation of assessment methods for head impact applied in appraisal of brain injury hazard, in particular in head impact on winshields. In: 1970 International automobile safety conference compendium, pp 1084–1112
Trosseille X, Tarriere C, Lavaste F, Guillon F, Domont A (1992) Development of a F.E.M. of the human head according to a specific test protocol. In: Proceedings of the 36th Stapp car crash conference. SAE paper no. 922527

Chapter 13
Urban Transport Planning in the Age of Global Warming

Hermann Knoflacher

13.1 Introduction

13.1.1 48 Years After "Global Warning"

"Limits to Growth" was the first report to the Club of Rome analyzing physical parameters of human activities by comparing them with the available resources of the globe. (Meadows et al. 1972) The report was translated in all major world languages. Deane and Dennis Meadows have shown that the existing exponential growth of population, use of resources, growing air pollution cannot continue in a limited world. It was the first global warning for humankind that the continuation of resource use and energy wasting of the industrial society cannot be sustainable (Fig. 13.1).

Radical changes in the system were recommended, but not accepted by politicians, experts and the society. If the politicians had accepted the report and acted accordingly, a sustainable future of the globe would have been possible after 1972. But this did not happen. Today, we know that there is no sustainable future for the human society any more. The global system is in a state of overshoot and collapse. Professor Dennis Meadows explained in many lectures and publications that changes in the next 20–30 years will be much bigger than the changes we had during the last 100 years. Through the exploitation of fossil energy, mankind has succeeded in exceeding the ecological carrying capacity of the planet. The ecological footprint as an indicator for human activities is 30% over the carrying capacity of the globe (Wackernagel and Beyers 2010) (Fig. 13.2).

This is only possible due to the enormous use of fossil fuel. The industrial civilization of today is depleting every year an amount of fossil energy stocks which

H. Knoflacher (✉)
Institute for Transport Planning and Traffic Engineering, Technical University of Vienna, Vienna, Austria
e-mail: hermann.knoflacher@tuwien.ac.at

G. Tiwari and D. Mohan (eds.), *Transport and Safety*, Springer Transactions in Civil and Environmental Engineering,
https://doi.org/10.1007/978-981-16-1115-5_13

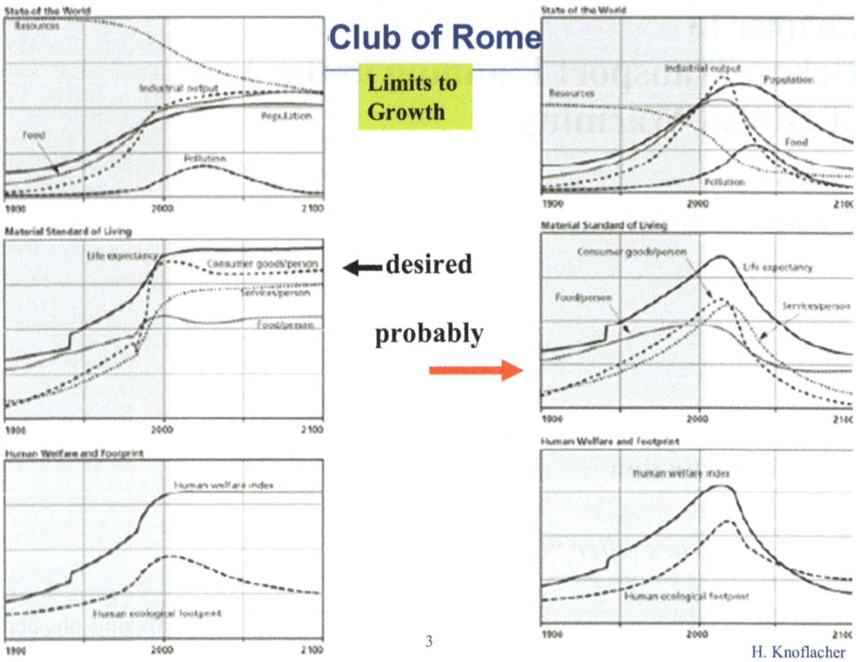

Fig. 13.1 Human and global indicators for different scenarios, analyzed by Dana and Dennis Meadows. Left side: sustainable development, if all recommendations for system change would have been implemented after 1972; Right side: the most probable future under prevailing conditions (Meadows et al. 1972)

have been stored in the globe for hundred million of years. The amount of energy released to the atmosphere as heat is greater than the release of energy as heat from the atmosphere to the outer space (Fig. 13.3).

Carbon dioxide, methane and other gases contribute to the global land–ocean temperature anomaly since the 1950s. Studies from the IPCC give the evidence of changes in the climate system. The mean surface temperature anomaly between 2001 and 2005 has never happened during the history of mankind as world wide effect. Land precipitation has changed significantly. Increases of precipitation in the north and decreases of precipitation in the south, where it is mostly needed, have been observed in the last decade. Since 2007, the surface temperature of the earth has continued to rise (IPCC 2019) (Fig. 13.4).

13.1.2 The City

The city is a man-made environment, developed over the last 7000 years of human history. The precondition for a city is a developed social system among a group of

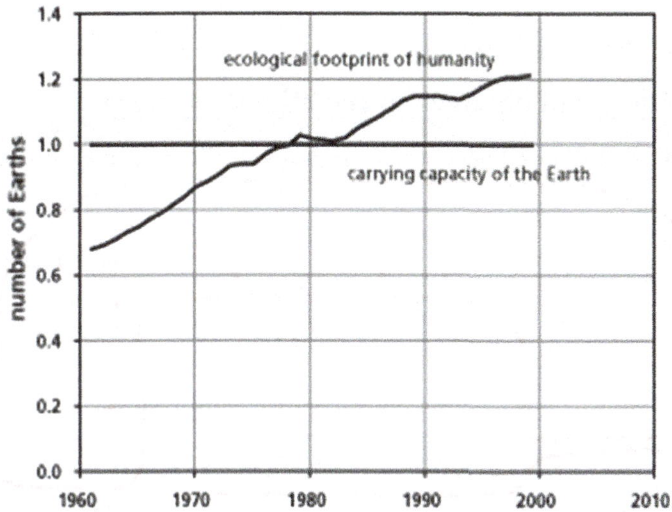

Fig. 13.2 Ecological footprint has exceeded the theoretical maximum carrying capacity of the global system about 30 years ago and is now 30% above the carrying capacity of the Earth (the real carrying capacity has gone down due to human activities) (Wackernagel and Beyers 2010)

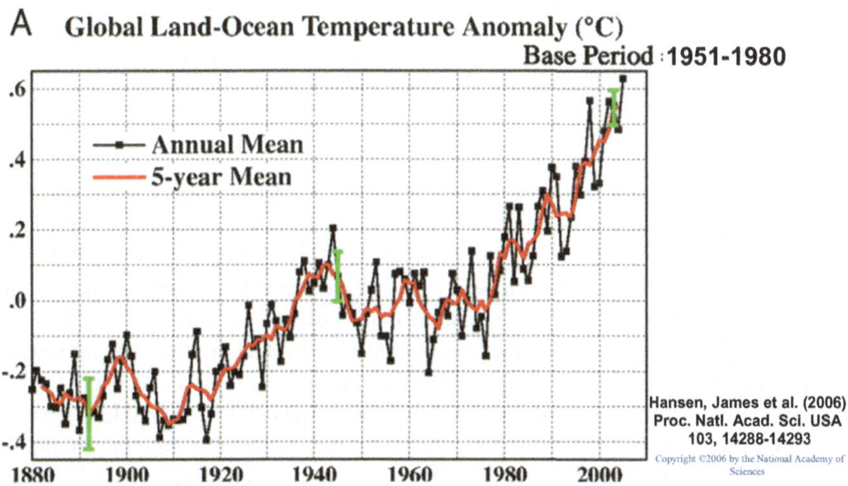

Fig. 13.3 Global land–ocean temperature 1880–2004 (The NASA Goddard Institute for Space Studies 2009). Source: Hansen James et al. Global Temperature Change. Fig. 1. Proceedings of the National Academy of Science of the United States 2006

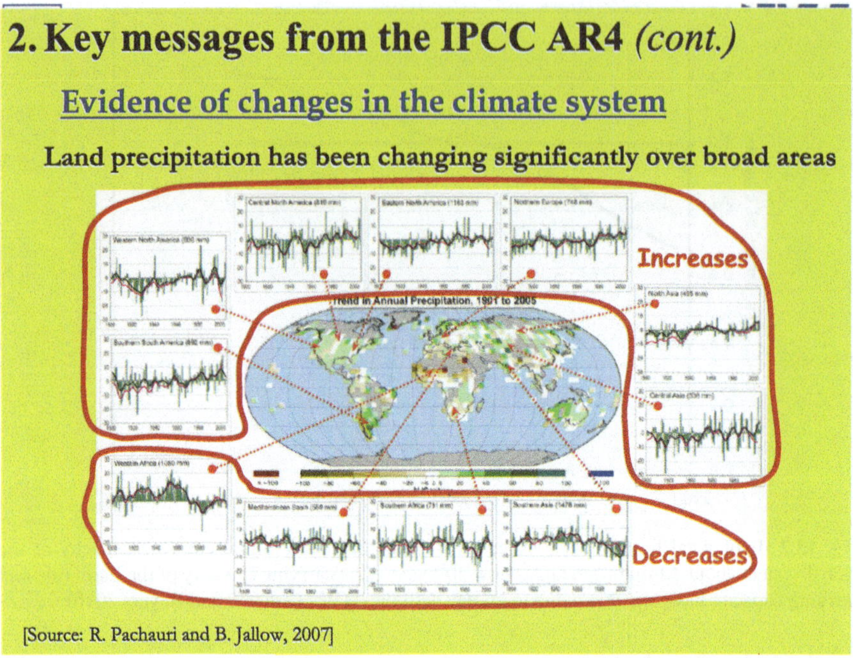

Fig. 13.4 Changes in precipitation (IPCC 2007). Source: National Center for Atmospheric Research Staff (Eds). Last modified 27 Feb 2020. The Climate Data Guide: GPCC: Global Precipitation Climatology Centre

people much greater than a family, a tribe or a clan, the urban society. Whenever we look at a historical urban structure, we find similarities in the pattern of the urban fabric, (Fig. 13.5 shows the city map of Venice with the street network), the built environment, the network of public space determined by the scale of the pedestrian.

It is the network of streets and places, the expression of human scale, whether the cities are in Asia, Europe or an other continent, where urban civilizations appeared. Maps from cities built before engine-driven transport systems have been implemented, show a similar scale in their network of streets and comparable distances between places (Koch 1990). It is also interesting that during this time the wealth of the society was distributed equally worldwide. There were no extremely rich and no extremely poor regions or states till 200 years ago.

Figure 13.6 shows an image of Gent city in the middle age.

There have always been large differences between poor and rich, as in any society, but globally, the mean value of wealth was fairly evenly distributed until the nineteenth century. The major part of the population lived in rural areas, and only a minority was living in cities.

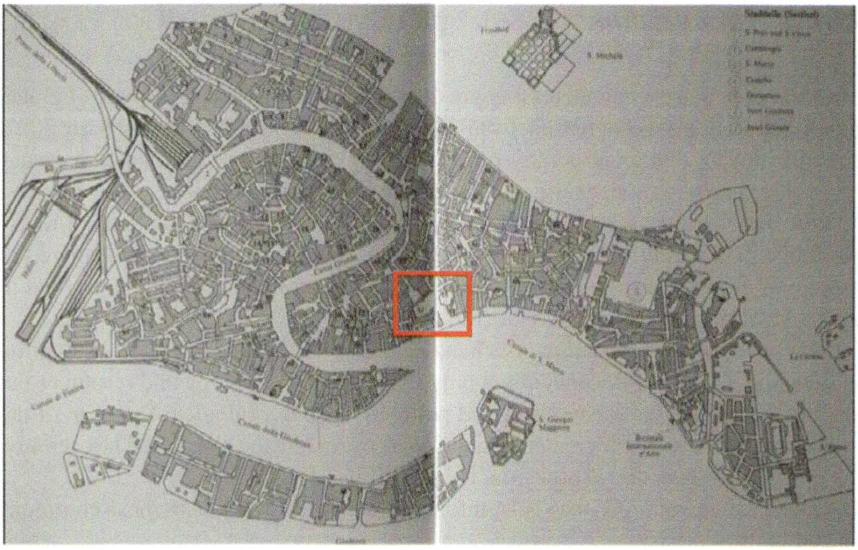

Fig. 13.5 Map of Venice, the car-free city since many hundred years

Fig. 13.6 Human scale city: image of Gent in the middle age

13.1.3 Urbanization

The last century shows an increasing growth of cities worldwide. The share of global urban population has increased from less than 30% to more than 50% in 2007. Urbanization of this scale is only possible, if the population in the cities has easy and cheap access to food, energy, freshwater and clean air. But this requires efficient, reliable and fast means of transport that provide access to resources that are not in the immediate vicinity of the cities. These transport chains may be endangered in the age of global warming or other influences, a question that has not been asked in urban and transport planning so far (Fig. 13.7).

This process of urbanization goes hand in hand with the total energy consumption of mankind. Cities are dependent not only on the inflow of energy and resources but also on the outflow of waste water and polluted air. The ecological footprint of the urban population is much greater than the area covered by the city. In the last century, the energy use shows an exponential increase (Figs. 13.8 and 13.9).

Global urbanization is in a self-reinforcing interrelationship with these enormous energy consumption.

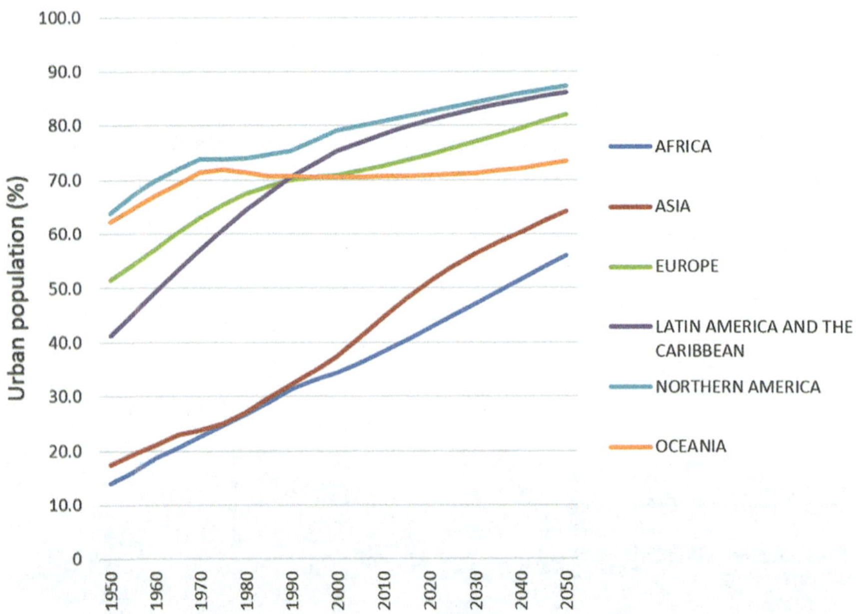

Fig. 13.7 Rapid increase of urban population since 1950 (Norton 2016). Source: Levels and trends of urbanization in selected countries (Population Facts UN Department of Economic and Social Affairs. Nr. 1/2018)

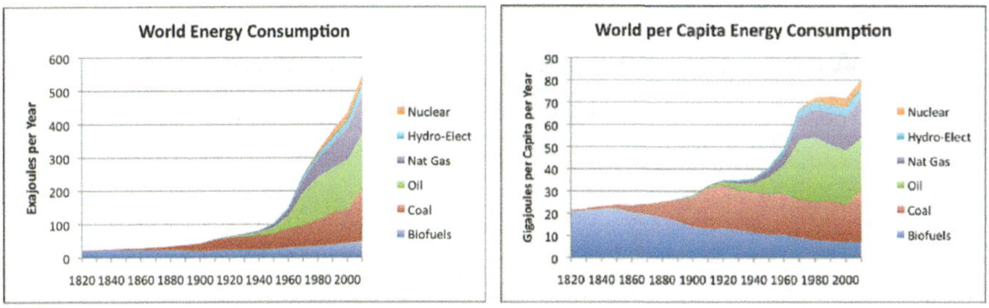

Fig. 13.8 World energy consumption total, left and per capita and year right (BP Statistical Data and Maddison Data) https://ourfiniteworld.com/2012/03/12/world-energy-consumption-since-1820-in-charts/. World Energy demand and outlook till 2050. (Source: Namit Sharma, Bram Smeets, and Christer Tryggestad McKinsey Quarterly April 2019

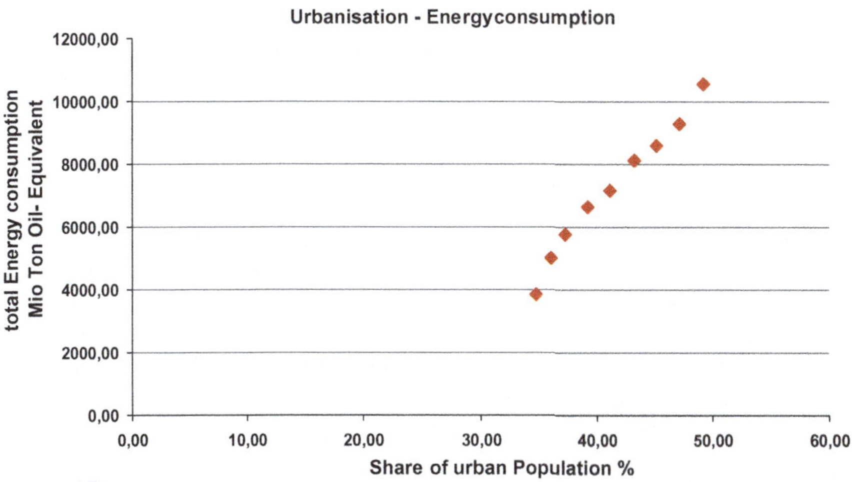

Fig. 13.9 Share of urban population goes hand in hand with fossil energy consumption (chart: author). Source: Graph Author, Data https://en.wikipedia.org/wiki/World_population

13.1.4 Political Indicators and Their Effects on the Globe

For national politicians, an indicator has become central importance since about 50 years: the Gross National Product, expressed in money terms as US Dollars. The global economy is driven by fossil fuel in the industrial age (Fig. 13.10).

Understanding systems, money is nothing else than energy, GDP is highly correlated with energy use of the society. It is therefore not astonishing that global temperature increases with the GDP. The politically targeted increase in GDP has therefore led to the emission of CO_2 in the atmosphere, which can no longer be absorbed in the natural cycle. One consequence of this is global warming, to which transport directly contributes about one third.

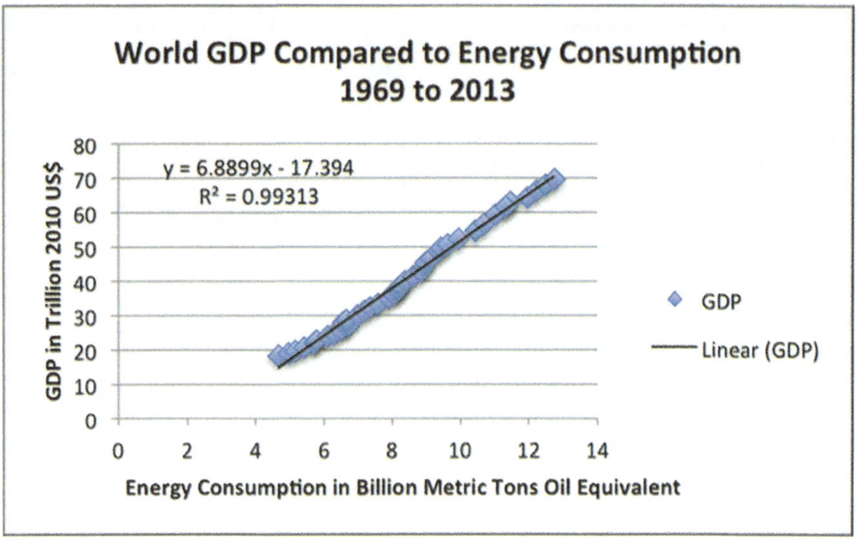

Fig. 13.10 Global energy consumption—global GDP (BP 2014). Source: Correlation of Energy Consumption and GDP. European Environmental Agency Feb. 2015 https://www.eea.europa.eu/data-and-maps/figures/correlation-of-per-capita-energy

The correspondence of Fig. 13.11 with the global energy consumption (Fig. 13.8) shows the effect on climate change. If politics continues to orient itself exclusively toward GDP growth without decoupling it from the consumption of fossil energy, this will inevitably lead to further global warming. Due to the massive urbanization since the last century, cities and thus urban transport play a central role.

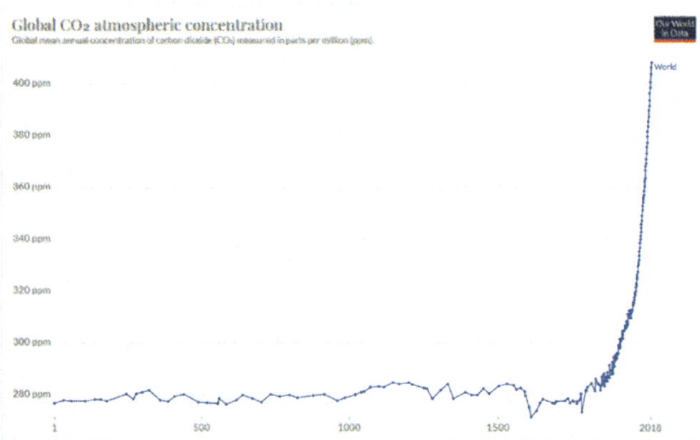

Fig. 13.11 CO_2 and green house gas emissions (Ritchie and Roser 2017)

Fig. 13.12 GDP and happiness. Figure taken from Layard (2005)

13.1.5 What is the Main Goal of Economic Activities?

The overall goal of each economic activity can only be the satisfaction of peoples need, the "happiness" of people. Capitalism assumes that happiness is increasing with the amount of available money. This is true only for the subsistence level and not further (Layard 2005, 2020) (Fig. 13.12).

The relationship between GDP, the scale for economic activities and happiness of people is logarithmic. The happiness indicator of the countries within a wide range of individual income (GDP/capita) is not significantly different, showing that income is determined by other variables from a certain value on. Indian people are as happy as people in Spain, although their relative income is much lower. Bangladesh, Nigeria, Brazil and China have happiness indicators in the range of southern European countries. Peru and South Africa have much lower happiness indicators compared to countries mentioned above due to their political system and the local situation. It is not GDP what matters, and it is happiness of people—at least over a certain minimum level of income.

13.1.6 The Transport System

Is the close relationship between income and car ownership a fate? With increasing income in most countries of the world, an increase of car ownership can be observed. One of the core hypotheses of traditional transport is based on this relationship (Kopits and Cropper 2003). This seems to be a kind of "law", if there were no exceptions. But there are two city states which demonstrate that the increase of income can be decoupled from the increase of car ownership (Fig. 13.13).

A very low level of car ownership income has not prevented people from Hong Kong and Singapore to have a high level of income, comparable with the rich industrial countries. In both "city states", the administration was able to organize a set of rules to keep the car ownership low and to provide the society with other means of mobility like public transport, in accordance with a good land use policy in order to save money, space and energy. These two city states prove that the linear relationship between wealth (GDP) and motorization is not a law of nature and must therefore be questioned. Due to the uncontrolled development of road transport, this sector uses more than two third of oil in the USA (Fig. 13.14).

The contribution of the transport system to global warming is not only energy to operate vehicles but also fossil energy to build, operate and maintain the transport infrastructure as well as the residential and service sector. Due to the effects of car traffic on the urban structures, urban maintenance cost, services for road lightning and other parts of infrastructure are increasing more than proportionally. Today, we are in a situation that oil discovery rates are much less that the rate of use (Fig. 13.15).

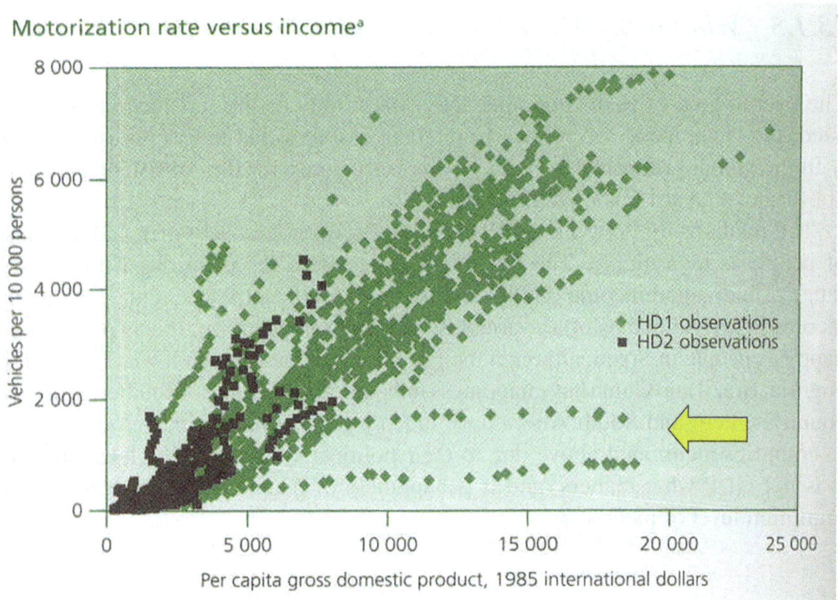

Fig. 13.13 Per capita GDP—motorization (Kopits and Cropper 2003)

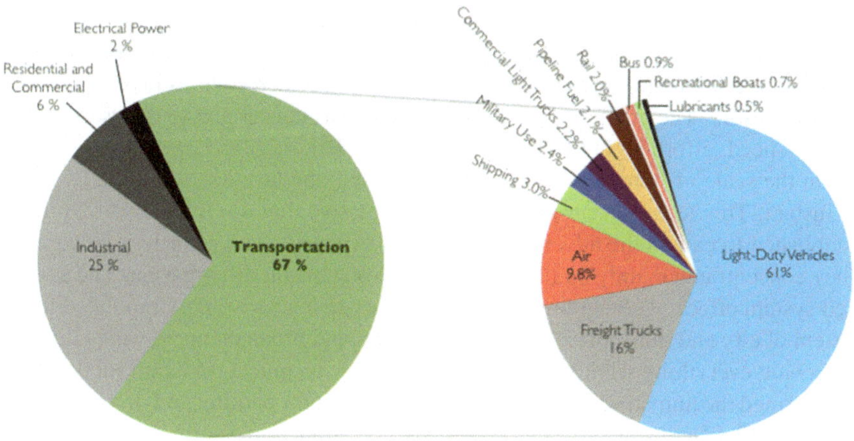

Fig. 13.14 US fuel consumption by sector 2004. Transportation uses more than 2/3 of oil in the USA (data from Ratner and Glover 2014)

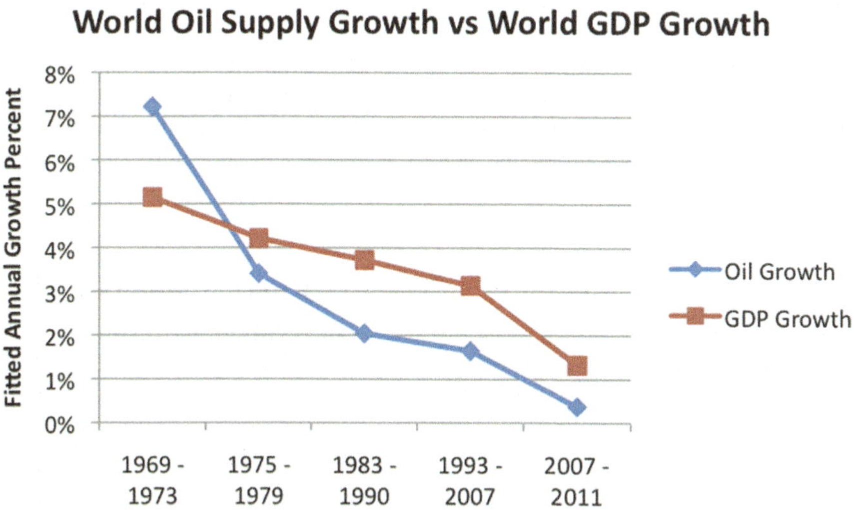

Fig. 13.15 Increasing difference between discovery and oil consumption (Tverberg 2012)

13.1.7 From Fascination to Addiction

In order to understand the effects of technical means of transport on people, their evolutionary experience with spatial mobility must be considered. For six million years, humans have been mobile as pedestrians, an energy-intensive form of transport. Few have been able to afford the luxuries of riding and draught animals in the last 2000 years. The desire for effortless mobility was therefore a human dream.

Experts, politicians and society were fascinated by the opportunities offered by the new technical modes—the rail and especially the car. The cities got new opportunities to reach the countryside and therefore human and material resources in big quantities as never before. The breakthrough for the society was the car giving the users access to high speed, effortless at the same time (Fig. 13.16).

On the scale of human evolutionary experience, the last 200 years are less than an instant. This technical progress was so impressive that disciplines were quickly established to extend the advantages of spatial expansion and effortless movement. They concentrated mainly on the car. Mobility soon meant driving a car. The associated system effects were ignored, and personal experiences were extrapolated to the system of city and transport. Traditional transport and urban planners were educated to provide everything what was needed for the new technical transport mode. They transformed the human scale city into an environment for parking and driving private cars (Fig. 13.17).

Transport and urban planning was not made any more for people, and transport and urban planning was made for vehicles, machines—first the rail-based public transport system and in the twentieth century for the car. Car traffic became the center of planning. Observations of car traffic and personal experiences became basic assumptions for planning.

The research was mainly interested in diligent traffic and road capacity and concerned with the description of car traffic manifestations, developed models and the design of roadways. System effects beyond these technical limits were only considered when serious problems such as accidents and later noise and exhaust pollution arose.

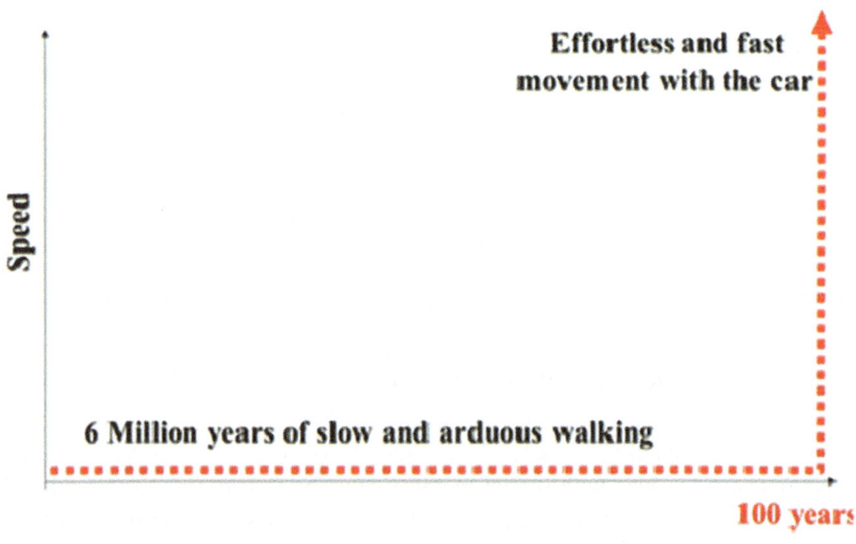

Fig. 13.16 On the scale of evolution of human being, the increase of travel speed happened in an instant during the last 100 years. Source: Drawing Author=Knoflacher

Fig. 13.17 What was the reason that people convert the public space into a place for cars, when motorization begin? (city Pristina 1995, photo: author)

"Growth of mobility", *"time savings through speed"* and *"freedom of choice of transport mode"* were taken for granted and not questioned. In Europe, but also in the USA, there were increasingly contradictions between the expectations raised and the real system behavior that needed to be investigated. Data from household surveys showed no increase in the average number of trips with increasing motorization. Worldwide, this figure is around three trips per person per day. It was also found in the 1970s that neither the travel time budget nor the distribution of travel times of the modes of transport provided evidence of a reduction in mobility time (Zahavi 1974; Knoflacher 1986; Schaffer et al. 2000). The scientific study of the system behavior of city and traffic proved that there is no increase in mobility due to more car traffic if it is defined in a meaningful and purposeful way. Because every road always has a purpose! There was also no evidence of a decrease in mobility time worldwide (Schafer et al. 2000). When the speed increases, only the distances increase because the structures change (Knoflacher 1986).

The effects are known and visible: urban sprawl on the one side and concentration of economic activities—mostly outside of the city (Fig. 13.18).

High speeds of the individual transport system created the unlimited city, the city without borders, with a plenty of space for cars, but not people on the streets. Public space was converted into carriageways cutting the urban living structure into pieces. The advantages of this development for the residents were obvious, living in a private

Fig. 13.18 Urban sprawl and high rise building characterize the car-city. Source: Photos author

home. The disadvantages came later, because apart from private homes there were no complementary functions nearby. Today, one begins to recognize this undesirable development and strives for improvement.

13.1.8 Future Outlook

The future outlook concerning global warming can be at least estimated and is not at all encouraging. To prevent a horrible future for human kind, it will be necessary to reduce the amount of fossil energy in industry, transport and households. We will also be forced not only to reduce the destruction of natural habitats through urban sprawl and deforestation, but also to revitalize them.

13.1.9 Consequences for the Urban Transport System

The choice in scientific terms is not difficult: Nonmotorized transport system users are more than hundred times more energy efficient than car users. All kinds of urban public transport are at least ten times more energy efficient than cars (Figs. 13.19 and 13.20).

Pedestrians, cycling and public transport also use public space one to two orders of magnitude more efficiently than car traffic.

In view of global warming, urban transport must significantly reduce its energy consumption and land requirements.

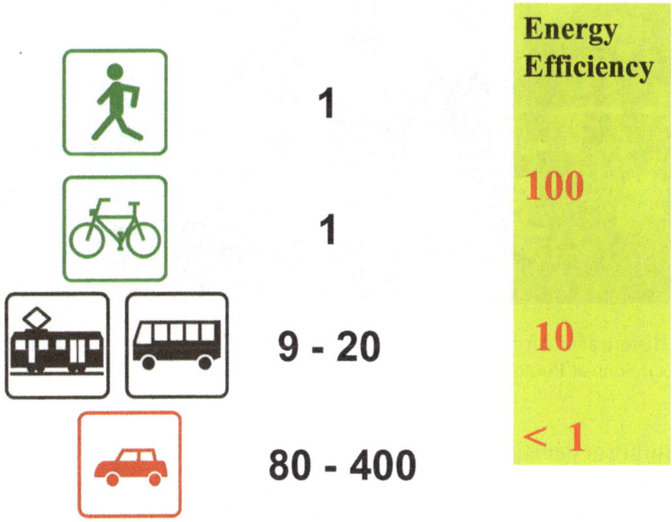

Fig. 13.19 Energy efficiency of existing urban transport modes. Source: Drawing Author

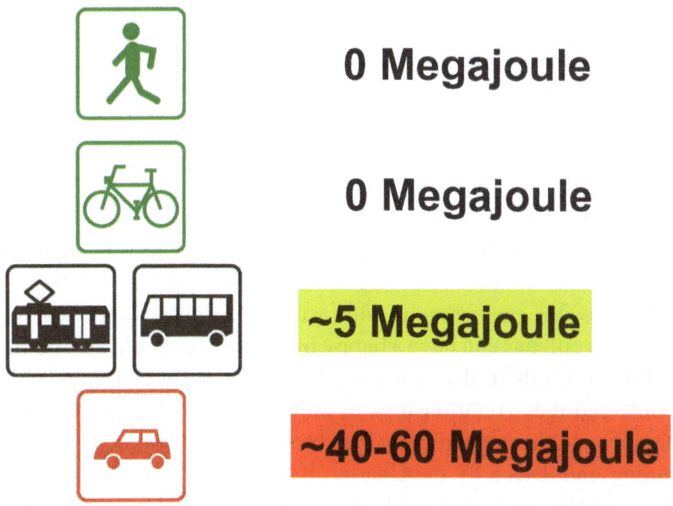

Fig. 13.20 Fossil energy use of common urban transport modes (megajoule per trip). Source: Drawing Author

13.1.10 *The Big Advantage of Urban Economy in Developing Countries*

The economy in Delhi in the walled city is probably one of the most efficient in the world. The turnover in money terms per square meter in the small shops is at least as

Fig. 13.21 Historical urban center of Delhi, the walled city is a model for a sustainable city (photo credit: author). Source: Photo Author

big as the turnover per square meter in western shopping centers. Since the transport system in Chandni Chowk is mainly based on pedestrians and cyclists, the efficiency is more than hundred times better than in the western car-oriented societies. This kind of economy exists since many hundreds of years. The shops in this city center still correspond to the ideal ideas of a fair functioning market (Fig. 13.21).

Only the most efficient urban transport system pedestrians, cyclists and rickshaws in combination with public transport are sustainable in the long term. It is the prototype for the future economy of urban structures as studies show also in western cities.

The energy consumption in transport per unit of goods in car-oriented shopping centers is ten to twenty times more, compared to local shops in urban street. These are results from a western society with a degree of motorization of 600 cars and more per thousand inhabitants (Fig. 13.22).

In order to cope with global warming, the city must offer its inhabitants protection from high temperatures and security against extreme weather events. A good city must provide a secure environment with a dense social network where children, women and senior citizens can have a happy live. The city should also be competitive and satisfy human needs in the best way. The demand for space and the speed of car traffic hindered these relationships in the existing historical parts of the cities, and the complex and diverse structure was broken down into monofunctional units. Urban planning, influenced by the Charter of Athens, divides the city according to monofunctional units into residential areas, commercial zones, traffic areas in two dimensions, which used to be three dimensional.

If the city is to be prepared for climate warming, the planning task is to redesign the transport system to make it more climate-friendly and to give preference to modes of transport that do not have a negative impact on the climate. Pedestrians, cyclists and public transport must once again be given priority over cars. Car traffic especially in the age of global warming has to be necessary only for special services.

For 7000 years, all city streets were pedestrian streets. In the last 100 years, the functions of public space were converted into priority space for fast moving vehicles.

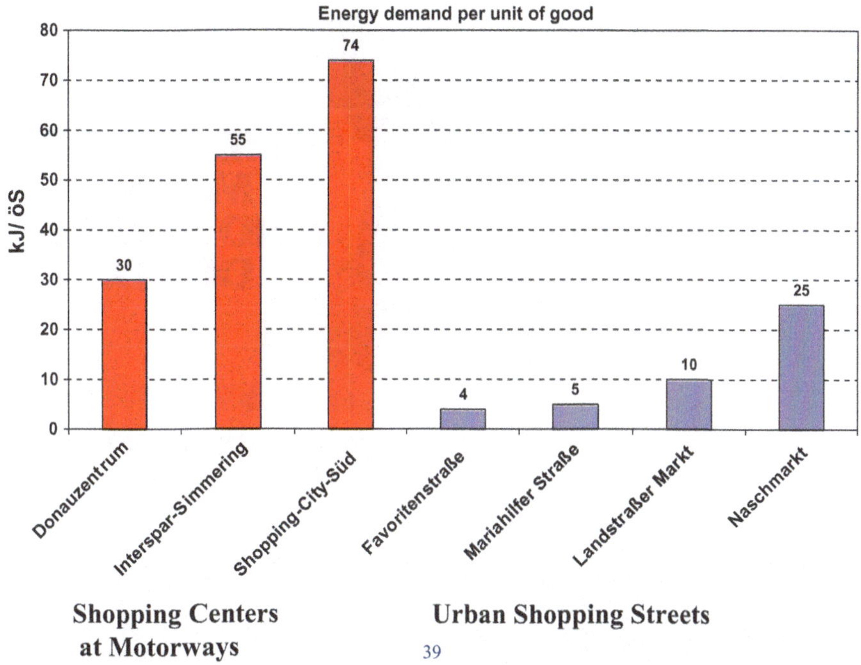

Fig. 13.22 Shopping centers outside of the city, compared to comparable shops in urban streets, need a much high transport energy input per item bought. Source: Author survey. Knoflacher 1997 Untersuchung der verkehrlichen Auswirkungen von Fachmarktagglomerationen Wien, MA 18 1997

In a place in which one false movement can kill the people, probably nobody would like to live (Fig. 13.23).

Is this the city you like to live in? Is this the place in which your children would like to live? Is this the place for your age time? Is this the result of what is called "progress"?

But this is exactly the public space of today. Different from other challenges such as health or education, urban transport has not improved with economic development, and it became even worse (Fig. 13.24).

These and many other examples—in Vienna, for example, a church was demolished to make way for car traffic—raise the question of what happened that led to this change in urban and transport planning. Car traffic has cut the city into parts and destroyed the urban fabric and the urban functions. People living in a street with less than 200 vehicles per peak hour have 7.6 acquaintances and three friends on the other side. With 2000 vehicles per hour, they have less than one friend on the other side and not more than three acquaintances on the other side of the street (Fig. 13.25).

Fig. 13.23 Traffic safety campaign poster. Source: Adams John, RSA Lecture Hypermobility too much of a good thing 2001

Fig. 13.24 Buildings of the old city of Shanghai were cleared away to make way for this traffic system. The former center of Shanghai had a dense street network (Map of Shanghai 1933). Source: Photo Autor

13.1.11 Unintended Effects of Car Traffic on Human Behavior

Donald Appleyard has demonstrated that the destruction of the internal relationships is vital for a vital city and that is created even by small amounts of car traffic. *Car traffic is cooling social relationships by heating up the atmosphere.* A further effect of global proportions, to which car traffic, which hinders active forms of mobility in cities, has contributed, is the obesity of the motorized population: *Car cities produce fat people* (Fig. 13.26).

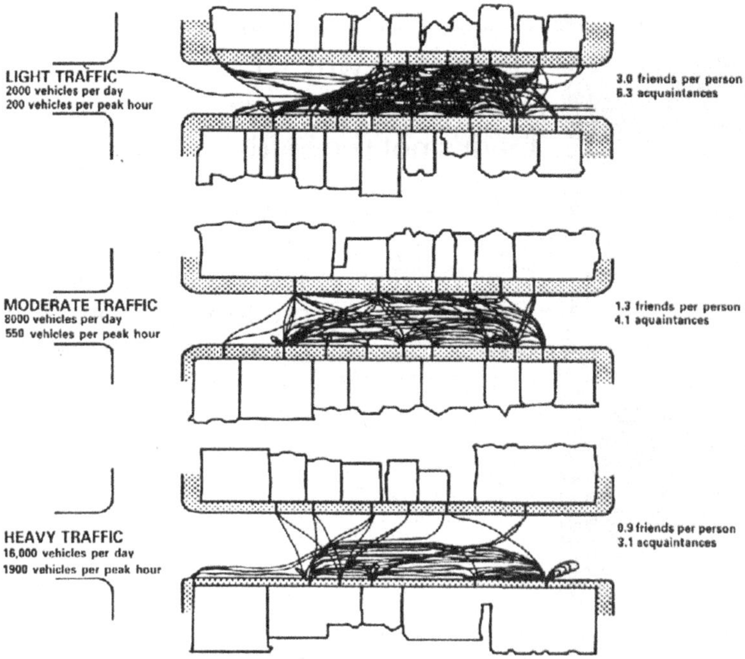

Fig. 13.25 Increasing car traffic—less friends and less acquaintances on the other side of the road (Appleyard 1981)

Decades of uncontrolled suburban sprawl conceived around the motorcar have left the people unable to walk, even if they wished to go. If the social network in the cities is destroyed, crime will grow. This has been proved by analyzing car increase and crime increase in Great Britain between 1903 and 1995 (Whitelegg 1994) (Fig. 13.27).

The rate of crime increased in accordance with the rate of car ownership. When human's eyes and ears do not control the public space since it has become an area of air pollution and noise, public space becomes unsafe. The society is going in an Orwellian society. As car traffic grows, so does crime.

Elevated motorways and flyovers are transport structures to damage the urban fabric and devaluate urban life. They are promoting the car-city and a driving force for global warming (Fig. 13.28).

The isolated treatment of car traffic without due consideration of the effects on other road users and on the city as a whole is shown by calculations based on car units. Starting from the elementary transport equation, the flow of cars per unit of time is determined without considering the causes of these car trips (Fig. 13.29).

The basic formula, to calculate traffic flow, is valid in the cross section of a road. This calculation is used to justify the need for the construction and upgrading of roads. It only takes a few years until a city highway, which was built to avoid traffic

Fig. 13.26 Car mobility makes people immobile, not only in the USA. Source: Adams John, RSA Lecture Hypermobility too much of a good thing 2001

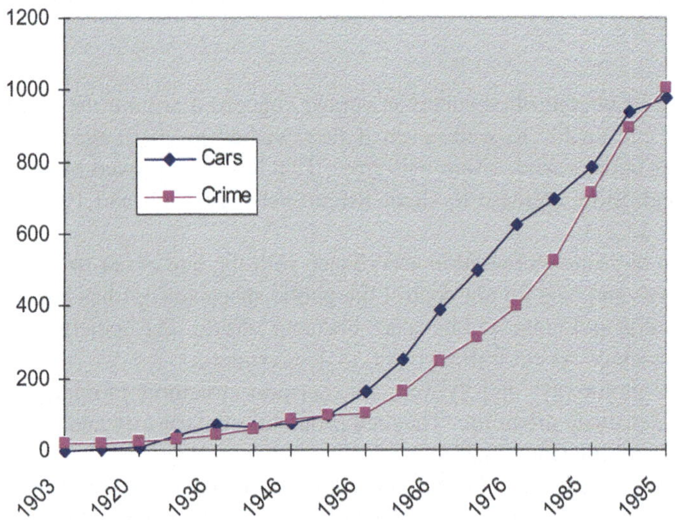

Fig. 13.27 Cars and crime in Great Britain 1903–1995 (1951 = 100). Source: Adams John, RSA Lecture Hypermobility too much of a good thing 2001

Fig. 13.28 Elevated motorways and flyovers damage the urban fabric, create social degradation and cut the urban body into pieces. Source: Photo Author

Measure to remove a local Bottleneck

$$Q = D \times V$$

short term

Fig. 13.29 Basic formula to calculate traffic flow (Q measures number of vehicles per hour, D is number of vehicles per km, and V is km traveled per hour). Source: Author

jams, does not only not prevent traffic jams, but itself becomes the source of an even bigger traffic jam (Fig. 13.30).

In the system, the same formula is valid. If local density is reduced, the car system becomes more attractive. This creates more car users, driving with higher speed in the system, which finally creates more car kilometers and ends up in congestion (Fig. 13.31).

This is how the problem treatment becomes the cause of the problem. As a side effect, these measures damage the urban fabric and finally the city. Global warming as a consequence is inevitable.

The urban transport system in the age of global warming must be much more intelligent and less energy consuming. Traffic engineering must use problem-solving, instead of problem-creating measures (Fig. 13.32).

Pillars from elevated motorways like in Shanghai and other cities will be wondering what the society who has created these structures thought. It will be

Fig. 13.30 Since decades, one tries to solve the traffic jam problem by even more road construction and therewith, one only creates the mega jam in the cities from Los Angeles to Beijing, from Stockholm to Bangkok

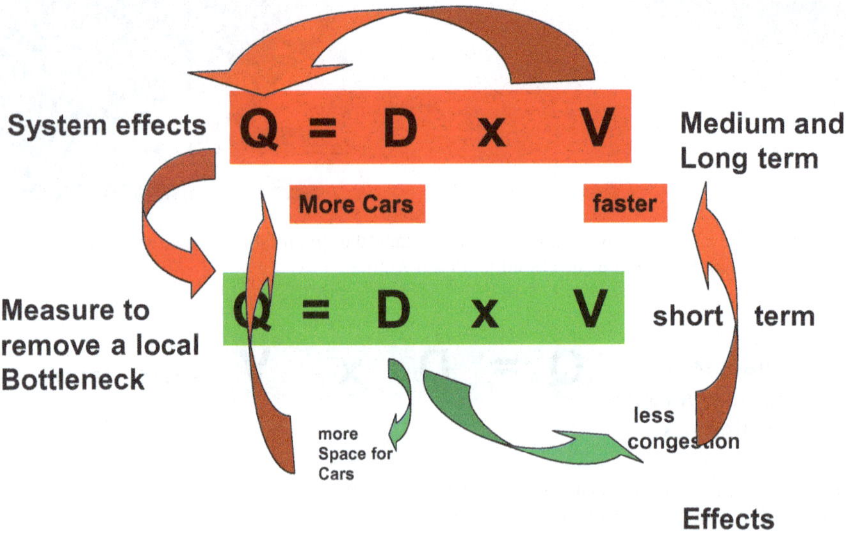

Fig. 13.31 Feedback effects from local measures on the transport system, which create increasing congestion. Source: Author

Fig. 13.32 What will people think, when they uncover pillars from this infrastructure 500 years from now? Source: Photo author: https://www.bing.com/images/search?q=osterinsel+photos&id

a view on our society as we have it today to the Eastern Island monuments. The Eastern Island society was not able to understand the natural environment in which they lived. They cut palm trees faster than they could regenerate. Finally, they were trapped on an Island without being able to escape from it. The Eastern Island civilization and the culture disappeared—as much as we know in a terrible degradation. Considering the exponential acceleration of climate change, it is urgently necessary to prepare the existing urban and transport structures for the future in a more climate-friendly way. Walking, cycling and public transport are better suited to this scenario than private car traffic because of their characteristics.

Concerning spatial efficiency, these three modes pedestrians, cyclists and public transport use a similar amount of public space. Cars are also in this respect totally inefficient (Fig. 13.33).

Pedestrians, cyclists and public transport users have a high social connectivity. Car users are isolated. The transport system for cities in the age of global warming must serve a dense city with at least 150–300 people per hectare. A city of this structure has not enough space for car traffic.

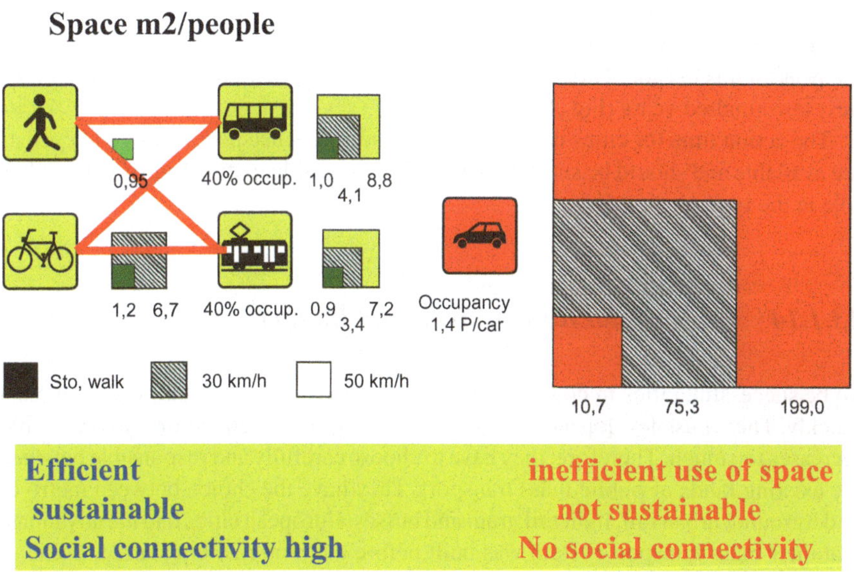

Fig. 13.33 From the point of rational science, the answer is simple. The modes on the left side are much closer to sustainability than the car on the right side. Source: Author

13.1.12 The Competitive Advantage of Developing Countries

The so-called developed countries and their cities have built an extended infrastructure for a transport system, which does not fit into the future. The degree of motorization is far too high and therefore not sustainable. Since the city-damaging effects of car traffic are known in European cities, they try to restrict car traffic and promote pedestrians, cycling and public transport. Countries with a low degree of motorization have the big competitive advantage if they can prevent fundamental mistakes happened in the western society by copying the American transport system. But there is a danger for the cities in the developing world: the exponential increase of car ownership. Cars and motorized two-wheelers are increasing exponentially, also in Delhi. The forecast for different regions of the world shows that Asia has the expected greatest growth of cars.

13.1.13 The Race Against Time

Fossil energy use started in 1930 in the transport sector in a big scale and has reached the peak in this decade. Global warming is now on the way and will increase in the next few hundred years (Fig. 13.34).

The action time for cities is therefore very limited. Since pedestrians and cyclists are available and should be supported as much as possible, public transport has a key role in the transition period.

13.1.14 The Crucial Role of Available Budget

To be successful, cities in countries with a low degree of motorization should react quickly. They must develop an efficient public transport system before car ownership increases too much. Therefore, they have to choose carefully and responsible between the existing kinds of public mass transport. They have the choice between metro or underground, urban rail, light rail, tram and busses. European cities had the advantage that their mass transport system was built before the motorization started. Some of the cities made the mistake to demolish the public transport system in the era of blind car enthusiasm imported from the USA. Since they have recognized that this was a fundamental mistake, they invest no into new tramlines, bus priority lines and restrict car traffic (Fig. 13.35).

Concerning construction costs of urban public transport, metro is the most expensive mode, and busses are the cheapest. It is therefore of strategic importance to choose the right mode to win the race against increasing car ownership. Compared with the metro, a bus rapid transit system is 20–100 times cheaper and therefore faster to implement. Modern rapid bus systems, if well planned and operated, have the same

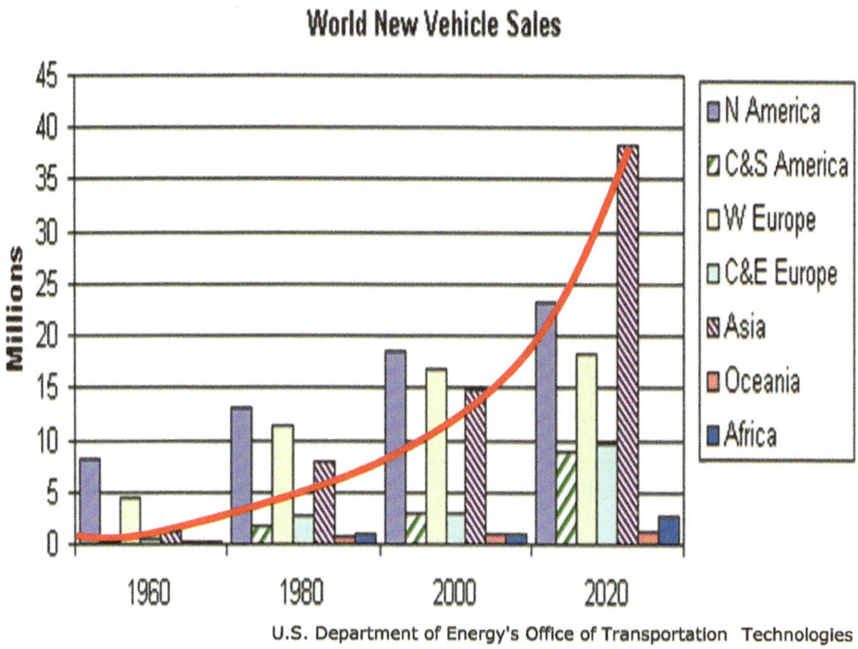

Fig. 13.34 Danger for cities in countries with low degree of motorization: the fast increase of car ownership

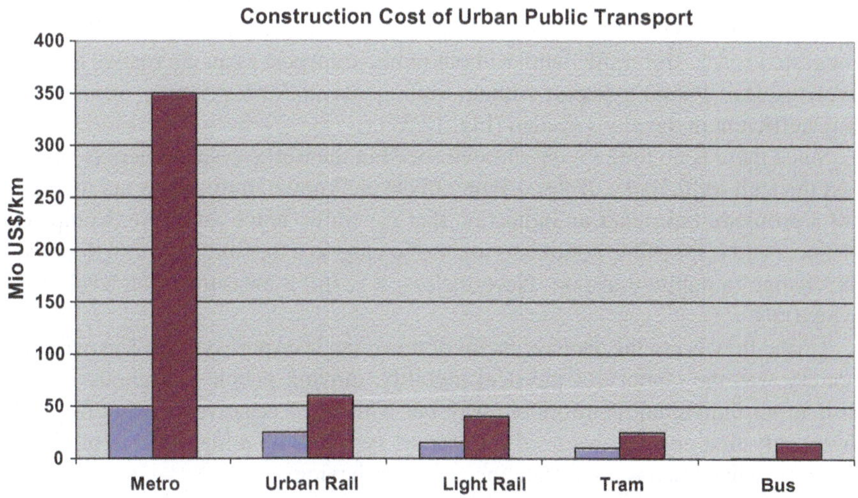

Fig. 13.35 Construction cost of different urban public transport modes (data from finalized projects, min. left, max. right)

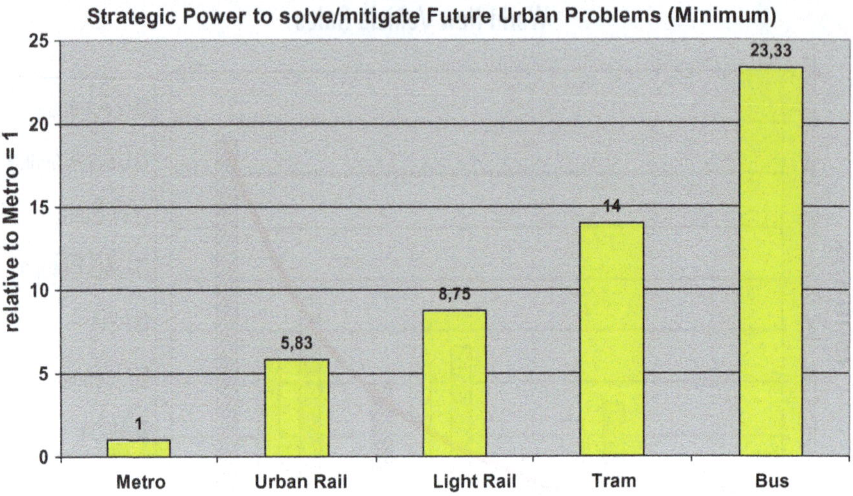

Fig. 13.36 "Strategic efficiency" of public urban transport modes (metro = 1). Source: Data and graph author

capacity as the metro. They only require the courage of planners and politicians to remove the traces of car traffic (Fig. 13.36).

If we compare the cheapest metro with the rapid transit system, BRT is 23 times cheaper and therefore faster to be implemented than the metro. The data from Delhi gives a figure from 1 to 25. To make public transport efficient, car traffic needs congested roads. But traffic jams without public transport as an alternative are relatively useless. Public transport without traffic jams and/or restrictions on car traffic is not efficient in the city's system (Fig. 13.37).

Since there is no time saving through speed in the traffic system, there is no time loss through traffic jams. If the system effects are known, traffic jams are therefore not a problem, but rather an indication that car traffic is too attractive. Competent planners and responsible politicians use traffic jams as a measure, as a tool to support the change to public transport. Nevertheless, it is still a measure at the level of the symptoms.

Traffic flow is not the decisive factor of the urban transport system. The twentieth century was the century of physical mobility, moving people and goods from A to B in an individual, comfortable way but with huge negative system effects. In the twenty-first century, we need a transport system with a high degree of mental mobility, which means understanding system behavior.

The history of mankind is characterized by energy limitations, which force the people to find clever solutions. Social mental mobility was therefore necessary, the precondition for the life in villages and in cities. Social systems are based on social mobility. This reached its peak before the industrial age. With technical transport systems, spatial mobility became so easy that it was no longer necessary to think about solving the problem locally, because the solution could always be found somewhere

Fig. 13.37 Congestion in car traffic is the environment for a good operation performance of public transport on separate lanes (e.g., Bogota). Source: Bogota BRT-System Photo Enrique Penelosa 2005

far away. Either the problems were shifted to faraway places, like exhaust fumes or garbage, or the resources needed were obtained from distant areas. The car was a welcome means of getting rid of the problem, at least in the short term. But the effects that cars had on our behavior were not considered. They are also not to be found in the disciplines that deal with the planning of cities and technical traffic systems.

13.1.15 The Cause of the Urban Traffic Problem: The Bond Between People and the Car

At the beginning of motorization, there was initially an interrelationship between car-using people and the car, which spread to the whole population with full motorization and changed life and the cities. Through the car, the human being, who as a pedestrian has to expend 0.1–0.2 hp of body energy for mobility, gets the feeling of being superhuman through the enormous power of the engine of 200 hp and more, so that he does everything to maintain this state. Associated with this is the feeling of power and speed with simultaneous effortlessness. An experience never before experienced in the history of human evolution. Even science and planners could not escape this fascination, nor could society. They rebuilt the world for the needs of the car (Fig. 13.38).

This has not only effects on the transport system; this also changes the inner structure of people totally. This can only be understood in the context of evolutionary epistemology. Cars are the invention of the twentieth century, the continuation of

Fig. 13.38 Unification of man with the engine of the car was not understood. Source: Author

human evolution into a technical evolution. But when people start to use the car, the car gets in contact with their oldest level of evolution, the level of body energy. (I use here the description as it was developed for a more vivid perception by the biologist, evolutionary researcher and epistemologist Rupert Riedl (1925–2005) with whom I could work together for many years.) (Figs. 13.39 and 13.40).

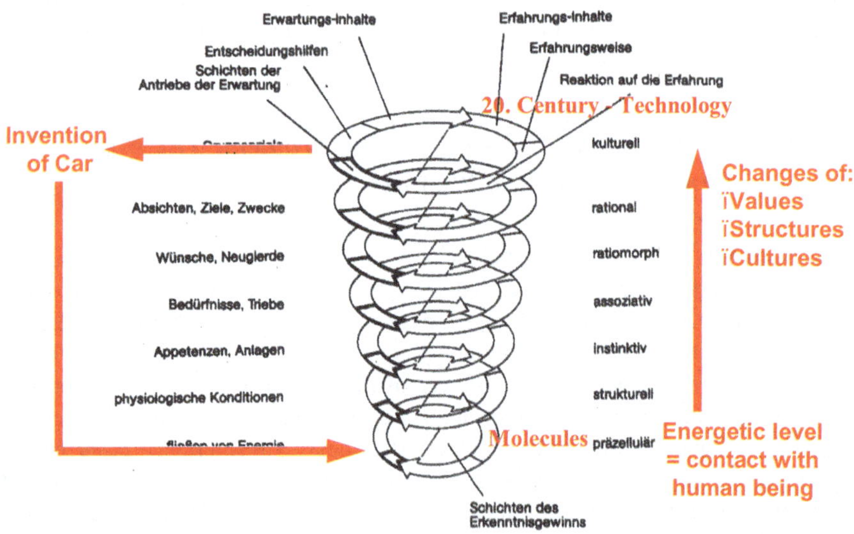

Fig. 13.39 Car as an invention of the newest level of human evolution gets in contact with the oldest level of human evolution—this changes everything. Source: After Riedl (1981)

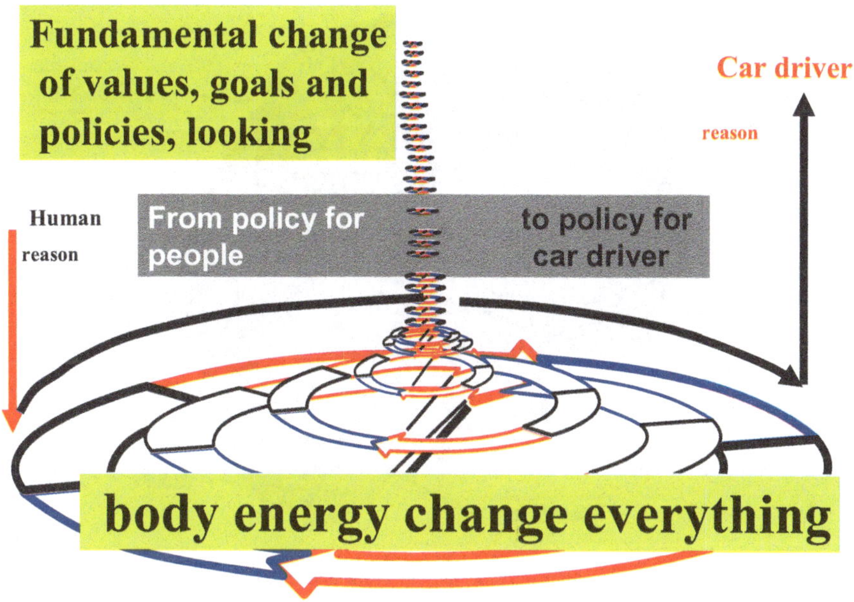

Fig. 13.40 Deep subconscious level, in which the car becomes active, changes everything above. It even changes perception of people, their values and their culture. Source: Author

This level of body energy is twisted toward the needs of cars, which are in general opposite to the needs of people.

Since the oldest levels of evolution are the strongest ones, they are moving everything above them. The outcome of this effect can only be understood in the interdisciplinary view (Riedl 2019). Since this also happens to politicians, they make a policy for car drivers and not a policy for people any more (Fig. 13.41).

Cars have therefore a far reaching effects on the society and the people. Through the deep-seated effect of the car on the level of body energy, it also changes perception in such a way that the abnormality of behavior is considered normal. The following photos are intended to make it easier to understand and evaluate the use of public "urban space" (Fig. 13.42).

This effect of the car on our behavior shows that the flow of traffic is only a symptom and the cause of the problems dealt with here lies where we come into direct contact with the car: the car park.

If the parking of cars is organized in the same way as before, so that the car gets a parking space at each starting point and each end point of a path, a physically based connection to the car is created. This bond is stronger than to all other things in the environment and surroundings, which can be observed in all cities.

This also explains why people also travel very short distances by car. Research also shows that in a car-free environment the acceptance of footpaths increases significantly (Knoflacher 1981, 1986, 2007) (Fig. 13.43).

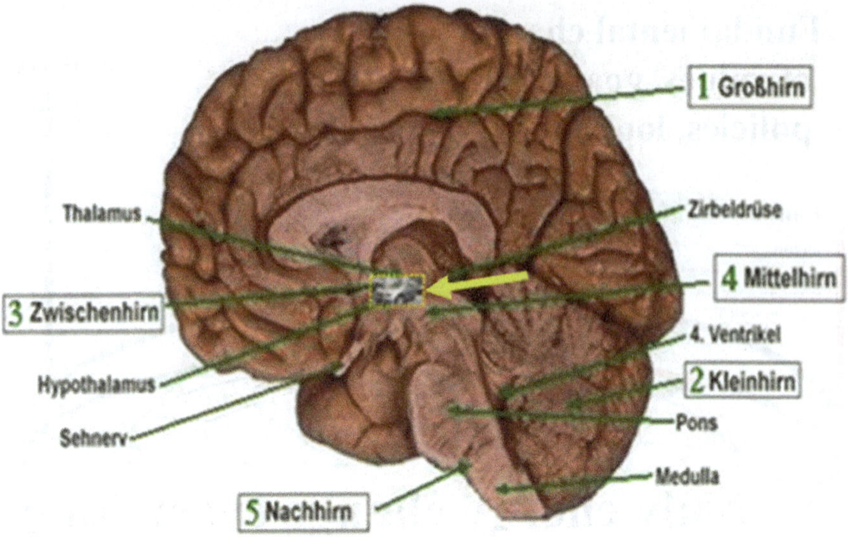

Fig. 13.41 Car goes directly deep into the human brain and controls human behavior. Source: Graphic author

The organization of car parking spaces determines the modal split and thus the climate impact of traffic.

If cars are parked at home, the city falls apart into monofunctional structures. Working, shopping, recreation and housing can be separated from each other. The human habitat, the urban fabric is thus changed in such a way that it is no longer adapted to the scale of the people, but to the car. There is no chance for public transport in this kind of cities (Fig. 13.44).

The necessary change in the transport infrastructure is therefore a reorganization of parking. *Cars can only be parked in specific places, which have to be organized in such a way that the walking distance to the parking places or parking garages is at least as long as the walking distance to the public transport stop.* If you only want to have the same chance to choose a car and public transport, the way to and from the parked car must be longer than the way to the public transport stop.

This is the minimum requirement for the free choice of transport for the population. In practice, however, the car-free city should be achieved as quickly as possible, with cars parked on the outskirts of the city and public transport or bicycles being used from there (Fig. 13.45).

Cars have to be parked at least as far away as the next public transport stop. This has to be implemented at the origins and destinations (Fig. 13.46).

The ways to the practical conversion are possible and different, however, must not lose sight of the goal of a car-free city.

In practice, more and more cities are starting to set up car-free neighborhoods. This development began in the 1960s with pedestrian zones and is now widely used

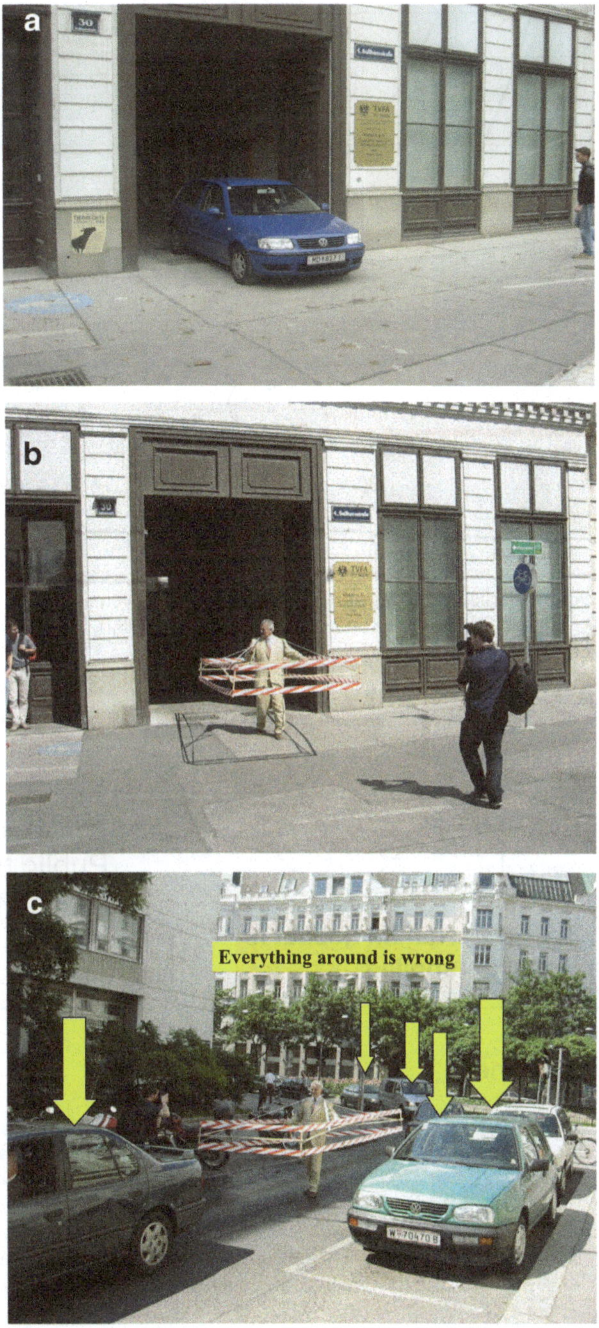

Fig. 13.42 **a** This seems to look like a totally normal situation for our society. **b** But if a pedestrian (with all the expected cultural background of mankind) uses the same amount of space, it looks rather crazy. **c** What is "normal" for our car-driven society is totally strange for a human society. Source: Photo author

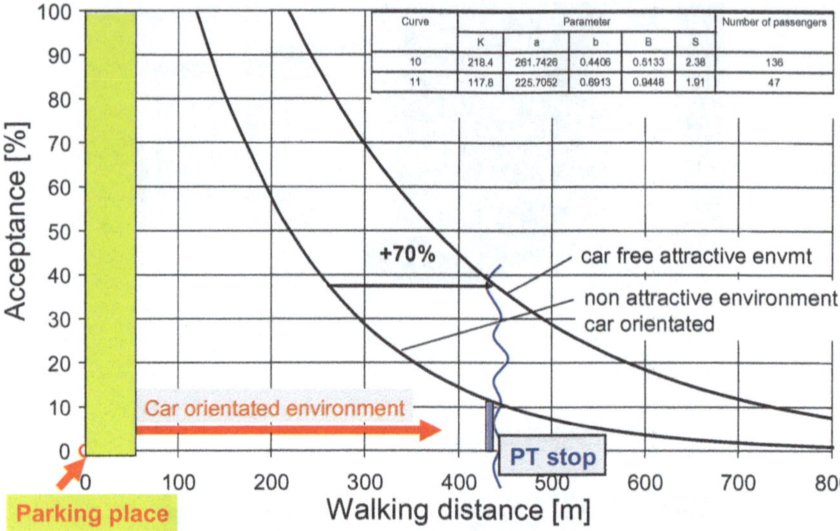

Fig. 13.43 Acceptance of walking distances in European cities in a car-oriented and a car-free environment. If cars are parked at the origin and destination, the acceptance is 100%—and public transport far away will not be accepted (Knoflacher 1981; Peperna 1982). Source: Drawing Author after Peperna (1982)

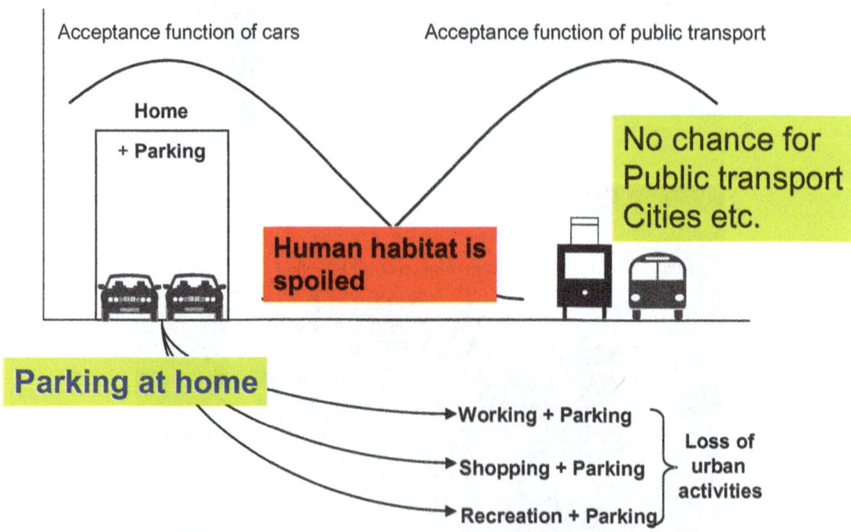

Fig. 13.44 This is the cause for all transport problems and most of the urban problems of today: parking the car at home damages the city. Under these circumstances, public transport has no chance. Source: Drawing author 1985

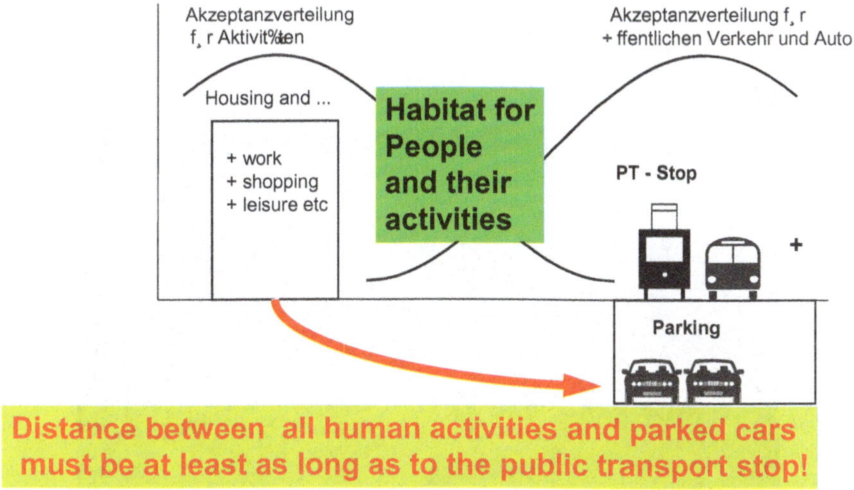

Fig. 13.45 Minimum precondition for a sustainable city for the future is car-free areas. Source: Drawing author 1985

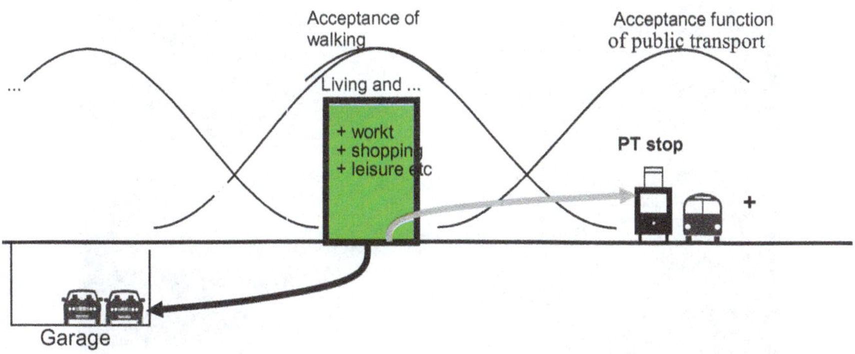

Fig. 13.46 Cars have to be parked at the outskirts of the city. Source: Drawing author 1985

in Europe for the planning of new development areas. Where good public transport is available, more and more lanes are being taken away from car traffic. If park and ride facilities are offered on the outskirts of cities, this has an accelerating effect on the desired choice of transport.

13.1.16 To Read the City—From Outside and from Inside

Architects read the city in general from outside (Fig. 13.47).

a
b
..from inside
c

◂**Fig. 13.47** **a** City and landscape are often considered as opposites, although they are closely interrelated. **b** The "living inner system" of the city has been designed over thousands of years according to the scaleg of the people. There were only a few 8–20 m wide streets (e.g. Chandni Chowk). **c** The secondary street network has the same-human dimension everywhere on the globe (left: Tokyo, right: Venice). Source: Internet and Photos by the author

A sustainable city can be recognized from the outside by its integration into the landscape and the local climate. The internal structures of all cities by human standards are similar to each other and have the same patterns of paths, squares and streets as Chandni Chowk or Venice. The greater roads have a width between eight and twelve meters. The secondary road system is in general not more than three meters wide, whether it is in Chandni Chowk City or in Tokyo or in an historical European city. The main streets have arcades. These constructions are characterizing elements of high level urban structures. Arcades protect people from sun, rain, wind, whether it is in Chandni Chowk, in Praha or Rome (Fig. 13.48).

The urban buildings have to be connected with people and the landscape like the famous city of Venice. We find this urban fabric full of pedestrian. Throughways, roads or canals have the similar size, the human scale, where children can play on public space and people can rest in pedestrianized areas like in the again pedestrianized city center of Vienna (Figs. 13.49, 13.50 and 13.51).

In an age of global warming, we must cool down the urban environment.

Fig. 13.48 Arcades protect people from sun heat, rain and wind—everywhere in the world. Source: Photo author Old Delhi

Fig. 13.49 City center of Vienna after pedestrianization. This place was used before by 80.000 cars per day and was totally congested. Source: Photo author

Fig. 13.50 Trees are the best and cheapest form of air condition in cities. Source: Photo author

Fig. 13.51 Shelters for people indicate mature urban planning. Source: Photo author

Therefore, buildings should not be higher than the trees grow in this area (Rainer 1948).

In the age of global warming, we cannot stop sunshine, but we can make cities, which create a temperature, convenient for human beings in a dense populated area with a mixture of functions, pedestrianized parts with trees as the best form of air condition. Intelligent architecture can also help to make the temperature in cities more pleasant, for example, with wind towers in hot climates.

But the city is only a part of the global network and has the obligation to take care on its environment. The rural landscape and the healthy of the rural landscape are therefore crucial elements for the city in the age of global warming. The wisdom of local engineers has created excellent solutions for the protection of transport system users in the past. The three alleys along streets give shadows to people and animals, stabilize the ground and protect the transport system users from wind (Fig. 13.52).

One of the most wonderful examples can be found on streets in India, where huge trees can cover the full road space, giving shadow and moisture to the users. What is happening on rural Indian roads is terrible today. With the support of World Bank money, traffic engineers are widening the road to build a motorway for a society with a motorization of less than 50 cars per 1000 people. While some Western European countries already have laws to protect natural monuments, elsewhere natural monuments are being willfully destroyed for traffic that damages the climate (Fig. 13.53).

Wrong educated transport engineers are cutting this treasures which would become a natural monument in Europe only for a questionable improvement of

Fig. 13.52 City in the age of global warming must take care also on the rural environment. Trees along streets indicate the understanding of local traffic engineers from the past (Indian rural road). Source: Photo author

alignment of a carriageway, based on questionable assumptions about traffic growth. This is unfortunately not an engineering achievement to be proud of, but, from the perspective of future climate change, downright criminal.

13.1.17 Is This Model City of the Future an Illusion?

It is not necessary to develop a modal city for the future: It exists. These are the medieval urban structures everywhere in the world where urban societies existed in the past (Figs. 13.54 and 13.55).

The right scale to design a city for the age of global warming, urban planners must learn what principles were used to build those cities that have survived and remained vital for hundreds of years, such as Chandni Chowk or Venice and other historic city centers and try to understand the structure of an organic living system, which has survived over hundred of years and has kept its vitality till today (Mumford 1961). If they can find a system, which is more effective in economic terms, more effective in variety, more effective in capacity and flexibility of the transport system and if they find a city which is more sustainable than the existing Chandni Chowk, they might prefer it. But as long as this is not the case they should admire the existing structures of our middle age cities, especially the walled city of Delhi, and try to discover their

Fig. 13.53 Cutting such a tree to make the carriageway wider is an act of barbarism and absolute ignorance about system behavior. Source: Photo author

secrets to be able to build cities for people again after they have been lost in the desert of urban structures of the twentieth century.

13.1.18 It is All About Reducing the Speed

If we reduce the speed to the pedestrian speed in the urban structure, harmony between the city and the transport system can be restored. The same is also happening with the highest speed in nature and the speed of light. Trees are reducing the speed of the sun beam to zero to use the energy for living matter (Fig. 13.56).

If we are able to apply the same principles to use the sun heat to heat the water, to use the energy of sun to make electricity on the roofs of the Chandni Chowk City, we will create a sustainable urban structure. The advantages of such cities are manifold. One of them protects the city from energy wasting transport systems.

Fig. 13.54 "The city" of cites, Venice is attracting people worldwide. Is secret: human scale and free of cars. Source: https://life4up.blogspot.com/2013/07/a-terra-vista-do-ceu.html

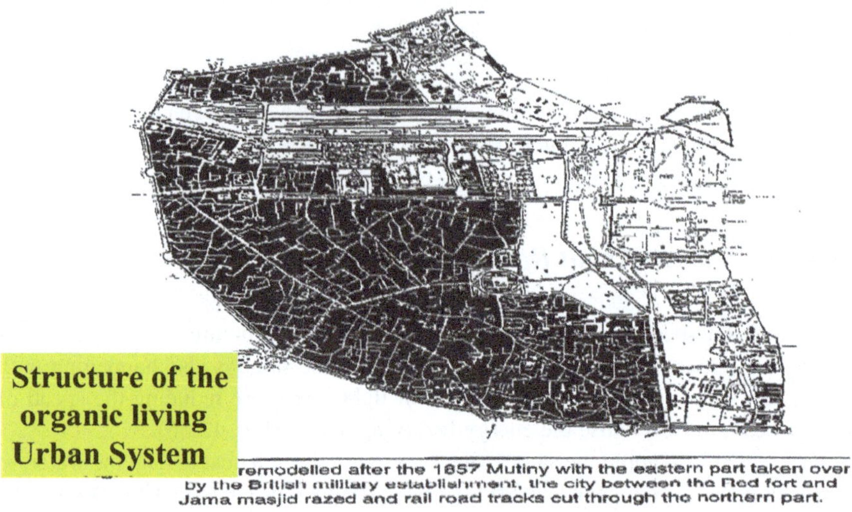

Fig. 13.55 If urban planners need a model how to design a city, they have to learn from the richness a variety of the walled city in Delhi

Fig. 13.56 Upgraded and equipped with solar energy facilities, our historical cities are the "best practice" for the city in the age of global warming

13.1.19 The Effect of TRIPP International Seminars

One of the students of the TRIPP international seminar has already set his knowledge into action and has introduced a pedestrian street in a small Indian city, called Fazilka close to the Pakistan border. He has analyzed the effects and demonstrated the advantage of the new approaches to transport and urban planning as they are taught in the TRIPP seminar (Figs. 13.57 and 13.58).

Fig. 13.57 **a** Boring and uninteresting road in Fazilka was converted into a pedestrian street temporary. **b** The street became alive and the city and its citizen also. Source: Photo by Asija Navdeep

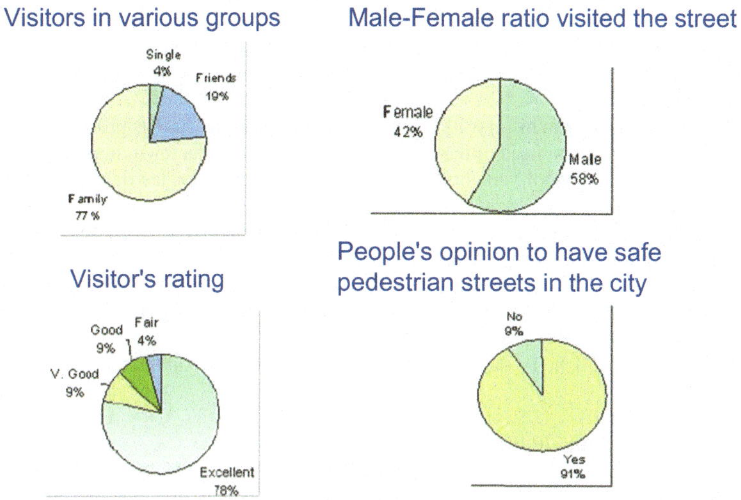

Fig. 13.58 As the colleague has learned in the courses of trip: everything has to be carefully checked and evaluated. His example was a contribution for a better climate, but also a contribution to more happiness and hope for the people of the city. Source: Graphs by Asija Navdeep

References

Appleyard DS (1981) Livable streets. University of California Press, Berkeley

BP (2014) Statistical review of world energy. Twerberg G (2015). https://ourfiniteworld.com/2015/05/06/why-we-have-an-oversupply-of-almost-everything-oil-labor-capital-etc/. Accessed 02 Apr 2020

Charter of Athens (1933). https://en.wikipedia.org/wiki/Athens_Charter

Gail T (2012) Evidence that oil limits are leading to declining economic growth, our finite World. https://ourfiniteworld.com/2012/07/13/plan-for-lower-growth-in-real-gdp-going-forward/

IPCC (2007) In: Core Writing Team, Pachauri RK, Reisinger A (eds) Climate change 2007: synthesis report. Contribution of working groups I, II and III to the fourth assessment report of the intergovernmental panel on climate change. IPCC, Geneva, Switzerland, 104 pp

IPCC (2019). https://www.ipcc.ch/site/assets/uploads/sites/4/2019/11/03_Technical-Summary-TS.pdf

Knoflacher H (1981) Human energy expenditure in different modes: implications for town planning. In: International symposium on surface transportation system performance, Washington DC

Knoflacher H (1981) Zur Frage des Modal Split. Straßenverkehrstechnik, Heft 5/1981, Sept./Okt

Knoflacher H (1986) Kann man Straßenbauten mit Zeiteinsparungen begründen. Internationales Verkehrswesen 38, Heft 6/1986, Nov/Dez

Knoflacher H (2007) Grundlagen der Verkehrs- und Siedlungsplanung. Band 1 Verkehrsplanung. Böhlau Verlag, Wien-Köln-Weimar

Koch Ch (1990) Platzfolgen und ihre Wechselbeziehung zum Fußgänger. Diplomarbeit Institut für Verkehrsplanung, Technische Universität Wien

Kopits E, Cropper M (2003) Traffic fatalities and economic growth. World Bank policy research working paper 3030

Layard (2005) Happiness: lessons from a new Science. London: Penguin Books

Layard R (2020) Can we be happier? Pelican Book

Map of Shanghai (1933) U.S. Army map service, 1945. University of Texas Perry-Castañeda Library Map Collection. https://www.lib.utexas.edu/maps/ams/china_city_plans/
Meadows D, Meadows D, Randers J, Beherens W (1972) Limits to growth. Potomac Associates Universal Books
Mumford L (1961) The city in history: its origins, its transformations, and its prospects. New York
Navdeep A, Pedestrianisation need of the hour: a case study of Fazilka town. https://www.ivv.tuwien.ac.at/fileadmin/mediapool-verkehrsplanung/Diverse/Allgemein/pedestrianisation_fazilka_india_full_paper.pdf
Norton A (2016) Facing an increasing urban world. Environment and urbanization. https://www.environmentandurbanization.org/facing-increasingly-urban-world
Peperna O (1982) Die Einzugsbereiche von Haltestellen öffentlicher Nahverkehrsmittel im Straßen-und Busverkehr. Diplomarbeit am Institut für Verkehrsplanung Hrsg. Wien: Technische Universität Wien
Rainer Roland Städtebauliche Prosa. Praktische Grundlagen für den Aufbau der Städte. Wasmuth Verlag 1948
Ratner M, Glover C (2014) U.S. energy: overview and key statistics. CRS report, Washington. https://fas.org/sgp/crs/misc/R40187.pdf
Riedl R (1981) Biologie der Erkenntnis. Verlag Paul Paray Berlin Hamburg
Riedl R (2019) Structures of complexity. A morphology of recognition and explanation. Club of Vienna 2019
Ritchie H, Roser M (2017) CO_2 emissions and green house gas emissions. Our World in data. https://ourworldindata.org/co2-and-other-greenhouse-gas-emissions
Schafer A, David G, Victor DG (2000) The future mobility of the world population. Transp Res Part A 34:171±205
The NASA Goddard Institute for Space Studies (2009) Global annual mean surface air temperature change, 2009. The NASA Goddard Institute for Space Studies, New York City, viewed 11th Jan 2010. https://data.giss.nasa.gov/gistemp/graphs/
Wackernagel M, Beyers B (2010) *Der Ecological Footprint. Die Welt neu vermessen*. Europäische Verlagsanstalt, Hamburg 2010, ISBN 978-3-931705-32-9
Whitelegg J (1994) https://worldcarfree.net/resources/free.php
Zahavi Y (1974) Travel time budgets and mobility in urban areas. DOT, Washington DC

Chapter 14
The Energy Glut: Transport and the Politics of Fatness and Thinness

Ian Roberts

14.1 Cars Are Dangerous to Pedestrians and Cyclists

Twenty years ago as a junior hospital doctor working in paediatric intensive care in Auckland, New Zealand, I discovered a public health problem that horrified me, and having just become a parent, frankly terrified me. I discovered what road traffic "accidents" really meant for children and for the people who care for them. Of course I knew, in an abstract knowledge sort of way, that being hit by a car was the leading cause of death in children in Britain, as indeed it was in most of the highly motorised countries of the world. I knew that children were continually being urged to take care whilst crossing the road. Indeed, I myself had grown up with the Green Cross Code.

What I had not appreciated, what I had not felt before, was the reality of the devastation, the horror and the real, raw human suffering, that being hit by a car entailed. I had known about it before but now I felt it. The experience of caring for seriously injured children dug deep groves in my soul that later channelled strong emotion onto problems that many people, most people actually, could not give a damn about. Most of my medical colleagues were interested in cardiology or endocrinology, areas where young doctors could really make a name for themselves, where they could get on. But having seen what energy can do a child's body what I wanted to know was how we could let that happen to children and why. I gave up being an intensive care doctor and studied epidemiology which is the science of disease causation. I started by working with transport engineers to look for risk factors for child pedestrian death.

The Oxford Dictionary defines an "accident" as an event that is without apparent cause or that is unexpected. The use of the word accident to describe child road deaths could not be more inappropriate. More is known about when, where and why

I. Roberts (✉)
Public Health, London School of Hygiene and Tropical Medicine, London, UK
e-mail: Ian.Roberts@lshtm.ac.uk

G. Tiwari and D. Mohan (eds.), *Transport and Safety*, Springer Transactions in Civil and Environmental Engineering,
https://doi.org/10.1007/978-981-16-1115-5_14

Table 14.1 Risk factors for child pedestrian death and injury-data from the Auckland child pedestrian injury study

Variable	Adjusted OR (95%CI)
Traffic volume (vehicles/h)	
<250	1.00
250–499	4.52 (2.04–9.98)
500–749	7.29 (3.09–17.20)
≥750	14.30 (6.98–29.20)
Mean speed (km/h)	
<40	1.00
40–49	2.68 (1.26–5.69)

child pedestrian-motor vehicle collisions occur than for almost any other disease in childhood. It would make much more sense to talk about a case of accidental leukaemia. Calling road deaths accidents implies that no one is responsible and no one is to blame. The child either made a bad judgment or was just unlucky. But parents are responsible for their children and are not meant to put them in a position where poor judgment or bad luck can be fatal. Blaming a child victim really means that the parents are held responsible. Many bereaved parents live out their lives in silent desperation (Table 14.1).

When the experiences of the hundreds of children killed and injured on Auckland's roads were aggregated, however, a very different picture emerged. Most of the children were injured close to home, often in the street where they lived. When we compared the traffic characteristics of the streets where injured children lived with a group of non-injured children selected from the general child population, we found that the main determinants of injury risk were the volume and speed of the traffic. The injury risk increased particularly steeply with rising traffic volume (Roberts et al. 1995). Children living in the busiest streets were fifteen times more likely to be injured than children living in the quietest streets.

Traffic and not erratic jaywalking children are the cause of child pedestrian injury. Children get hit by cars because the cars are there. There is one street, and it is either a place for children or a place for cars. Mixing the two, at least at average urban speeds, does not work without bloodshed. A vehicle driving down a residential street at forty miles per hour packs more destructive energy than a bullet. If a child is unlucky enough to be hit, a single shot can kill and crossing a busy main road is like making a dash through machine gun fire.

Epidemiological studies reveal associations. In this case, the association is between the volume of traffic and pedestrian injury risk. Whether this is a cause–effect relationship is a matter for judgement. It seems very likely but perhaps we would like to confirm it with a natural experiment. What would happen to road deaths if the volume of traffic actually fell? Because the history of western motorisation is one of almost relentlessly increasing traffic, there has been little opportunity to answer this question. However, the Middle Eastoil crises of 1974 and 1979 provided a rare insight into what happens when traffic volume falls. Despite its clean green image, New Zealand is one of the most heavily motorised countries in the world and the 1974 energy crisis, which was accompanied by a fourfold increase in the price of petrol, hit hard. The government responded by introducing "car-free days" when every car had to be off the road for one day each week. There was also a weekend ban on petrol sales which lasted until August 1980. During the period of high petrol prices, child pedestrian death rates plummeted. Between 1975 and 1980, child pedestrian death rates fell by 46% (Roberts et al. 1992). But when the oil started flowing and traffic volume resumed its upward trajectory, the number of children killed and injured on the roads increased along with it.

The oil price shocks of 1974 and 1979 also coincided with reductions in road death rates in the USA and Britain. The oil crises had revealed link in the chain of causation, the link between the price of fossil fuels and the amount of danger on the roads. When petrol prices rise, fewer children die; when they fall, more children die. To a physicist, this connection would seem obvious. Petroleum is chemical energy, and the petrol in the fuel tank is the source of the energy that kills and injures. The tens of thousands of controlled explosions that drive the pistons that spin the wheels are fuelled by a steady stream of petroleum, and whatever obstructs this flow of chemical energy, whether it is war in the Middle East or hurricane damage to oil refineries in the Gulf of Mexico, there will be less road danger and fewer road deaths as a result. However, the oil crises were just a temporary blip, and for the next thirty years, traffic volumes would soar on the back of reliable supplies of cheap petroleum. Had nothing else changed, the bloodbath on the streets would have made current road death statistics seem trivial. But something else did change. The pedestrians and the cyclists got out of the way.

14.2 Road Death Rates Fall as Roads Get More Dangerous Because People Get Out of the Way

A vehicle travelling at forty miles per hour down a residential street is an obvious threat. Due to its mass and velocity, it contains enough kinetic energy to break bones and tear flesh. There is an equation in the physical sciences that tells us that the kinetic energy in a moving object is equal to one half of its mass times its velocity squared. This means that a car weighing 1500 kg driving at forty miles per hour (eighteen metres per second) has $\frac{1}{2} \times 1500 \times 18 \times 18 = 243{,}000$ J of energy. A car moving twice as fast has four times as much kinetic energy. We do not need to have been hit by a car, or to have seen someone else being hit, to understand this. Evolution has hard-wired this knowledge into our brains. In the street, might is right. The person with the power is the person behind the wheel of the car. They hold the power to kill or disable, and if they do kill or disable, then they face little or no legal sanction. When faced by an assailant with a dangerous weapon, you have two choices: to run or to fight. Picking a fight with raw kinetic energy is stupid and futile. The normal

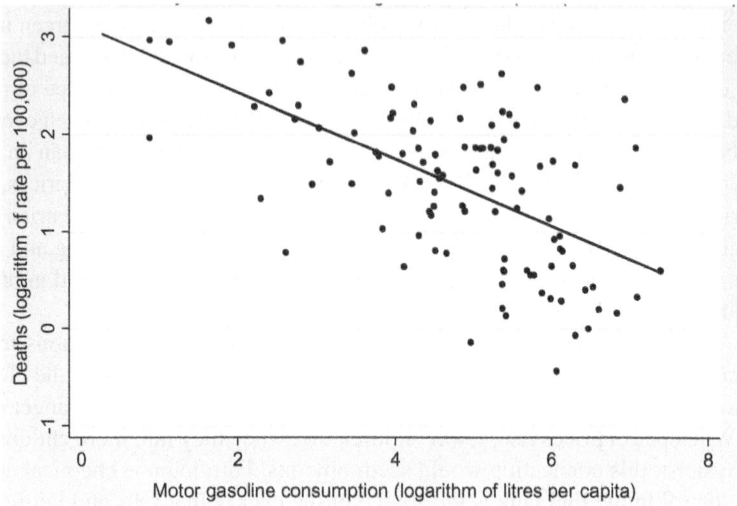

Fig. 14.1 Pedestrian and cyclist deaths and motor gasoline consumption (107 countries, 2008)

human response is to get out of harm's way and make sure that your children are out of harm's way too.

Figure 14.1 shows, in general, as per capita motor vehicle gasoline consumption increases, pedestrian and cyclist death rates fall. Child road deaths in Britain and the USA have been falling for decades. Death rates fell more steeply when oil prices were high but the fact that the overall trend was down, despite rising road danger from increased motor vehicle traffic, meant that something else was going on. As the volume of road traffic increased and the streets became rivers of lethal kinetic energy, the pedestrians got out of the way. Parents kept their children indoors and those who could afford a car started driving rather than walking, even for short distances. The body counters at the Ministries of Transport of course claimed that death rates were falling because the traffic planners and police were doing a great job. Transport ministers proudly claimed responsibility for the fall in the number of deaths. No one bothered to count how many live people there were out on the streets (Fig. 14.2).

There is another common response to an assailant with a dangerous weapon apart from getting out of the way. Obtain such a weapon for yourself. You would prefer that you and your children were safe inside a vehicle, rather than vulnerable on the outside. A survey of parents of primary school aged children in north London found that most were very worried about road danger on the daily journey to school and that most would drive their children if they had access to a car (DiGuiseppi et al. 1998). And so begins the motorised arms race which drives the downward spiral of walking and cycling. If what I am saying is true, we would expect more pedestrian deaths among poor people who cannot afford to use a car and fewer pedestrian deaths among the wealthy people who can. Every death in England is recorded along with the occupation of the person who died. For dead children, the occupations of their

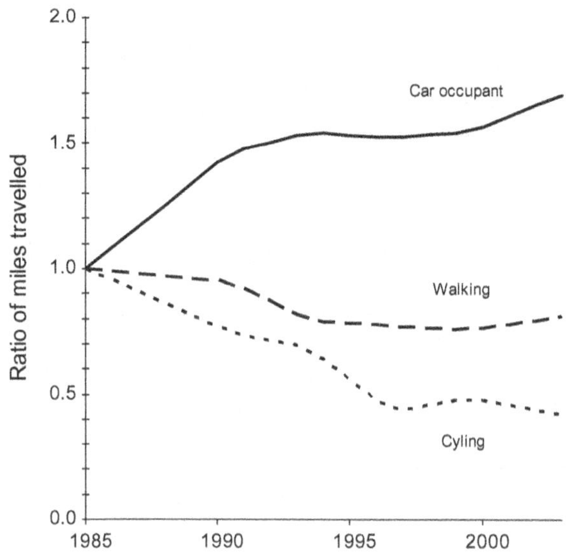

Fig. 14.2 Walking cycling and car use trends 1985–2005

mothers and fathers are recorded. When we looked between 2001 and 2003, there were 205 children killed as pedestrians in road crashes in England and Wales. Of these, seventy-one children had parents who were "long-term unemployed", sixty-two had parents who were in "routine occupations" and twenty-five children had parents in "managerial and professional occupations". When taking into account the number of children in each group, the risk of road death for a child in the lowest social group was five times that of a child in the highest social class (Edwards et al. 2006). Walking surveys show that children from families without a car walk much more often than children in car owning families (Roberts et al. 1997). Poor children are outside the car because they cannot afford to get inside.

Increasing road traffic has decimated walking and cycling in Britain, the USA, and most other highly motorised countries. Data from the UK National Travel Survey shows that the average distance walked per person per year fell from 255 miles per person per year in 1975 to 192 miles in 2003. Over the same period, the average annual distance cycled fell from fifty-one miles per person to thirty-four miles. The distance walked by children has fallen by almost a quarter. Children today walk less than ever before in the history of humanity (Hillman et al. 2000). It is likely that by 1975, when the UK Department of Transport first started measuring walking and cycling, most of the decline had already taken place. We can get an indication of the extent of the changes from old photographs and paintings. Lowry, the Manchester artist famous for his matchstick figures, painted Salford streetscapes in the 1920s and 1930s before the hay day of the motor car. His sombre skies show the pollution of industrial England, but his streets scenes show a vitality that is completely absent today. The streets Lowry painted belonged to the people living in them. Now, they belong to the car. The same process is currently underway in India.

Traffic is not the only danger people consider when deciding whether to put on their walking shoes. For the past ten years, conducting clinical trials with trauma doctors on four continents, I have visited some of the most violent cities in the world. Many of these are in Latin America where the juxtaposition of conspicuous wealth and obscene poverty elevate routine urban violence to war-like proportions. There are forty homicides per 100,000 people each year in Medellin, a murder rate over ten times higher than in London.

When it comes to violence, even though it is people we fear, we feel most afraid when there are few people about. We feel safest in the peripheral vision of other people's awareness. We don't want to be stared at but we do want to be seen. In Medellin, as soon as I turn into an empty street my heart starts racing. In her book "The Death and Life of the Great American Cities" Jane Jacobs wrote how "eyes on the street" help to keep the street safe (Jacobs 1992). The more eyes the better, especially if those eyes can summon for backup if there is cause for concern. She describes a brief urban drama in her neighbourhood. A man is seen dragging a young girl up a side street. The child is resisting, crying and shouting. A crowd quickly gathers. The eyes on the street did not like the look of what was going on. It turns out that the man is her father, the child is having a tantrum and that there is nothing amiss. The crowd disperses. On this occasion, there was no cause for alarm, but people were concerned and came to help. According to Jacobs, this collective concern is what keeps communities safe.

Traffic takes eyes off the street. It divides the street. Interactions between people on the other side of a busy street are less likely to be noticed, voices might not be heard and the mood of interpersonal situations might not be understood. Yes there are eyes inside the cars; but when travelling at speed, sight lines are polarised ahead, along the road and not on the pavement. Would someone stop if someone was being attacked? Just as for road danger, the response to stranger danger is to get off the street and into a car, another vicious circle. Traffic makes a street seem hostile. This leads to more traffic and more hostility. The only people left on the street are the very poor people.

Research shows that when considering whether to let their children walk to school, London parents fear stranger danger more than road danger. But are these parents worried that someone would drag their child screaming and yelling up a busy street or that they would be pulled into a passing car and whisked away? The latter seems far more likely. Enrique Peñalosa the former mayor of Bogotá mused that children are like an indicator species for urban safety in the way that the presence of fish indicates whether or not a river is polluted. By taking children off the street, traffic increases our fear of violence. A colleague told me a story about a street party in England, where the street was closed to traffic for the day, as a demonstration project for a safe street initiative he was involved in. The children from the street were playing outside. One of the residents said that it was an excellent event but asked my colleague "where did you get all the children from?"

14.3 The Resulting Decline in Human Movement Moves the Population BMI Distribution Upwards

The human body is a vehicle perfectly designed for personal transportation. It will keep running until the day we die. Whether moving or not, it requires energy but provided that we fill it up regularly with enough food, and it will meet most reasonable transportation demands. It has an important special feature. If the amount of food energy taken in is surplus to requirements, it will store the excess energy as fat. This energy can be called on later if needed, allowing the body to keep running for weeks, even on an empty tank. Body fat accumulates when the amount of energy we eat as food exceeds the amount of energy we use moving around and keeping warm. Fat is stored energy.

Most of our body fat is stockpiled under our skin and around our internal organs as adipose tissue, which is mostly made up of fat cells, also known as adipocytes. Fat cells contain a large droplet of lipid. This lipid, which has the consistency of margarine on a hot day, pushes the nucleus and everything else to the edges of the cell. The lipid droplet plumps up the cell, rather like a silicone implant plumps up a breast and the bigger the droplet the fuller the cell becomes. As we get fatter, the number of fat cells in our body increases, and the fat cells that we already have expand as the lipid droplets they contain get bigger. Energy intake need only be slightly higher than energy output for fat to accumulate. Doctors gauge how fat we are by working out our body mass index (BMI), which is our weight in kilograms divided by our height squared (height in metres multiplied by height in metres). Doctors diagnose "overweight" if a person's BMI is between twenty-five and thirty, and "obesity" if it is thirty or more.

If we measured the BMI of all the people in a particular country and plotted on a graph how many people there were at each BMI value, we would find a wide spread of values. This spread of values is called a distribution. There would be some really skinny people with a very low BMI and some really fat people with a very high BMI, but most people would have a BMI close to the centre of the distribution.

As average levels of human movement have decreased, and the average BMI has increased with it. Figure 14.3 shows the BMI distribution in Canada in 1978 and 2004. It can be seen that the entire distribution has shifted upwards. Everyone has become fatter—the whole of the Canadian population.

The same trend is evident all around the world. Increasing body fat is not a problem limited to a particular country. Wherever you happen to be living, you can be reasonably sure that the population is getting fatter. Canada is not as fat as the USA, but the Canadian BMI distribution is slowly sliding upwards. In 1978, the average BMI in Canada was twenty-five. In 2004, it was twenty-seven. China has a relatively low-average BMI, but it has already started drifting upwards. Between 1989 and 2000, the average BMI of Chinese men increased from 21.3 to 22.4 and that of Chinese women increased from 21.8 to 22.4. It is the same story in Europe, Asia and South America. In 2006, the World Health Organisation reported that there are 300 million obese adults living on planet earth, and there will be many more in the

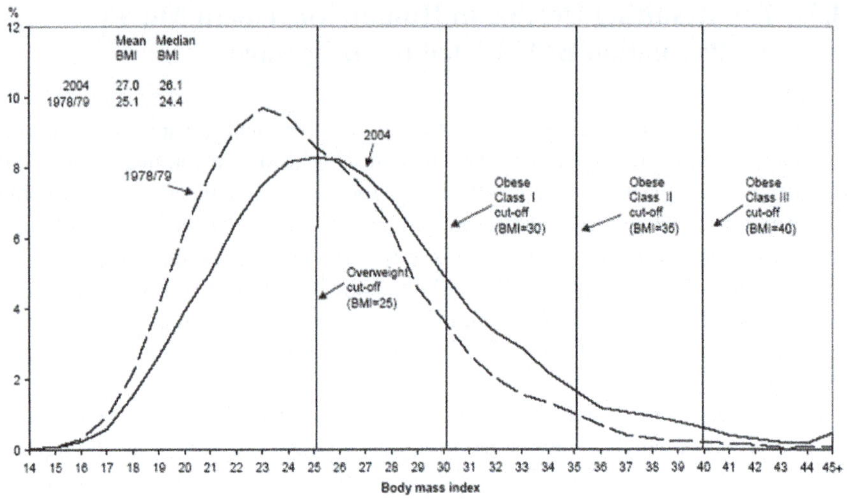

Fig. 14.3 BMI distribution in Canada in 1978 and 2004

future. Epidemics that affect the whole world are called pandemics. The USA may be the epicentre, with very large increases in the weight and size of the population (see Fig. 14.4), but no country is immune.

In brief, the entire population distribution of BMI is shifting upwards, in almost every country in the world. Governments internationally tend to portray the increasing

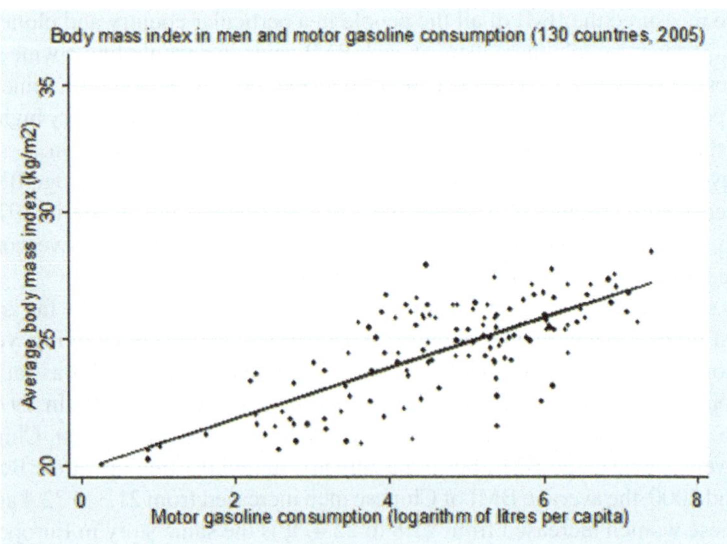

Fig. 14.4 Abdominal adiposity in US adults: prevalence and trends, 1960–2000

in population fatness as a personal problem, highlighting the number of people in the upper "obese" tail of the distribution. The British government, for example, recently launched a campaign to tackle obesity, arguing that the British people were developing bad habits. This is of course nonsense. If this was a habit problem, it would imply that almost everyone on earth had fallen into exactly the same unhelpful habits, a bad habit pandemic on an unprecedented scale! Decades of health research have shown that population fatness is an environmental problem and not a personal weakness. Our tendency to fall into unhelpful habits is the same as it ever was. What has changed is that thanks to motorisation and mechanisation, there were fewer opportunities to move our bodies than ever before, whilst at the same time we are besieged by a food industry that uses the best marketing brains in the world to sell us mountains of cheap energy dense food.

Figure 14.5 shows the relationship between the average BMI and motor vehicle gasoline consumption per capita for the 130 countries for which the necessary data were available. Each dot represents a country, and the solid "regression" line is

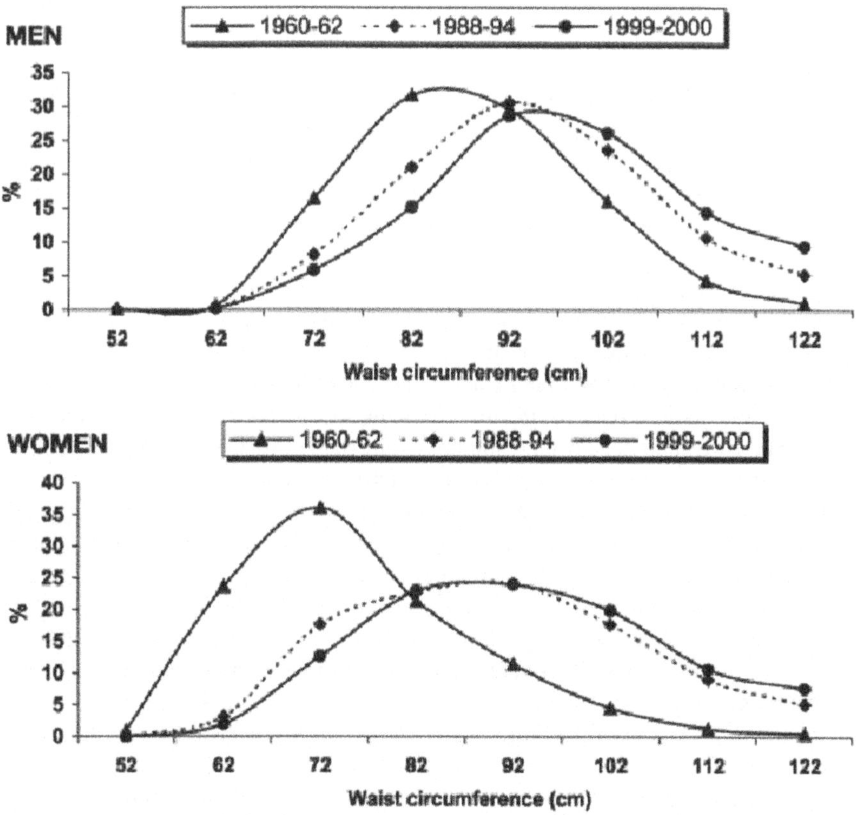

Fig. 14.5 Relationship between the average BMI and motor vehicle gasoline consumption per capita for the 130 countries

chosen, so that it comes as close to the points as possible. This line provides an indication of how average BMI varies as motor gasoline consumption increases. There is a clear strong relationship between average BMI and motor vehicle gasoline use. Where motor vehicle gasoline consumption is low, average BMI is also low, and where gasoline consumption is high, average BMI is also high. The association is present in both men and women, although strongest among men.

The current panic in the West about rapidly rising obesity levels is a consequence of forgetting that the whole BMI distribution was moving upwards. The slow but steady movement of the BMI distribution could be likened to the New York marathon, with the obesity threshold as the finishing line. Once the race starts, the whole population of runners starts moving towards the finish. For the first few hours, no one crosses the finishing line but then the front runners start crossing in dribs and drabs. An hour or so later, the main lump of competitors reaches the line, and the number of finishers increases rapidly. In most wealthy countries, the main lump of the population is now crossing the obesity threshold, and so, obesity is rising very rapidly. But there is little point searching frantically to find out what is happening now to explain the increase. The race towards obesity started decades ago. Here, however, the analogy ends, because unlike runners in a marathon, on our increasingly fat planet, the runners just keep going.

Doctors and health scientists have been astonished by the sudden increase in obesity and have only recently started conducting studies looking for the causes. Most of these studies compare the food intake and daily activity levels of obese and non-obese people within a particular population at one point in time. The problem is that because we are now all exposed to the same transportation system, physical activity levels are at a record low across the entire population and the factor that most differentiates the fat from the thin is food intake. By analogy, if everybody smoked and to exactly the same extent, and we compared people with and without lung cancer, smoking would not differentiate those with and without cancer. But this would not mean that smoking was not a cause of lung cancer. When it comes to fatness, scientists have been asking the wrong question (a focus on obesity rather than on average BMI) and have been using the wrong data (a focus on what is happening now rather than what has happened over the past half century).

I will not go into detail here about the political reasons for the neglect of motorisation as a cause of population fattening, other than to say that scientists tend to have a blind spot when it comes to gradual societal transitions and that they tend to limit their consideration of causes to those things that they believe are amenable to change. As we will see later, the notion that the de-motorisation of society is either possible or desirable has effectively been barred from consciousness by a sustained bombardment of industry and government propaganda. Road danger is a key environmental determinant of declining physical activity levels even though it scarcely given the time of day in contemporary public health literature. Entire populations do not passively give up walking and cycling because they are tempted to do so by the speed, comfort and glamour of motorised travel. On the contrary, they are driven off the street by deadly force, with a barrage of propaganda being required for its justification. The forceful eviction of people from urban public space is currently underway in many low and middle income countries, in particular in India and China. Many politicians call it progress.

14.4 The Car Industry Says the Main Problem Is a Lack of Road Safety Education

The huge increase in road death and injury that results from the forceful eviction of the walking public from the streets space is politically problematic for the car industry. The problem is too big to ignore but at the same time tackling it at source would be bad for business. They understand the political threat posed by the great mass of the walking public. Currently, they are busy manoeuvring themselves into pole position in road safety politics. They have to ensure that efforts to improve global road safety do not adversely impact on car sales. In 2006, the Federation Internationale de l'Automobile set up a commission for global road safety with a remit to "examine the framework for and level of international cooperation on global road safety and to make policy recommendations". At its inception, the commission was chaired by former UK Defence Secretary, Lord Robertson, and had eight commissioners, one from each of the G8 group of wealthy nations.

If you wanted to represent the interest of motoring classes, you could not put together a more able group of commissioners. Canada was represented by an Executive Director at General Motors, Japan by a Board Member of the Bridgestone Corporation, the major transnational tyre maker. Russia was represented by the president of the Russian Automobile Federation and Italy by a former president of the Automobile Club of Italy. Michael Schumacher represented Germany, and France was represented by Gerard Saillant, a doctor who works on the medical aspects of Formula One. The UK commissioner was the chief economist at Lehman Brothers, the US investment bank whose later collapse precipitated the perfect storm of global economic chaos. The commission's Patron is Prince Michael of Kent, a former racing driver, now a member of the British Racing Drivers Club and the Bentley Drivers Club. Lord Robertson himself was then deputy chairman of the Board of TNK-BP, a Russian oil company. According to the House of Lords' Register of Interests, which shows that the FIA paid Robertson to attend the 2006 Monaco Grand Prix, the commission holds its meetings at the race track (Roberts 2007).

Working through the commission, the FIA and the car lobby are taking a lead role in global road safety. They would like to set the policy agenda for road safety and thus gain considerable influence in global transport policy. They do not want to fund road safety efforts but to dictate how other organisations spend their money, and in particular, how development money is spent. Former World Bank President Paul Wolfowitz was eager to confirm the bank's willingness to implement the commission's recommendations, and the former UK secretary for State for International Development Hilary Benn welcomed the Commission's proposals.

Although most people in poor countries will never own a car and most of the victims of road traffic crashes are pedestrians, the commission has worked hard to ensure that the views of the motoring elites dominate transport policy decisions. Unelected, with only token representation from developing nations, the car lobby wants to dictate how poor countries' governments spend the development loans that their impoverished people will repay for decades.

The car lobby's favourite road safety policy is pedestrian education. Despite decades of evaluation research, safety education has never been shown to reduce road injury rates, a point emphasised by the WHO in the World Report on Road Traffic Injury Prevention (Peden et al. 2004). Road user education is favoured by the car lobby because it places the responsibility for road traffic injury squarely on the victim and has no impact on industry profits. Its primary purpose is ideological. It sends the message that the road space belongs to drivers and that pedestrians and cyclists must look out or die. This also applies to children by the way, who account for 300,000 of the 1.2 million road deaths each year. Awareness campaigns are another favourite (Duperrex et al. 2002). The commission promoted the "Think Before You Drive" campaign, supported by the Bridgestone Corporation, which reminded drivers to use child seats and seatbelts and to check their tyres. Sensible suggestions they may be, but such exhortations have no discernable effect on road safety. On the other hand, the campaign may improve Bridgestone's corporate image.

At first sight, there appear to be many different stakeholders in the global road safety policy arena, but careful examination reveals otherwise. In 1999, the World Bank—arguing that a partnership between businesses, NGOs and governments can deliver road safety improvements in poor countries—established the global road safety partnership (GRSP), a business partnership that includes the automotive giants General Motors, Ford, Daimler Chrysler, Volvo and drinks multinationals such as United Distillers and Bacardi-Martini. General Motors was represented on the GRSP and on the Commission for Global Road Safety.

A 2006 study compared the frequency of use of different road safety related words in GRSP road safety reports and in the World Report on Road Traffic Injury Prevention, a report that was prepared relatively independently of business concerns by the World Health Organisation. In the GRSP reports, there was a clear lack of reference to pedestrians and cyclists. In the WHO report, "speed limit" occurred seventeen times in every 10,000 words; in the GRSP reports, just once. "Pedestrian" was used sixty-nine times by the WHO, and fifteen times by the partnership; "buses" and "cyclists" were mentioned thirteen and thirty-two times, respectively, by the WHO but not once by the partnership (Roberts et al. 2006).

14.5 Structural Reasons for the Current Situation

The main reason why car travel is annihilating human movement and putting our planet in peril is that motor vehicle travel is highly subsidised. Motorised transportation depends on three essentials: roads, oil and vehicles. Although most of the world's population will never own a car, road building is invariably funded by public funds in "developed" and "developing" countries alike. Road transportation is 95% oil dependent and ensuring a steady supply of cheap oil also involves massive public expenditures. Then, there is the automobile industry, which received billions of dollars of taxpayers' money at a time when thousands of small businesses were going to the wall. In December 2008, US car makers went cap in hand to Congress seeking a $34 billion bailout package and received the best part of it.

And it does not stop there. Motorised transport causes a mountain of suffering. Who pays for the road traffic crashes that kill 1000 children per day and permanently disable ten times as many? Who pays for transport related air pollution and the cardiac and respiratory diseases it leads to? Who pays for physical inactivity and the obesity, diabetes, heart disease, stroke and cancer it causes? And who will pay for climate change? These are the real social and environmental costs of motorised transport, but it is the public and the global environment, and not the people who benefit from motor vehicle use who pay.

Table 14.2 shows the top ten corporations in the world according to the Fortune 500 annual ranking for 2008. Eight of the top ten are oil companies or car makers. The corporations that sell oil and cars pack enormous economic and political clout with revenues higher than the gross domestic product of many developed countries.

As regards roads, the wealthy world needs poor countries to build more roads, so that the car makers can remain profitable. The market for cars in high-income countries is nearly saturated. In the year 2000, there were 769 cars per 1000 people in the USA and 441 per 1000 people in UK. Although there is some turnover, as old and damaged vehicles are taken out of stock, a process that was recently given a boost with a generous injection of public funding, the main prospect for a growth in sales is in Africa and Asia. Nigeria has eleven cars per 1000 people, and India has seven per 1000. To survive, the car industry must sell more cars, and to make sure that it can, poor countries must build the roads to accommodate them.

Table 14.2 Top ten corporations in the world according to Fortune 500 annual ranking for 2008

Rank	Company	Product	Revenue ($ millions)	Profit ($ millions)
1	Wal-Mart Stores	Supermarket	378,799	12,731
2	Exxon Mobil	Oil	372,824	40,610
3	Royal Dutch Shell	Oil	355,782	31,331
4	BP	Oil	291,438	20,845
5	Toyota Motor	Cars	230,201	15,042
6	Chevron	Oil	210,783	18,688
7	ING Group	Finance	201,516	12,649
8	Total	Oil	187,280	18,042
9	General Motors	Cars	182,347	−38,732
10	ConocoPhillips	Oil	178,558	11,891

Getting impoverished countries to spend public money on road building requires some serious propaganda. Poor countries have lots of pressing problems to deal with. The most pervasive misinformation is that road building is good for development. In 2005, the British Prime Minister Tony Blair launched the Report of the Commission for Africa. The commission's objective was to diagnose African woes and make a prescription for a better future. Its conclusion was that Africa needed more roads. More important than health care, AIDS prevention, security or better governance and road building it was argued would jump-start the stalled economy of a continent mired in misery for decades.

The commission's analysis was simple. Africa is poor because its economy is not growing. Improving its transport infrastructure would make its goods cheaper, allowing Africans to break into world markets and trade their way out of poverty. Of the estimated $75 billion needed to implement the commission's recommendations, 27% would be spent on infrastructure, mainly road building, with 13% spent on AIDS and 10% on education.

If reducing the costs of getting African goods to western markets is really the cause of African poverty, as Tony Blair and the Commission for Africa claim, Britain could help Africa's poor by reducing the transportation costs for African goods once they reach Britain. Like many wealthy countries, Britain has high levels of fuel taxation. In most of Africa, fuel is not taxed but subsidised. In 2004, a litre of super gasoline in Nigeria retailed for US$0.40. The corresponding cost in the UK was US$1.56. British politicians would not contemplate reducing fuel taxes since these are such an important source of government revenue. Indeed, the money raised from fuel sales in Britain helps to pay for the huge state apparatus required to service a car-based transportation system. This includes the police needed to enforce road safety laws, a judiciary, a system of pre-hospital and hospital care and a social safety net for injured victims. African fuel prices do not even cover the costs of road maintenance. According to the World Health Organisation, the economic loss associated with road traffic injuries in poor countries is around 2% of GDP, nearly US$100 billion, twice the amount they receive in development aid (Peden et al. 2004). Reducing the cost of road transportation in Africa might be good for trade, but not so good for most Africans.

In 2006, the UK Department for International Development commissioned transport expert Professor David Banister at University College London to collate the scientific evidence on the link between road building and development (Banister and Wright 2005). Although he did find a statistical link between the road infrastructure and the size of the economy, he could find no evidence that the former caused the latter. It is no surprise that rich countries have more roads than poor countries. Wealthy countries have more cars, and so, there is a higher demand for roads. There are more swimming pools in wealthy countries, but no one would claim that swimming pools are "central to development" which is what the World Bank claims for road building. However, the report did point out that the congestion resulting from rapid motorisation hampers economic productivity and that the poor bear the lion's share of the negative impacts of road transport. At the Department for International Development, the report was given a hasty burial.

If roads are the cure for African poverty, we have learned nothing from history (Rodney 1981). For centuries, Africa's roads have led to its impoverishment. In his economic history of Africa, Walter Rodney describes the role of Africa's transport infrastructure in this way: "means of communication were not constructed in the colonial period, so that Africans could visit their friends nor were they laid down to facilitate internal trade in African commodities. There were no roads connecting different colonies or different parts of the same colony to meet Africa's needs and development. All roads and railways led down to the sea. They were built to extract gold or cotton and to make business possible for the trading companies and for white settlers".

So, who does benefit from reducing the transportation cost of transnational trade? Global business revolves around resources, factories and markets. Raw materials are transported to factories where workers produce manufactured goods. These goods are then transported to markets, so that consumers can buy them. If consumers are willing to pay more for the finished goods than it cost to produce them, the business will make a profit. And making a profit is what business is about. Cheap transport is good for profits because it reduces the costs of production and the costs of getting goods to markets. It also enables companies to take advantage of the lower wages of workers in poor countries. Indeed, cheap labour is one of the main reasons why the captains of industry are so excited about transnational trade. It is more profitable to set up factories in poor countries where wages are low than in rich countries where workers enjoy decent wages and standards of living. But poor people cannot afford to buy expensive manufactured goods, and so, the goods have to be transported back to markets in high-income countries. All this depends on cheap transport which is bad news for road safety, physical activity and climate change, but good for profits (Roberts 2004).

The globalisation of trade leads to more freight, longer journeys, more road danger and more greenhouse gas emissions. In the poor countries that bear the brunt of the road death epidemic, trucks are responsible for the majority of crashes. In India, trucks are involved in half of crashes in cities and two thirds of crashes on highways. The victims are mostly pedestrians and cyclists. Their experiences are part of the real social cost of international trade. And will small farmers in Africa compete with the subsidised grain from US agribusiness? Transnational trade is a new name for an age old activity.

For centuries, countries with greater economic and military power sought access to the resources, and markets of weaker countries and millions died in the process. Rich countries claim that "trade" benefits both rich and poor, but the historical record suggests otherwise. Walter Rodney believed that Africa's roads were built, so that white settlers could make themselves rich at the black continent's expense (Rodney 1981). The Uruguayan journalist Eduardo Galeano came to the same conclusion (Galeano 1973). In his book "The Open Veins of Latin America", Galeano wrote how the continent's transportation infrastructure was developed to drain its wealth into the ports and then out to the colonial economy. Nevertheless, publicly funded road building is only one of the three essential elements of a profit centred transportation system. Cheap transport runs on cheap oil, and keeping oil prices within profitable limits entails huge public subsidies and masses of misery for a great many people.

14.6 The Good News and Why We Should Act

So far, this paper has argued that the increase in global motorisation has profoundly important consequences for the health of the population. First, there is an epidemic of death and injury on the roads. This is followed by a decline in pedestrian and cycling activity and an increase in average population fatness, followed in turn by an epidemic of chronic disease. However, because transport runs on fossil fuel, it also has important consequences for the health of the planet. Climate scientists predict an average global temperature increase of between 1.5 and 6 °C by 2100, depending on the extent of future emissions, possibly reaching a temperature not experienced in the past 100,000 years. This will have dire consequences for plant and animal life and for our health. Even cautious scientists are talking in apocalyptic terms about famine, disease and environmental refugees. That climate change is real, and man-made is not in doubt. All that is uncertain is how bad it will turn out to be.

However, continuing our dependence on fossil fuel powered transportation presents a serious ecological threat to the future of our civilisation, and reducing it presents unrivalled opportunities for improving population health and well-being. Policies to reduce greenhouse gas emissions have the potential to bring about large reductions in heart disease, respiratory illness, cancer, obesity, diabetes, depression, and road deaths and injuries. They could improve food security, reduced inequalities, strengthen communities, build a resilient and sustainable economy and reduce the threat of large-scale violence.

The health benefits arise because climate change policy necessarily affects two of the most important determinants of health: human nutrition and movement. Although health professionals increasingly recognise the benefits of policies to address climate change, they are not widely appreciated by public policymakers. As regards climate policy, the existence of these health benefits implies a dramatic reduction in the net cost of taking strong action to mitigate climate change—which means that failure to understand their importance could have serious environmental consequences.

In 2009, as part of a research project to estimate the health effects of reducing fossil fuel energy use, researchers from the London School of Hygiene and Tropical Medicine and the Indian Institute of Technology in Delhi estimated the health effects of transport policies that would meet greenhouse gas emissions reduction targets (Woodcock et al. 2009). They compared business as usual 2030 transportation scenarios (without policies for reduction of greenhouse gases) for both London and Delhi (as examples of British and Indian cities), with more sustainable transportation scenarios. In both settings, meeting greenhouse gas emissions targets in the transport sector would require modest increases in walking and cycling and reduced car use as compared with the business as usual scenarios.

Based on the epidemiological evidence linking transportation patterns and health, the resulting increase in physical activity would dramatically reduce rates of chronic disease. In India, which currently faces a major epidemic of chronic disease, there would be large reductions in ischaemic heart disease (11–25% of total ischaemic heart disease burden), stroke (11–25% of total stroke disease burden) and diabetes (6–17% of total diabetes disease burden). There would also be reductions (27%) in road traffic injuries. The increase in physical activity from more sustainable transport policies would also improve mental health, with an estimated 6% less depression. There would be additional mental-health benefits from more green spaces, less noise pollution and improved physical fitness.

The experience of Cuba in the 1990s confirms the health benefits of societal reductions in fossil fuel energy use. During the Cuban energy crisis, the proportion of adults were physically active more than doubled (Franco et al. 2007, 2008). The population average BMI fell by 1.5 units with a halving in the prevalence of obesity from 14 to 7%. Deaths from diabetes fell by 51%, from heart disease by 35% and from stroke by 20%. No one starved, because Cubans recognise food as a human right and not an economic commodity to be rationed according to the ability to pay. Cuba ranks seventh on the New Economics Foundation Happy Planet Index for 2009. Its neighbour, the USA, ranks 114th, next to Nigeria. Cuba shows that weaning off oil can be achieved whilst maintaining high levels of sustainable well-being.

Sustainable transportation policies would reduce the population fatness. World-wide, a total of 1.5 billion adults are either overweight or obese. In the USA, more than one-third of the population is obese, and UK scientists predict that the United Kingdom will be "a predominantly obese society" by 2050. The prevalence of obesity in India is very low but set to increase with increasing motorisation. The fact that motor vehicle use contributes to population fatness is well recognised but the fact that it can also contribute to under-nutrition has received much less attention. In April 2008, Evo Morales, the president of a poor and increasingly hungry Bolivia, pleaded for "*la vida primero, los autos segundos*" (life first, cars second), exhorting the wealthy world to stop burning food every time they drive—a reference to Western governments' policies on bio-fuels. However, as was discussed earlier, motor vehicle use and food prices were linked long before bio-fuel policies.

Car use drives up food prices through its influence on the price of oil, indeed, whereas agricultural economists reject the theory that the rising demand for food from in India and China was partly responsible for 2008 food price crisis, and they do accept that the rising demand for oil in India and China is likely to have played an important role. Reducing oil use in the transport sector will help to prevent starvation. Until agriculture unshackles itself from dependence on oil, petrol tanks and stomachs will be competing to be filled. The decarbonisation of transport will contribute importantly to improving food security and reducing global hunger.

Reducing oil demand through sustainable transport policies would also help to reduce hunger by promoting the sustainable economic growth that is needed to lift people out of poverty. Rising oil prices threaten economic growth by increasing inflation. In February 2011, India's prime minister warned that the country's rapid economic growth is under "serious threat" from inflation. It is worth noting that almost every major economic recession in the USA has been preceded by an oil price rise, and every oil price rise has been followed by a recession. Indeed, the sudden rise in oil prices in 2008 is believed to have been a key causal factor in the current global recession. Prior to this period, low petrol prices had helped to keep inflation down which in turn kept interest rates low. However, between January 2004 and January 2006, US oil prices surged from thirty-five to sixty-eight dollars per barrel. This set off a burst of inflation which pushed interest rates up from one to five percent. Suddenly paying the mortgage and the interest on the loans for the two family four by fours became a lot more difficult, quite apart from the costs of filling them up at the pumps. Food prices also went up. Then, the bubble burst, the banks went bust and the economy went into recession.

When the financial meltdown began, the immediate response from western governments was to get unsustainable consumption back on the roads. In both Europe and the USA, car makers were given cheap loans that cost the taxpayer billions. European car makers claimed that the loans would be used to manufacture greener vehicles. In 1998, the car makers had promised to cut the greenhouse gas emissions of their vehicles voluntarily. By 2006, it was clear that they had lied and the European Commission announced that it would set compulsory standards. By 2009, they were claiming that they could only go green with massive public subsidies.

Propping up the car industry through public loans and by subsidising the purchase of new cars is not a long-term solution. What is needed is an ambitious decarbonisation programme that will cut across all the major areas of fossil fuel energy use. This would include the decarbonisation of our energy supplies, increasing the energy efficiency of homes, the creation of an urban infrastructure for safe walking and cycling and the greening of our cities.

Renovating towns and cities for walking and cycling will require architects, artists, arboculturists, builders, carpenters, engineers, ecologists, educators, planners, planters, street performers and urban farmers. Their job description would be to ensure that walking and cycling provide the most enjoyable, the most satisfying, the safest and the most direct means of getting around. These will be socially useful jobs that build the foundations for better health, safer and stronger communities and a sustainable economy. Resuscitating a carbon-based economy is a short-term fix that can only fail. As soon as the economy starts growing and the demand for oil increases, petrol price rises will choke it back into a recession. The decarbonisation of transport and society more generally is not the path towards austerity—it is the only way to avoid it.

References

Banister D, Wright L, Transport Research Laboratory (2005) The role of transport in supporting sub-national growth: an overview study. DfID

DiGuiseppi C, Roberts I, Li L, Allen D (1998) Determinants of car travel on daily journeys to school: cross sectional survey of primary school children. BMJ 316(7142):316, 1426-8

Duperrex O, Bunn F, Roberts I (2002) Safety education of pedestrians for injury prevention: a systematic review of randomised controlled trials. BMJ 324:1129

Edwards P, Roberts I, Green J, Lutchmun S (2006) Deaths from injury in children and employment status in family: analysis of trends in class specific death rates. BMJ 333:119

Franco M, Ordunez P, Caballero B, Granados T, Lazo T, Bernal J (2007) Impact of energy intake, physical activity, and population-wide weight loss on cardiovascular disease and diabetes mortality in Cuba, 1980–2005. Am J Epidemiol 166:1374–1380

Franco M, Ordunez P, Caballero B, Copper R (2008) Obesity reduction and its possible consequences: what can we learn from Cuba's special period? CMAJ 178:1032–1034

Galeano E (1973) Open veins of Latin America. Monthly Review Press, New York

Hillman M, Adams J, Whitelegg J (2000) One false move.... study of children's independent mobility. Policy Studies Institute, London

Jacobs J (1992) The death and life of great American cities. Random House Inc., New York

Peden M, Scurfield R, Sleet D, Mohan D, Hyder A, Jarawan E, Mathers C (2004) World report on road traffic injury prevention. World Health Organization, Geneva

Roberts I (2004) Injury and globalisation. Inj Prev 10:65–66

Roberts I (2007) Formula one and global road safety. J R Soc Med 100:1–3

Roberts I, Marshall R, Norton R (1992) Child pedestrian mortality and traffic volume in New Zealand. BMJ 305

Roberts I, Norton R, Jackson R, Dunn R, Hassall I (1995) Effect if environmental factors on the risk of injury of child pedestrians by motor-vehicles: a case control study. BMJ 310:91–94

Roberts I, Carlin J, Bennett E, Bergstrom B, Guyer B, Nolan T, Norton R, Pless I, Rao R, Stevenson M (1997) An international study of the exposure of children to traffic. Inj Prev 3:89–93

Roberts I, Wentz R, Edwards P (2006) Car manufacturers and global road safety: a word frequency analysis of road safety documents. Inj Prev 12:320–322

Rodney W (1981) How Europe underdeveloped Africa. Howard University Press

Woodcock J, Edwards P, Tonne C, Armstrong BG, Ashiru O, Banister D, Beevers S, Chalabi Z, Chowdhury Z, Cohen A, Franco OH, Haines A, Hickman R, Lindsay G, Mittal I, Mohan D, Tiwari G, Woodward A, Roberts I (2009) Public health benefits of strategies to reduce greenhouse-gas emissions: urban land transport. Lancet 374:1930–1943

Chapter 15
'How Can Transport Contribute to Other Urban Agendas?'

David Satterthwaite

15.1 Introduction

The focus of this lecture is on whether and how transport can serve and support other local and global agendas in urban areas. There are important contributions that a well-functioning city transport system can make to a range of goals—including improving housing (and lowering housing costs), reducing poverty (including increasing income-earning opportunities and lowering transport costs), reducing disaster risk, adapting to climate change and climate change mitigation. But to what extent do the transport components of these strategies overlap or conflict? And, perhaps as importantly, is it possible to implement what is needed, especially in regard to land-use management that supports these goals.

I was very flattered to be invited to give this lecture—but a bit puzzled in that I know little about transport. But I work on urban agendas where transport has considerable importance. So it provides me with an opportunity to consider what different agendas want or expect from transport—including provision for the different modes of transport.

15.2 How Does Transport Serve Four Urban Agendas

As someone who works on urban poverty reduction, I want transport systems that cut housing costs and increase the access of low-income groups to income-earning opportunities. Of course I also want transport systems that are safe and keep down time and monetary costs for users. Most of this lecture is devoted to this issue.

D. Satterthwaite (✉)
International Institute for Environment and Development (IIED), London, UK
e-mail: david.satterthwaite@iied.org

G. Tiwari and D. Mohan (eds.), *Transport and Safety*, Springer Transactions in Civil and Environmental Engineering,
https://doi.org/10.1007/978-981-16-1115-5_15

But as someone who also works on disaster risk reduction, I want transport systems that are resilient to extreme weather or other potential catalysts of disasters and that allow those exposed to high risks in their homes or workplaces to move to safer locations if needed. Also, transport systems with the capacity are to recover quickly from disruptions; this has particular importance for low-income groups for whom disruption in their incomes of only 2 or 3 days presents them with challenges to their nutritional status. Avoiding, for instance, the disruptions to livelihoods and the very functioning of the city was evident in Mumbai in the floods in 2005.[1] And of course, transport systems may have particular importance in getting people away from a site particularly at risk from an approaching storm or flood.

As someone who works on climate change adaptation, transport has importance for reducing disaster risks that are linked to climate change and for building resilience and redundancy within all the infrastructure and service networks that are important for cities. And given how long most transport infrastructure lasts, building into new infrastructure investments new safety margins so they can cope with more intense or frequent extreme weather events and other (likely or possible) impacts from climate change.

There is much common ground between poverty reduction, disaster risk reduction and climate change adaptation. All three have a strong focus on reducing risks, especially for low-income groups and/or those living in informal settlements that lack risk-reducing infrastructure and services. But these three different urban agendas tend to focus on different sets of risks. However, the infrastructure and services that are so important for poverty reduction (good quality, regular water supplies piped to homes, good provision for sanitation, drainage and solid waste collection, health care and emergency services, schools, rule of law/policing and accountable local government) are also key to disaster risk reduction. They (and the institutional and financial systems that underpin them) also provide a valuable base from which to address climate change adaptation.

As someone who works on climate change mitigation, transport has obvious importance. Greenhouse gas emissions per person in any city are much influenced by urban form and by the quality of provision for public transport (and of course provision for public transport also influenced urban form). It is worth recalling the seminal work of Newman and Kenworthy showing how many of the European cities with the highest standards of living had per capita gasoline use that was a quarter that of many US cities.[2] Barcelona that has a high-quality public transport system and a historic city core that is oriented to pedestrians has much lower greenhouse gas emissions per person than (for instance) Atlanta. Curitiba in Brazil with its well-known use of bus rapid transit and its land-use planning and feeder bus system has a less carbon intensive transport system than most other Brazilian cities.

[1]Revi (2005).

[2]Newman and Kenworthy (1999).

If dangerous climate change is to be avoided, a city's quality of life needs to be delinked from high greenhouse gas emissions per person. There are also many examples of cities in low-income nations with very low emissions per person.[3] But we also have some examples of cities that have relatively low greenhouse gas emissions despite high per capita incomes—as in Barcelona or Curitiba. Porto Alegre, a Brazilian city with a very high quality of life is reported to have a tenth of the greenhouse gas emissions per person when compared to most cities in the USA. Oslo, Copenhagen and other European cities with a very high quality of life also have per capita greenhouse gas emissions far below those of cities in the USA. However, part of this is explained not by transport systems but by the origin of their electricity supplies; countries (and cities) with much of their electricity generated by hydropower have an obvious advantage over those that rely on thermal power stations in regard to greenhouse gas emissions per person.

Although the climate change literature points to 'win-wins' between different urban agendas, there is not much overlap between adaptation and mitigation—although cities with high levels of ambient air pollution can find that lowering greenhouse gas emissions can contribute to lower levels of air pollution. But in the longer term, unless the world's governments agree to greenhouse gas emission reduction on a massive scale to avoid dangerous climate change, even cities with strong adaptation programmes will be unable to protect themselves.

15.3 Making Sense of City Growth

One difficulty facing any city government is not really knowing what is happening in their city. Today and everyday, in any city, many individuals or households are thinking of moving. So take Delhi. There are thousands of people who are moving or thinking of moving to Delhi from another city, town or rural area. There are also many Delhi residents moving or thinking of moving to somewhere else within Delhi—perhaps looking for more space or to find somewhere cheaper or better located in regard to income-earning opportunities. Meanwhile hundreds of businesses are considering whether to expand or move within the city or move elsewhere. Then there are many businesses outside Delhi who are considering investing here. All these individual, household and enterprise decisions influence demand for transport—but no city government has a record of these decisions. Every ten years, censuses tell us about how populations have changed for cities and their surrounds and the changes in population within cities. But censuses so often bring surprises—as the city in which I live has a larger or smaller population than expected—but we only learn about this every ten years.

We know that migration flows to a city where there are new investments and economic opportunities. Cities can be seen as labour markets (and as Alain Bertaud

[3]Hoornweg et al. (2011).

notes, without a functioning labour market there is no city[4]). This is a relief because people are moving to where there are more job or livelihood opportunities. But a large part of the migrants and the 'already in the city' residents have low incomes. All need to find accommodation that they can afford and that provides them with access to income-earning opportunities. Here, it is worth looking not only at cities as labour markets and how those with labour seek to insert themselves in this market (or to move to a more advantageous position in it) but also at what this means for their housing and their transport.

15.4 Do All Low-Income Groups Really Live in 'Slums'?

There is much discussion in India and elsewhere about 'slums' and about how many people live in them. Bhan discusses how the size of Delhi's slum population (and the proportion living in 'slums') is much influenced by what definition is used and notes that many of Delhi's most vulnerable poor live in makeshift shacks or sleep on the street and remain uncounted in any assessment of who lives in 'slums'.[5]

Of course, how many people live in 'slums' also depends not only on how they are defined but also on how accurately they are measured. The United Nations produces statistics for the proportion of the urban population living in 'slums' in most nations and globally[6] but there are serious doubts as to the accuracy of these 'slum' statistics. First, there are the criteria used for defining 'slum' households. A household is defined as a slum household if it lacks one or more of the essentials like 'improved' water, 'improved' sanitation, durable housing or sufficient living area. But a large proportion of households with 'improved' water or 'improved' sanitation still lack provision to a standard that meets health needs (or, for water, what is specified in the Millennium Development Goals as sustainable access to safe drinking water).[7] If there were the data available to apply a definition as to who has provision for water and sanitation to a standard that cuts down health risks and ensures convenient and affordable access, the number of 'slum' dwellers would increase considerably in many nations. If we had statistics as to who had water piped to their premises that was of drinking water

[4]Bertaud (2014).

[5]Bhan (2009).

[6]See UN Habitat (2012).

[7]For water, improved provision includes piped water into dwelling, yard or plot, public tap or standpipe, tubewell or borehole, protected dug well, protected spring or rainwater collection. For sanitation, improved provision includes use of use flush or pour-flush toilets to piped sewer system septic tank or pit latrine, ventilated improved pit latrine, pit latrine with slab or composting toilet. These cover such a wide range of types of provision and will include many that are very deficient in urban contexts. See UNICEF and WHO (2012), Progress on Drinking Water and Sanitation; 2012 Update, Joint Monitoring Programme for Water Supply and Sanitation, UNICEF and WHO, New York and Geneva, 60 pages.

quality and regular and classified households without these as 'slum' households, the number of slum dwellers would increase dramatically.[8]

A second reason for concern as to slum statistics' accuracy is that they show very large drops in the proportion of urban dwellers living in 'slums' in some nations for which there is so little supporting evidence. For instance, the United Nations Human Settlements Programme (UN-Habitatl) states that the proportion of the urban population living in 'slums' in India has fallen from 54.9% in 1990 to 29.4% in 2009. If this is true, then India has had one of the world's most successful programmes in reducing slum populations. For Bangladesh, the proportion of the urban population living in 'slums' is said to have fallen from 87.3 to 61.6% in this same period. Where is the supporting evidence for this? It may be that most of the apparent fall in the slum population globally between 2000 and 2010 was simply the result of a change in definitions—as a wider range of (inadequate) sanitation provision was classified as 'improved'. It is also not clear where UN-Habitat gets its annual figures for the proportion of the urban population in 'slums' yet these figures are so widely used and quoted.

But perhaps the issue is not to improve the definition and measurement of 'slums' and 'slum' households but to better understand the ways in which city residents get accommodation in a range of housing sub-markets through which they buy, build, rent or otherwise get to use housing. Then it becomes possible to consider how transport does or could better serve their inhabitants. Some housing sub-markets may be assessed as very poor in terms of housing quality but very good in terms of access to income-earning opportunities. Each individual or household has their own particular priorities in terms of location, size, quality (of building, infrastructure and services) and price. For those with limited incomes, a lot of these are in informal settlements or in overcrowded rental accommodation. It is a big mistake to label all these as slums as this gives no sense of their diversity. What we need to understand is how low-income individuals or households seek the accommodation that provides the best compromise for their multiple needs and how these can be supported.

There was a low-income settlement I saw in Montevideo that was called Barrio Nicol. Barrio means neighbourhood. But why Nicol? It was an abbreviation for 'ni-colectivo'—neighbourhood where there is no bus. For the residents to name their settlement in this way shows a strongly felt need.

In cities, low-income groups usually face the greatest difficulties finding housing that is close to income-earning opportunities. Just as all of us do, they have to make choices that reflect trade-offs that balance location in relation to income-earning opportunities, size, quality, provision for infrastructure and services and tenure security. Many would like to get a plot of their own on which they can build their own house. But they have less than we do in regard to what they can afford. My many trips to Mumbai and what I have learnt from the National Slum Dwellers Federation, Mahila Milan and SPARC are clear examples of particular trade-offs by low-income groups. For instance, the pavement dwellers who prioritize locations that maximize their access to income-earning opportunities (most able to walk to where they work)

[8]Mitlin and Satterthwaite (2012).

end up getting very small, poor quality and usually insecure accommodation. As the census of pavement dwellers done by the pavement dwellers showed, these were not unemployed recently arrived rural migrants (which is what the authorities thought) but fully employed city residents. They lived on the pavement because there was no better option that they could afford.[9]

We can look at Dharavi as a slum.[10] Or as a source of cheap accommodation within walking distance of many jobs or income-earning opportunities. Or as a highly productive economy with very few traffic accidents that generate some US$400 million worth of goods and services in a year; or as a model low-carbon economy with very low private automobile use and lots of recycling and reusing of materials. Or as a place that may be the nightmare of any environmental health officer or occupational health specialist.

Many of the residents of Dharavi share rooms or even beds in rented accommodations which keep down costs and provide immediate access to income-earning opportunities—but this also means very little space, no security and inadequate or no provision for basic services in the home.

So instead of labelling all the housing sub-markets used by low-income groups as slums, let us consider the different housing sub-markets through which low-income groups buy, build, rent or occupy accommodation, who lives in them and how these are structured around transport options and costs.

Box 1: Housing sub-markets used by low-income groups[11]

- Renting rooms in inner-city tenements or subdivided housing (where the structures are legal but with high levels of overcrowding and shared facilities; these are usually well-located in relation to labour markets).
- Renting rooms in other formal housing structures (including public housing).
- Renting a room or a bed in cheap boarding/rooming houses (these cheap boarding houses are often clustered in locations with income-earning opportunities).
- Renting a room or a shack in an illegal settlement (that range from those with relatively secure tenure to those with insecure tenure and from central to very peripheral locations).
- Renting a land plot on which a temporary shack is built (including rooftops).
- Renting space—e.g. in hot beds, cages, public sites, warehouses, workplace, graveyards… (mostly with quick access to labour markets).
- Employer-provided room (e.g. domestic servants).
- Building a home in an informal settlement (that range from those with relatively secure tenure to very insecure tenure).

[9]SPARC (1985).

[10]Patel and Arputham (2008), Lantz and Engqvist (2008), Patel et al. (2009), SPARC and KRVIA (2010).

- Building a home on illegal subdivision.
- Invading empty houses/buildings or part constructed buildings … (often with central locations).
- Building a home within a site and service scheme.
- Building house or shack in a temporary camp or on the pavement.

Box 1 lists different ways by which those with low-incomes buy, build, rent or otherwise obtain accommodation. Box 2 lists some different categories of low-income groups that influence the options that each has for getting accommodations—for instance households with low but steady (formal) employment have more possibilities of getting loans for housing than those without. Single persons will often choose poorer quality accommodation than households with children to maximize the amount they can save or send back to their family. Some low-income groups face more constraints than others in seeking accommodation—for instance those who suffer discrimination in getting access to land or housing or loans.

Box 2: Who are the low-income groups seeking accommodation

- Households with low but steady incomes (including many lower rank public employees).
- Households with low but fluctuating incomes.
- Single people/childless couples with low incomes (and/or seeking to save as much as possible...).
- Most students.
- Low-income older people (often with low/falling pensions).
- Temporary residents with low-incomes seeking to minimize what they spend on housing.
- Weekly commuters/seasonal or circular migrants.
- Those who suffer discrimination in getting access to housing, land or credit to build housing (single mother households? Particular ethnic groups? …).

Box 3 lists the factors that influence the choices made by individuals or households in regard to housing. This now allows a consideration of how transport influences or can influence these. Obviously, for almost all low-income individuals or households, access to income-earning opportunities (which means location and the time and monetary cost of getting this access) are important. Improving public transport can (in theory) reduce time and money costs and this contributes to poverty reduction if this benefits low-income groups. If Barrio Nicol got its colectivo, it would likely improve incomes and reduce time spent travelling. But improving public transport (and access) could also serve low-income groups that want to develop their own home if it helps increase the supply and reduces the cost of land for housing (whether this

[11] This and the other boxes draw material from Hardoy and Satterthwaite (1989).

is legal or illegal in terms of land occupation and formal or informal as to its land-use regulations).

Box 3: What are the priorities of each individual or household in regard to housing

- Location for income/livelihoods.
- Cost and how to pay.
- Size.
- Quality of house.
- Quality of site (and space for expansion).
- Quality and accessibility of services.
- Tenure/security.
- Permanency.
- Shelter as savings account.
- Extent to which they can help build the shelter.

Many low-income households that manage to get a plot of land on which they can build a home see this home as a savings account as well as a home. This savings account is increased as they incrementally build, improve and extend the house. Its value is increased if infrastructure and services are provided or improved—and if tenure can be secured.

Box 4: How low-income groups get housing

- renting
- leasing
- invading
- purchasing
- inheriting
- Informally using
- sharing
- getting from employer

A house, apartment, shack, room or open space

15.5 Housing Options

If low-income groups have managed to get accommodations that are well-located in relation to income-earning opportunities—however, poor the quality of their housing and insecure their tenure—they generally want to stay there. Indeed, this easy access was often the reason why these developed—and the inhabitants have to put up with the insecurity, poor quality housing and lack of infrastructure but they have a good location. What they want is basic services and avoiding eviction. For the de facto owner occupiers, they want tenure. There are many good examples of community-managed incremental upgradations which turn an informal settlement into a formal settlement. And usually, these are quite high in density and where a high proportion of trips are walking. This incremental upgrading can produce good quality accommodation with infrastructure and services and nice neighbourhoods to live in. These upgraded settlements actually fulfil many of the goals of the New Urbanism. Lots of shops and services within walking distance. High densities that mean high and concentrated demand for public transport. Good provision for walking and bicycling (and often many lanes that are only for pedestrians and bicyclists). And often public spaces that are heavily used and lots of work opportunities.

So it is very common for those living in informal settlements to want to stay there because they have a relatively good access to labour markets, and they also want to avoid the disruption to social networks that a move would entail. There is also a literature going back to the 1930s on how moving low-income households to 'better quality' accommodation in peripheral locations actually impoverishes them as this move reduces their access to income-earning opportunities.[12]

So incremental upgrading of informal settlements preserves the locational advantages while also addressing the other needs of the residents. This is perhaps best seen in the work of the Community Organizations Development Institute (CODI) in Thailand that provides loans and support to low-income community residents to develop their own upgrading plan and negotiate secure tenure from the owner of the land on which they live.[13] These have transformed the quality of housing and provision for infrastructure and basic services for tens of thousands of households.

Upgrading is not always the preferred solution; the pavement dwellers in Byculla (Mumbai) want better quality and more secure accommodation as long as it does not impose too high a monetary cost for the housing and services and too high a time and monetary cost getting to and from work.[14] The 20,000 households that lived along the railway tracks in Mumbai did not want to stay there if there were housing options that better suited their needs and priorities. They agreed to move although what was unusual in this move was the extent to which those who moved were themselves engaged in decisions about where to move to, when to move and how.[15]

---

[12]Turner (1976).

[13]Boonyabancha (2005), also Satterthwaite and Mitlin (2014).

[14]Patel (1990).

[15]Patel et al. (2002).

As a city grows and expands, so there is a need to upgrade infrastructure. For many cities, there is also a very large deficit in basic infrastructure that needs to be remedied. Many cities also have a high concentration of low-income groups living in areas at risk from extreme weather events—and often also from the increase in risks that climate change is bringing or will bring. So there will be some need for relocations. What stands out from the successful and the unsuccessful relocations is the key role of location and the quality of public transport. And as critically, who chooses the site for the relocation—and who manages the move (as in the example given above of those relocated from beside the railway tracks in Mumbai). Also, with low-income groups in informal settlements allowed to challenge official plans and limit those who have to move to a minimum.

15.6 Can Transport Costs Be Reduced?

For at least two decades, there have been studies showing the high proportion of household expenditure going to transport among urban populations—or the high costs facing particular urban poor communities. These include studies showing public transport costs representing a significant part of total household expenditure. Travel can take 20–25% of daily wages among the low-income population living in cities such as Delhi, Buenos Aires and Manila and up to 30% in Pretoria and Dar es Salaam.[16] Yet transport costs are often not even considered in the setting of poverty lines.[17]

In Buenos Aires, a 2002 survey found that the poorest quintile spent over 30% of family income on public transport.[18] In Sao Paulo, a 2003 survey showed low-income groups spending 18–30% of their incomes on travel; by comparison, wealthy residents spent 7% of their incomes and were able to travel far more frequently.[19] In Salvador (Brazil), a household survey in two peripheral low-income neighbourhoods found that transport expenditures averaged 25% of monthly expenditures.[20] In informal settlements in Nairobi, residents spend 8% of their income on transport[21] Although given the variety of locations for informal settlements in Nairobi in regard to how close they are to income-earning opportunities, there is likely to be considerable variation in this between different informal settlements. A study of average household expenditures across Zambia's urban population found that 12% was spent on transport.[22]

[16]UN-Habitat (2013).

[17]Mitlin and Satterthwaite (2012), op. cit.

[18]Carruthers et al. (2005).

[19]Ibid.

[20]Winrock International (2005).

[21]World Bank (2006).

[22]Central Statistical Office, Zambia (1998).

But averages across urban populations hide the great diversity in costs between different settlements and locations—and many low-income groups live in very poor quality and overcrowded accommodation in more central areas to get quicker, cheaper and easier access to income-earning opportunities. So those living in more distant poorly located settlements may be paying two or more times the average. In addition, what such figures do not show are the other consequences of high public transport costs. These include the time and energy burden of having to walk more. Some studies have shown how many low-income groups walk long distances to keep their transport expenditures down.[23] In a survey of Wuhan, China, the bottom quintile reported walking for almost half of their journeys; so too did the bottom quintile in Buenos Aires.[24]A survey in Nairobi showed the high proportion of trips done by walking for those living in 'slums'; women were more affected—as 67% of low-income women walked, significantly higher than the 53 per cent of their male counterparts.[25] High transport costs may also result in employment opportunities forgone because the time and monetary costs of getting there are too high. A 2003 survey in Wuhan, China, showed how prohibitively high transit costs resulted in low-income groups rejecting jobs far from their homes.[26] A study of eight informal settlements in Cairo noted that high travel costs were one reason why few children went to secondary school.[27]

15.7 What Could Transport Do for Housing Costs?

In theory, expanding road networks in and around a city should increase the supply and reduce the cost of land for housing. Expanding public transport services to a larger area in and around the city (including rail/light rail, bus-rapid transit and metro) should also do so. This seems to go against the very considerable literature on how improving any location's access to central city locations (or labour markets) increases land prices. But surely this is only for the locations most favoured by the improved access. I want to know what happens to land prices in locations that benefitted from the new road, BRT, rail or metro but that were further away from the stations or bus stops. When Curitiba developed and then extended its bus-rapid-transit system, land prices must have increased in and around the bus stops—but overall this bus system served to increase the land area in and around the city that was within (say) half an hour and a particular cost of the industrial area or the central city.

[23]See, for instance, Huq et al. (1996) for various cities in Bangladesh, and Barter (1999) for central Bombay/Mumbai and Jakarta.

[24]Carruthers et al. (2005), op. cit.

[25]Salon and Gulyani (2010).

[26]Carruthers et al. (2005), op. cit.

[27]Sabry (2010).

However, when I asked friends of mine who are transport specialists, they suggested that I was wrong on this. They pointed to how improving the access of peripheral areas to central city labour markets meant land speculation and real estate interests focussing on commercial developments or developments for wealthy households with very little new land available to low-income groups. They also pointed to how land-use regulations (and how they are applied) constrained any increase in the supply and reduction in the cost of land for housing. But I want to stick with my perhaps naïve hope that expanding the area in and around a city that has access to central city (or other clusters of) income-earning opportunities (and services) can also increase the possibilities of low-income households to get land on which they can build housing. So the key issue here is how improvements in transport can be linked to land-use changes that do increase the supply and reduce the cost of land for housing. Three examples are given here of different ways in which the supply of land for housing was increased and costs lowered.

The first is from the city of Ilo in Peru, an industrial city that was growing very rapidly because of a large copper smelter located closeby. The first Mayor to be elected there (in 1982) saw the difficulties that low-income households had in finding accommodation so the municipal government bought a plot of land in a good location in relation to the city and employment, plotted it, put in basic infrastructure and sold the plots for the equivalent of US$60 per plot. This opened up the possibility of legal housing for low-income groups even if they had to build their own homes. Here the key intervention was not on transport but on increasing the availability of low-cost housing plots well-located in relation to income-earning opportunities.[28]

A similar story in Namibia. Here, a national federation of homeless people had been formed by savings groups. Their members could not afford to buy legal housing plots that local governments developed and sold at cost. But this federation negotiated with local government to allow smaller plots (below the official minimum lot size) and lower-cost infrastructure. This reduced the cost of legal land plots for housing and widened the proportion of households able to afford these.[29]

A different story from Tunisia that was documented in the 1970s but for which I have found no more recent documentation. Here, a government agency purchased land in and around cities, subdivided it into plots, put in the basic infrastructure and sold it—including sales to housing companies. This was done on a rolling programme so the sale of the (now prepared) land generated the revenues to allow the agency to continue buying land. This did not bring direct benefits to low-income groups because this boosted the supply and reduced the cost of formal housing that they could not afford—but it increased options for lower-middle income groups who then were no longer in competition with low-income groups for existing housing.[30] There were also other government initiatives to improve conditions in informal settlements that would have brought more direct benefits to low-income households.

---

[28] Follegatti and Luis (1999).

[29] Mitlin and Muller (2004).

[30] Hardoy and Satterthwaite (1989), op. cit.

There are also some historic moments when a city government greatly increased the plotted area around it (as in New York and Barcelona) and what did this do for housing prices? How much did the expansion of public transport in London through suburban railways and the metro increase the supply of land for housing and reduce the cost (or were all the benefits concentrated in middle and upper income groups).

Of course, whether or not an increase in the area in and around a city with good access to income-earning opportunities brings cheaper housing also depends on the price and availability of building materials (and how these are influenced by laws, codes and regulations on building design and materials). It also depends on the cost of getting permission to build, extend, buy or sell and the cost implications of meeting land-use regulations. Like so many other aspects of urban development, it is influenced by the attitudes of politicians and civil servants—in this instance, regarding the best use of government land and attitudes of politicians and civil servants and elites regarding development of unused land for low-income groups. It depends on who can get credit for land purchase and housing and who can access it.

So how can we combine the expansion of transport systems with increasing the supply and reducing the cost of land for housing and the infrastructure it needs. But without urban sprawl. And without building into the urban form increasing car dependency.

15.8 Conclusions

There are some strong and important linkages within a good transport policy that can serve the different urban agendas. Public transport systems that are not only used by those with low-incomes that keep down time and monetary costs for users, that widen housing choices for low-income groups, that are pleasant and safe and that help keep down transport-related greenhouse gas emissions. And in doing so, reinforcing the comparative advantage of that city in retaining or attracting new investment?

Can we learn from the way that low-income groups and their own organizations have upgraded their settlements to show how a high quality of life can be achieved with high densities and a high proportion of all trips made by walking, bicycling and public transport?

How can more attention be paid to the five Ds that influence travel demand: Density, Diversity of land uses (jobs and homes, provisions for walking and bicycles as well as cars); Design (street layout that encourages pedestrians and bicycles), Destination accessibility (what can be reached within 30 min) and Distance to transit.[31]

Can all discussions of relocation now be done with those 'to be relocated' with particular attention paid to how this affects their access to labour markets and the time and transport costs they would face. Clearly, there are very large variations in upgrading programmes from serving those who live there to those that push them

[31] UN-Habitat (2013), op. cit.

out.[32] Clearly, settlements will be displaced as cities upgrade their infrastructure (and address the often enormous deficits in basic infrastructure). But the cost of doing relocation well (so that all those who are relocated are fully engaged in planning and managing this) is so often a very small fraction of the infrastructure investment budget.

So much of the above depends on changing the nature of the link between city government and those living in informal settlements. Do city politicians and civil servants understand the key contributions to the city economy made by those living in informal settlements and also understand their needs and priorities?

Are there forms of redistribution to which transport systems contribute? For instance, where the transport modes that low-income groups use are faster than using private automobiles? Where provision for bicycle use is so good that the middle-class also choose to bicycle (see some cities in Europe with 30–55% of trips made by bike).[33]

And perhaps the most difficult in any successful city: How to get control of land use and land-use changes so that these are in the public interest? And here with particular attention to increasing the supply and reducing the cost of land for housing in locations with good access to labour markets. As noted by Angel et al.,[34] few city governments are actively preparing for urban expansion. The planning horizon of a politician is too short. Many governments are still anti-urban and will not plan for fear that it encourages the 'rush to cities' even though there is no credible evidence that shortages of housing, roads, open space, drinking water or public facilities have any effect on rural–urban migration.[35]

References

Angel S, Sheppard SC, Civco DL (2005) The dynamics of global urban expansion. Transport and Urban Development Department. The World Bank, Washington DC, 205 p

Barter PA (1999) Transport and urban poverty in Asia. A brief introduction to the key issues. Reg Dev Dialogue 20(1):143–163

Bertaud A (2014) Cities as labour markets. Working paper no 2, Marron Institute, New York University, New York, 49 p

Bhan G (2009) This is no longer the city I once knew; Evictions, the urban poor and the right to the city in Millennial Delhi. Environ Urban 21(1):127–142

Boonyabancha S (2005) Baan Mankong; going to scale with 'slum' and squatter upgrading in Thailand. Environ Urban 17(1):21–44

Carruthers R, Dick M, Saurkar A (2005) Affordability of public transport in developing countries. Transport paper 3. World Bank, Washington DC

Central Statistical Office, Zambia (1998) Living conditions in Zambia 1998; the evolution of poverty in Zambia 1990–1996. Government of the Republic of Zambia, Lusaka

[32] Patel (2013).

[33] UN-Habitat (2013), op. cit.

[34] Angel et al. (2005).

[35] Ibid.

Follegatti L, Luis J (1999) Ilo: a city in transformation. Environ Urbanization 11(2):181–202
Hardoy JE, Satterthwaite D (1989) Squatter citizen: life in the urban third World. Earthscan Publications, London, UK, 388 p
Hoornweg D, Sugar L, Gomez CLT (2011) Cities and greenhouse gas emissions: moving forward. Environ Urban 23(1):207–227
Huq AT, Zahurul M, Uddin B (1996) Transport and the urban poor. In: Islam N (ed) The urban poor in Bangladesh, Dhaka. Centre for Urban Studies
Lantz M, Engqvist JH (eds) (2008) Dharavi: documenting informalities. The Royal University College of Fine Arts, Art and Architecture, Stockholm, 303 p
Mitlin D, Muller A (2004) Windhoek, Namibia: towards progressive urban land policies in Southern Africa. Int Dev Plan Rev 26(2):167–186
Mitlin D, Satterthwaite D (2012) Urban poverty in the global south: scale and nature. Routledge, London
Newman P, Kenworthy J (1999) Sustainability and cities: overcoming automobile dependence. Island Press, Washington DC, 442 p
Patel S (1990) Street children, hotels boys and children of pavement dwellers and construction workers in Bombay: how they meet their daily needs. Environ Urban 2(2):9–26
Patel S (2013) Upgrade, rehouse or resettle? An assessment of the Indian Governments' Basic Services for the Urban Poor programme. Environ Urban 25(1):177–188
Patel S, Arputham J (2008) Plans for Dharavi: negotiating a reconciliation between a state-driven market redevelopment and residents' aspirations. Environ Urban 20(1):243–254
Patel S, d'Cruz C, Burra S (2002) Beyond evictions in a global city; people-managed resettlement in Mumbai. Environ Urban 14(1):159–172
Patel S, Arputham J, Burra S, Savchuk K (2009) Getting the information base for Dharavi's re-development. Environ Urban 21(1):241–252
Revi A (2005) Lessons from the deluge: priorities for multi-hazard risk mitigation. Econ Pol Wkly 40(36):3911–3916
Sabry S (2010) How poverty is underestimated in Greater Cairo, Egypt. Environ Urban 22(2):523–541
Salon D, Gulyani S (2010) Mobility, poverty and gender; travel 'choices' of slum residents in Nairobi. Transp Rev 30(5):641–657
Satterthwaite D, Mitlin D (2014) Reducing urban poverty in the global south. Routledge, London
SPARC (1985) We the invisible. A Census of Pavement Dwellers, Bombay, 41 p
SPARC, KRVIA (2010) Reinterpreting, re-imagining, redeveloping Dharavi. SPARC and Kamla Raheja Vidyanidhi Institute for Architecture and Environmental Studies, Mumbai, 81 p
Turner JFC (1976) Housing by people—towards autonomy in building environments. Ideas in progress. Marion Boyars, London
UN Habitat (2012) State of the World's cities report 2012/2013: prosperity of cities. United Nations Human Settlements Programme, 149 p
UN-Habitat (2013) Planning and design for sustainable urban mobility. Earthscan Publications, London, 317 p
Winrock International (2005) Enabling urban poor livelihoods policy making: understanding the role of energy services: Brazil country report. Available at https://www.urbanenergy.utwente.nl/resources/reports/country_reports/rep_winrock.pdf
World Bank (2006) Kenya inside informality: poverty, jobs, housing and services in Nairobi's slums. World Bank, Washington DC

The manufacturer's authorised representative in the EU is Springer Nature Customer Service Centre GmbH, Europaplatz 3, 69115 Heidelberg, Germany. If you have any concerns regarding our products, please contact ProductSafety@springernature.com

Printed and bound by CPI Group (UK) Ltd, Croydon, CR0 4YY
30/09/2026
02225287-0011